AF333856

Control of Continuous Linear Systems

Control of Continuous Linear Systems

Kaddour Najim

First published in Great Britain and the United States in 2006 by ISTE Ltd

ISTE Ltd
6 Fitzroy Square
London W1T 5DX
UK
www.iste.co.uk

ISTE USA
4308 Patrice Road
Newport Beach, CA 92663
USA

© ISTE Ltd.

First South Asian Edition 2007

The rights of Kaddour Najim to be identified as the author of this work has been asserted by them in accordance with the Copyright, Designs and Patents Act 1988.

This edition is licensed for sale in India, Bangladesh, Pakistan, Sri Lanka and Nepal only. Not for export elsewhere.

Library of Congress Cataloging-in-Publication Data

Najim, K.
Control of continuous linear systems / Kaddour Najim.
 p. cm.
Includes bibliographical references and index.
  ISBN 13: 978-1-905209-12-5
  ISBN 10: 1-905209-12-6
  1. Linear systems--Automatic control--Mathematics. 2. Linear control systems. 3. Calculus, Operational. I. Title.
TJ220.N35 2006
629.8'32—dc22

2006009140

British Library Cataloguing-in-Publication Data
A CIP record for this book is available from the British Library
ISBN 10: 1-905209-12-6
ISBN 13: 978-1-905209-12-5

Printed in Brijbasi Art Press Ltd., I-72, Sector-9, Noida, U.P. India.

Contents

Introduction

Reader's guide

In this book, we present a method of teaching the theory of the control of linear continuous systems. This method consists of introducing some basic definitions, and then presenting the theory related to these systems in the form of solved problems while appealing to computer tools for the more difficult problems. This method has another advantage, in that students will be more involved in the educational process and will have to play an active and dynamic role that will be beneficial to their training.

The objective of this book is to provide the reader with problems and their solutions in order to aid them to acquire and deeply understand the fundamental notions related to the foundations of the control of linear continuous systems, and to help them to be able to implement control systems. Many problems can be solved using available software such as MATLAB. We have rejected this solution. The computer has become an essential tool, but we see very dangerous drift. In fact, students have blind confidence in this tool. They tend to lose their spirit of criticism and analysis. As teachers, we have to review our pedagogy. In other words, the primary purpose of this book is to help the reader to acquire a deep knowledge of the theoretical tools related to the control of linear continuous systems. For example, learning how to sketch a root locus is very important in order, among other things, to check the results of a simulation, and drawing a Bode diagram manually gives us an understanding of how the location of poles and zeros affects the shape of this diagram. This does not prevent the reader from using computer tools in order to obtain, for example, a more precise drawing of the root locus or a rapid study of the stability of the system. We recall the main definitions and theoretical tools at the beginning of each chapter.

The first chapter is dedicated to process modeling. It presents some modeling techniques for chemical, electrical and mechanical systems. A set of accurate models is presented. Taking into account developments in computer technology, phenomenological models can be used in order to support decisions that need to be taken online.

The reader should observe that, for a given process, many phenomenological models can be developed. They depend on the assumptions made about its behavior and the desired objective, i.e., for what purposes the model will be used. In a sense, the model designer can be considered as a photographer who get obtain, for the same subject, different photos with different zooms. The main objective of Chapter 1 is to help the reader to understand and to develop phenomenological models, or at least to be able to understand the main lines related to the kinds of models developed by engineers involved in the areas concerned (electrical, chemical, mechanical, etc.). The treatment presented in this chapter is not intended as a complete description of modeling techniques but merely as a basic introduction to the subject. This introduction may help automatic-control engineers to communicate easily with engineers involved in other areas.

The main core of Chapter 2 deals with the use of Laplace transforms for solving various kinds of problems. In particular, the derivation of transfer functions, as well as block diagrams and their simplifications, is considered. Laplace transforms significantly support the modeling of systems by providing simple rules for manipulating a set of interconnected systems. Of paramount concern in linear control theory is the transfer function, which leads to block diagrams. Block diagrams agglomerate all the available information concerning a given process. We end the chapter by presenting a general method for calculating the coefficients of the partial fraction expansion of a rational function.

Chapter 3 is devoted to the transient and frequency analysis of linear systems. Many examples are treated in order to illustrate such an analysis. We present a set of approaches to the statement of the frequency response of a given system. As a large class of systems can be modeled by a first – or second-order – system, we present a set of identification techniques based on the impulse and step responses of these systems, and simple ideas for distinguishing a high-pass filter from a low-pass filter. The chapter also presents an analysis of some commonly used filters (band-pass, notch, etc.). Chapter 4 is dedicated to stability and precision analysis. Stability is one of the most important challenges in the design of control systems. Before one optimizes the behavior of, for example, a chemical reactor where an exothermic reaction takes place, it is necessary to study its stability. Algebraic and graphical stability criteria are presented. A method based on integral phase evaluation is presented. This method allows one to check the stability of feedback systems without visual inspection of the Nyquist diagram. Examples illustrating precision and stability analysis are described. A set of examples should help the reader to draw easily the root locus of any system.

The primary purpose of Chapter 5 is to introduce a number of PID tuning techniques. We focus our attention mainly on the ideas behind the development of these tuning techniques. It is impossible to make an analytical comparison of the available PID tuning techniques because they are based on different model approximations, different control objectives and, sometimes, different PID parameterizations. Taking into account the control objective and the process model, the reader has to select the tuning method which yields the best control performance. In a thermal power plant, it may happen that a turbine is damaged, and it is necessary to heat the turbine at specific points to straighten it. These specific points, as well as the energy to be used for heating the turbine, depend on the know-how of the technician who carries out this job. This know-how cannot be found in books on heat transfer and materials. This job can be compared to the job of a PID designer (open-loop shaping by manipulating the gain, zeros and poles) who has to select the PID settings in order to modify the diagram (Bode, Nyquist or Nichols) of the uncompensated system such that the diagram (frequency-response curve) of the compensated system will correspond to the desired diagram (a curve which meets the control specifications). We can also compare the job of a PID designer to the operation done by an ophthalmic surgeon in order to correct the curvature of the eyes of a person with myopia using a laser. Several problems dealing with the design of transfer functions from specifications related to the desired dynamics of the controlled system are presented in detail. The chapter ends with a brief introduction to the integrator wind-up problem, and to the two main useful representations of non-linear systems, namely the Wiener and Hammerstein structures. These structures are very interesting in the sense that if they are connected to the inverse of the static non-linear part, any control strategy designed for linear systems can be implemented.

In each chapter, we recall the necessary mathematical tools in a very simplified and didactic manner. As signal processing and automatic control use the same tools, we study some commonly used filters.

The book ends with an Appendix which presents some mathematical and practical developments related to the impulse function (or Dirac delta function) and its relation to the residence time and the unit step, as well as some proofs concerning stability. We present proofs of the Nyquist and Routh–Hurwitz criteria, and a proof related to the asymptotes of the root locus. We present a rigorous statement of the formulae giving the intersection of the asymptotes of the root locus with the real axis. Some of these results are difficult to find in books dedicated to the control of continuous linear systems. These results are very important in the sense that:

1) for a given plant, stability is the main objective to be achieved, before optimizing its behaviour;

2) the proofs constitute good exercises in themselves.

Recent chemical disasters remind us, unfortunately, that stability is very important. For example, it is absolutely necessary to stabilize a chemical reactor where an exothermic reaction occurs, before optimizing its yield. The Appendix also deals with the quas-ilinearization of non-linear systems such as relays. This method is used to find the limit cycle (crossover frequency) and some other points of the frequency responses of systems, which are useful in some PID tuning methods.

In summary, the objective of this book is to provide the reader with a sound understanding of the foundations of the modelling and control of linear continuous systems. In other words, this book should provide the reader with depth and breadth of knowledge in this field. It contains more than 150 solved problems. This book is written in such a manner that students should be able to extend their knowledge to address new problems that they have not seen before. From a mathematical point of view, this book is self-contained. The book also can serve as a tool for students to test their knowledge.

I would like to thank my friends and colleagues E. Ikonen (University of Oulu, Finland), A. S. Poznyak (CINVESTAV, Mexico City) and P. Thomas (Université Paul Sabatier, Toulouse, France) for providing valuable comments on the manuscript.

Professor Kaddour Najim

Process Control Laboratory, ENSIACET, I.N.P. Toulouse, France

University of Oulu, Finland

Notation

Throughout this book, we use the following notation:

$1(t)$	Unit step
$F(s)$	Closed-loop transfer function
$G(s)$	Transfer function of the forward path
$H(s)$	Transfer function of the feedback path
OS	Overshoot
Q	Q factor
$R(s)$	Transfer function of the regulator
$T(s)$	Open-loop transfer function
T_u	Ultimate period
$W(s)$	Sensitivity transfer function
$d(t)$	Perturbation
$g(t)$	Impulse response
k	Static gain
k_c	Controller gain
k_{lim}	Critical (ultimate) gain
k_m	Gain margin
p_i	ith pole
$u(t)$	Process input
$v(t)$	Voltage
$y(t)$	Process output
$y_r(t)$	Desired output (set-point)

z_i	ith zero
$\Xi(s)$	Laplace transform of the error
$\delta(t)$	Dirac impulse (impulse function)
$\varepsilon(t)$	Error
ζ	Damping factor
τ	Time delay
τ_d	Derivative time
τ_i	Integral time
φ_m	Phase margin
w_c	Critical frequency
w_r	Resonance frequency

Indices

$i = \overline{1, N}$	$i = 1, 2, 3, \cdots, N$
r	Reference

A filled square ■ indicates the end of a proof.

Chapter 1

On Process Modeling

1.1. Introduction

Modeling is a common activity in many engineering areas [AGU 99]. A model can be considered as a mapping of input variables into output variables. This chapter presents a set of problems related to some fundamental notions about the representation of dynamical systems in the form of models. The most serious difficulty in implementing control strategies is the lack of adequate models. A myriad of models can be developed for a given process. The models obtained are in general complex (non-linear, high-scale, etc.).

Observe that the complexity of a system is not correlated with its scale. It is, for example, easier to derive a control policy for an industrial phosphate drying furnace 40 m long than for a rapid thermal system used in a semiconductor wafer fabrication process. Notice also that the complexity can be derived from multiple simple dynamic components that interact in varying and complex ways. For various reasons (improved conversion and selectivity, heat integration benefits, avoidance of azeotropes, etc.), chemical engineers are now concerned with process intensification [RAM 95], which generally leads to simple systems. For example, the manufacturing of methyl acetate is usually done in a plant consisting of a chemical reactor and nine distillation columns. This manufacturing can be done instead in a single reactive distillation. The resulting reactive distillation process is very simple and more economical, and it is easier to control it than to control a set consisting of a reactor, nine distillation columns, and many heat exchangers and pumps.

We can consider linearity as a view of our mind. There exists no general technique for the design of controllers for non-linear systems. This explains why linear models are used, because also the theory related to the control of linear systems is well established[1]. The model obtained is linearized around a given operating point.

We begin this chapter by reviewing the main approaches used in processes modeling, bearing in mind our objective: the development of a control strategy in order to achieve the desired control objective.

1.2. Model classification

Models can be classified into three main categories [IKO 02]:

1) phenomenological models;

2) input-output (black-box) models;

3) grey-box models.

Phenomenological models are developed on the basis of physical laws describing the behavior of a given system. In simple mechanical systems, setting force equal to mass times acceleration yields the equation of motion. The parameters of the model have physical significance: for example, the heat transfer coefficient in a thermal process. In chemical engineering, mass, energy and momentum balances are the basis of the development. Observe that the total derivative also plays an important role in the development of models for large-scale processes, where assumptions concerning the homogeneity of temperatures, concentrations, etc. along the three axes (x, y, z) cannot be done [NAJ 88]. The models obtained depend on the assumptions made about the behavior of the system. Indeed, changes in one or more assumptions lead to different models [NAJ 83]. In some sense, the model designer can be considered as a photographer who obtains, for the same subject, different photos with different zooms. For a given system, many phenomenological models can be derived. The model is developed according to the objective to be achieved and the use for which it is designated. The development of this kind of model is, in general, very time-consuming and necessitates a deep understanding of the phenomena (transfers, kinetics, fluid mechanics, etc.) involved in the process considered.

For the synthesis of black-box models, designers adopt a model structure (transfer function, state-space representation, Hammerstein structure, neural network, etc.),

---

1. Notice that the Wiener, Hammerstein and Uryson models are quite general representations of non-linear systems. Recall that the Hammerstein structure consists of a non-linear static system followed by linear dynamics. On the basis of the use of the inverse of the non-linear static system, any linear control strategy can be used for the control of this kind of system (see for instance [IKO 02]).

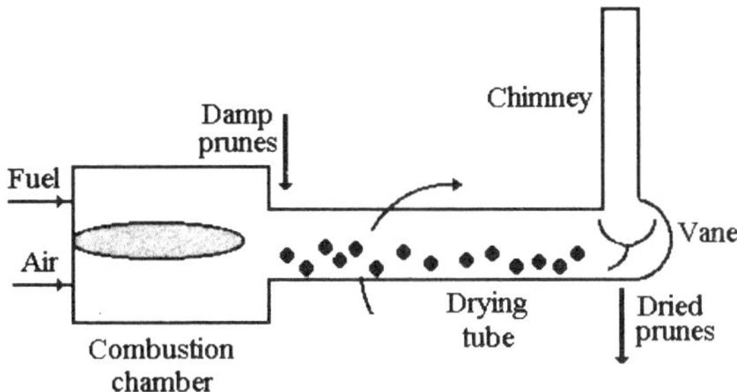

Figure 1.1. *Drying furnace*

and by making use of the available data, they identify the parameters of the structure. These parameters have no physical significance. Compared with the previous approach, the time savings of this approach are evident.

Gray-box models are a combination of the modeling approaches described above. For example, in an electric heating system, it is preferable to use the energy, which is related to the square of the voltage, as a control variable instead of the voltage itself. In other words, gray-box models are input-output models where physical insight into the process considered is included.

For many processes, the variation of the dynamics is usually due to changes of the operating point (feed flow rate, etc.). For these processes, another modeling approach can be used. This approach consists of building local models on the basis of a database (measurements) for a specific operating point when they are needed. This approach is called "model-on-demand" and has been studied mainly in the Division of Automatic Control and Communication Systems, University of Linköping, Sweden.

PROBLEM 1.1. Consider the prune-drying rotary furnace depicted in Figure 1.1. This dryer consists of a combustion chamber, a drying tube of length L, a vane and a chimney. Derive its block diagram.

SOLUTION 1.1. Let us first determine the list of the physical variables characterizing the behavior of this dryer. The behavior of this furnace depends on the following main variables: fuel flow rate, air flow rate, combustion gas temperature, flow rate of damp prunes, moisture content of damp prunes, ambient temperature, moisture content of dried prunes, and temperature and flow rate of dried prunes. These physical variables play different roles in the behavior of the furnace, and are classified as control and controlled variables, measured perturbations, and random perturbations as shown in

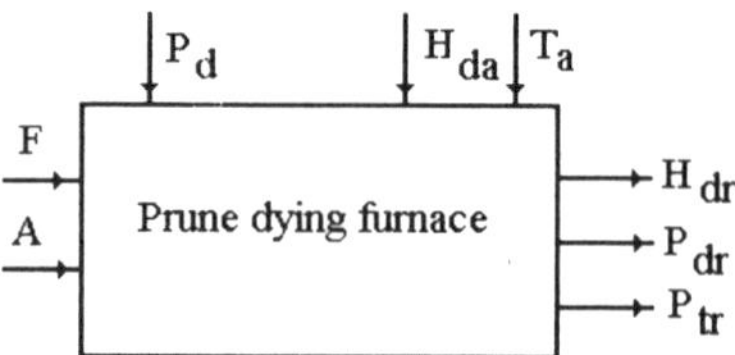

Figure 1.2. *Block diagram of the drying furnace*

the block diagram given in Figure 1.2, where the flow rate of fuel (F) and air (A) are the control variables, the flow rate of damp prune (P_d), the ambient temperature (T_a) and the moisture content of the damp prunes (H_{da}) represent the measured perturbation and the unmeasured[2] random perturbations. The moisture content of the dried prunes (H_{dr}), the flow rate (P_{dr}) and the temperature of the dried prunes (P_{tr}) correspond to the controlled variables. Notice that if the input flow rate of the damp prunes is not fixed at its nominal value, which corresponds to the capacity of the dryer, and can be varied, then it can be considered as a control variable.

REMARK 1.1. The establishment of a list of the physical variables conditioning the behavior of a given process and their classification is the first step for the gathering of knowledge about the behavior of the process. This step is fundamental in the framework of the development of control systems. The most valuable contribution of block diagrams is their ability to identify and categorize information about the controlled process.

REMARK 1.2. Figure 1.2 defines in a certain manner the border between the process considered and its environment: a system.

PROBLEM 1.2. Characterize the time delay associated with this drying furnace.

SOLUTION 1.2. There is a noticeable delay between the instant a change in the input (control variable) is implemented and when the effect is observed, with the process output displaying an initial period of no response. When a process involves mass or energy transport, a time delay (transportation lag) is associated with the movement. In this case, this time delay is equal to the ratio L/V, where V represents the velocity of the raw material (prunes).

---

2. In the case where the furnace is not equipped with sensors for the online measurement of the corresponding variables.

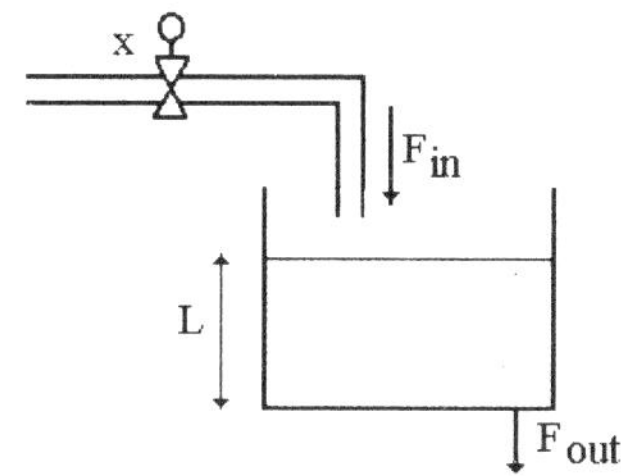

Figure 1.3. *Tank-level-regulation system*

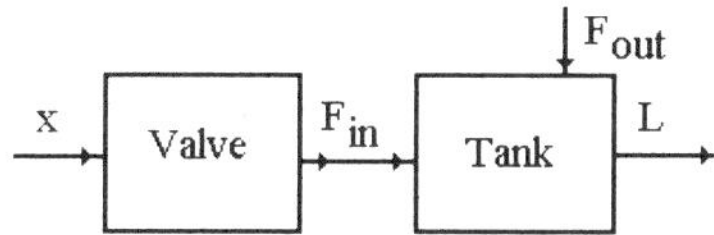

Figure 1.4. *Block diagram of the tank-level-control system*

The next problem shows that the role (control variable, output, etc.) played by a given physical variable depends on the system considered.

PROBLEM 1.3. Consider a tank-level-regulation system (see Figure 1.3). This consists of a manual valve and a tank. Determine a block diagram and a dynamic model of this level-control system.

SOLUTION 1.3. The block diagram is shown in Figure 1.4, where the rate of opening of the valve is x, the inlet flow rate is F_{in}, the tank level is L and the liquid leak rate is F_{out}.

In view of the previous remark, we observe that the inlet flow rate plays two roles: a controlled variable for the valve and a control variable for the tank. The dynamic model can be derived from a mass balance consideration. For an interval dt of time, we obtain:

$$F_{in}\, dt - F_{out}\, dt \quad = \quad \text{variation of the volume of liquid contained in the tank,}$$
$$F_{in}\, dt - F_{out}\, dt \quad = \quad A\, dL,$$

where A represents the cross-section of the tank. Observe that the accumulation of water in the tank is modeled by an integrator. The association of an integrator with a time delay permits us to model many chemical processes.

1.2.1. *Heat and mass balances*

The next problems concern the development of a phenomenological model of a set of systems. Let us first recall the main idea behind this development process. If, in a given system, mass and/or heat transfers take place, mass and/or energy balances yield differential equations governing the behavior of the process. In the framework of fluid mechanics, the force-momentum balances are also considered:

– *Mass balances* express the fact that the quantity of material entering the system minus the quantity of material leaving it is equal to the accumulation of material in the system.

– *Energy balances* express the fact that the heat (energy) supplied to the system is equal to the sum of the quantities of heat transferred to all the components of the system and its surroundings, plus the accumulation of energy in the system.

The heat transfer may occurs via conduction, radiation or convection. Radiation occurs at high temperature. The heat transferred by radiation is proportional to T^4, where T represents the absolute temperature expressed in kelvin. It remains negligible for temperatures less than $200 - 300°C$. Transfer by conduction is proportional to the temperature gradient. For example, if the outside temperature decreases, the loss of energy from a furnace increases. Convection corresponds to heat transfer by mass motion of a fluid such as air (heating in a building) or water (in a kettle) when the heated fluid, which carries energy with it, moves away from the source of heat. In processes involving mass transfer, non-linearities of product type appear. The quantity of a product A contained in a mixture is given by:

$$F_m(t) C_A(t)$$

where $F_m(t)$ and $C_A(t)$ represent the flow rate of the mixture and the concentration of the component A, respectively.

PROBLEM 1.4. Consider the system depicted in Figure 1.5. This consists of a feed system (valve), two tanks and two restrictions. Derive a mathematical model of this system.

SOLUTION 1.4. This system involves only mass (liquid) transfer. During an interval of time dt, the mass balances lead to the following equations:

$$F_{in}\, dt - F_{out1}\, dt \;=\; \text{accumulation of water in the first tank,} \qquad (1.1)$$

$$F_{out1}\, dt - F_{out2}\, dt \;=\; \text{accumulation of water in the second tank.}$$

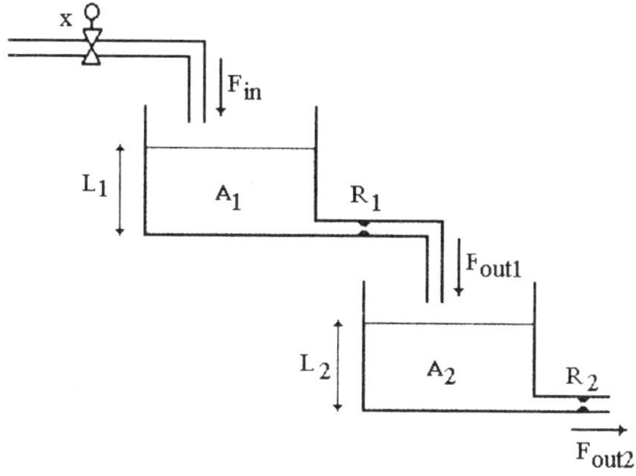

Figure 1.5. *System of two tanks*

Now let us calculate the accumulation of water (liquid) in the two tanks. This accumulation corresponds to the change in the volume of water contained in each tank:

$$\text{first tank: } A_1 \, dL_1,$$
$$\text{second tank: } A_2 \, dL_2. \tag{1.2}$$

where A_1 and A_2 represent the cross-sections of tank 1 and tank 2, respectively. From (1.1) and (1.2), we derive:

$$F_{in} - F_{out1} = A_1 \frac{dL_1}{dt}, \quad F_{out1} - F_{out2} = A_2 \frac{dL_2}{dt}. \tag{1.3}$$

According to Bernoulli's law,[3] the restrictions induce flow rates varying according to the square root of the level, i.e.:

$$F_{out1} = k_1 \sqrt{L_1}, \quad F_{out2} = k_2 \sqrt{L_2}. \tag{1.4}$$

Finally, we obtain the following non-linear model:

$$F_{in} - k_1 \sqrt{L_1} = A_1 \frac{dL_1}{dt}, \quad k_1 \sqrt{L_1} - k_2 \sqrt{L_2} = A_2 \frac{dL_2}{dt}. \tag{1.5}$$

---

3. In simple words, Bernoulli's law states that the output flow rate F_{out} is proportional to the square root of the level L of water in the tank considered, i.e.:

$$F_{out} = k\sqrt{L}.$$

This relation can also be used to model the relation between the output flow rate of a valve and its opening ratio.

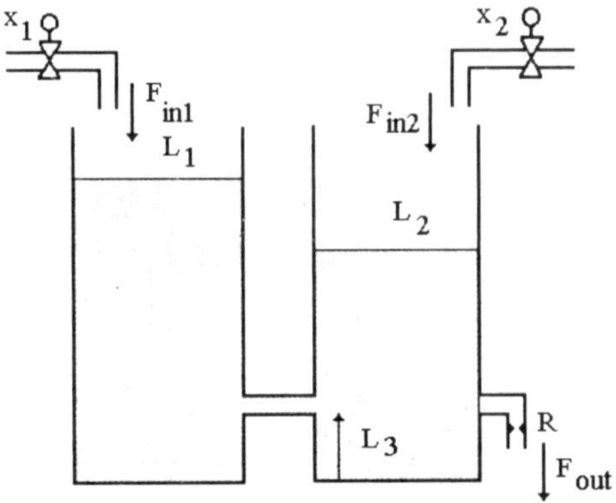

Figure 1.6. *Two communicating tanks*

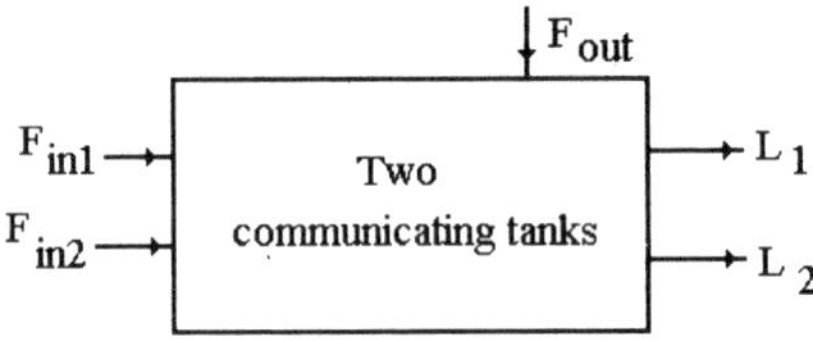

Figure 1.7. *Block diagram of two communicating tanks*

This model will be linearized later in this chapter.

PROBLEM 1.5. Derive the block diagram and the dynamic model of the system pictured in Figure 1.6. The cross-sections of these tanks are assumed to be constant and equal to A. We assume that the output flow rate represents a perturbation.

SOLUTION 1.5. This system is characterized by two control variables, two controlled variables and one perturbation. The block diagram of this system is given in Figure 1.7.

In order to derive a model of this system, let us consider an interval of time dt. During this interval of time, the variation of the volume of the liquid (water) contained in the first tank is equal to the volume of water poured into it, associated with the feed flow rate F_{in1}, minus the amount of water exchanged between this tank and the second one. For the second tank, during the same interval of time, the variation is equal to the volume of water poured into it, associated with the feed flow rate F_{in2}, plus the volume of water exchanged with the first tank minus the volume of water associated with the output flow rate F_{out}. Notice that the direction of the exchange of water between the two tanks depends on the sign of the difference in levels $(L_1(t) - L_2(t))$. In order to model these exchanges, we shall use Bernoulli's law, the relation between $L_2(t)$ and $L_3(t)$, and the output flow rate F_{out}. The mass balances express the following:

$$(F_{in1} \pm \text{flow rate between the two tanks})\, dt$$
$$= \text{variation of the volume of water contained in the first tank,}$$
$$(F_{in2} \mp \text{flow rate between the two tanks } - F_{out})\, dt$$
$$= \text{variation of the volume of water contained in the second tank.}$$

The considerations above lead to the following system of differential equations:

$$A\frac{dL_1(t)}{dt} = F_{in1} - sgn(L_1(t) - L_2(t))\, k_1\sqrt{|L_1(t) - L_2(t)|},$$
$$A\frac{dL_2(t)}{dt} = F_{in2} + sgn(L_1(t) - L_2(t))\, k_1\sqrt{|L_1(t) - L_2(t)|}$$
$$- k_2\sqrt{|L_2(t) - L_3(t)|}.$$

The terms $AdL_1(t)$ and $AdL_2(t)$ represent the variations of the volume of liquid contained in the first and the second tank, respectively.

PROBLEM 1.6. A tank is supplied with water via a serpentine cooler and a funnel (see Figure 1.8).

The serpentine cooler introduces a delay equal to $2s$. The flow rate is limited to $1l/h$ by the funnel. The accumulated inflow rate from 0 to t is denoted by $u(t)$ and is equal to zero for $t \leq 0$. The volume of water collected in the tank is denoted by $y(t)$, and $y(0) = 3l$. Determine the expression relating $y(t)$ to $u(t)$.

SOLUTION 1.6. If the inflow $u(t - 2)$ is less than 1, then:

$$y(t) = u(t - 2) + y(0) = u(t - 2) + 3. \tag{1.6}$$

If $u(t - 2) > 1$, then:

$$y(t) = y(t - 1) + 1. \tag{1.7}$$

Combining Equations (1.6) and (1.7) yields:

$$y(t) \leq \min\left(u(t - 2) + 3, y(t - 1) + 1\right).$$

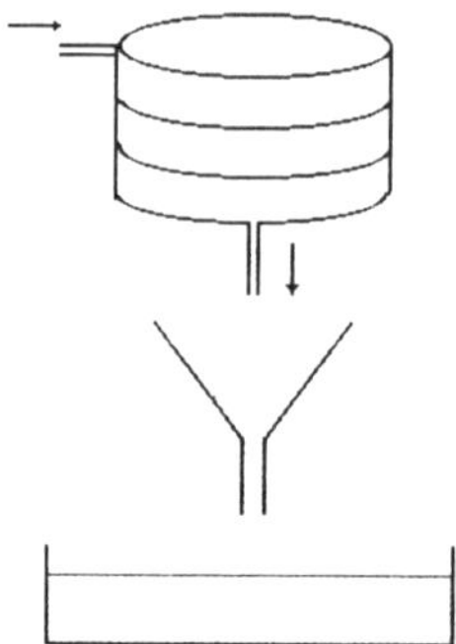

Figure 1.8. *System with delay*

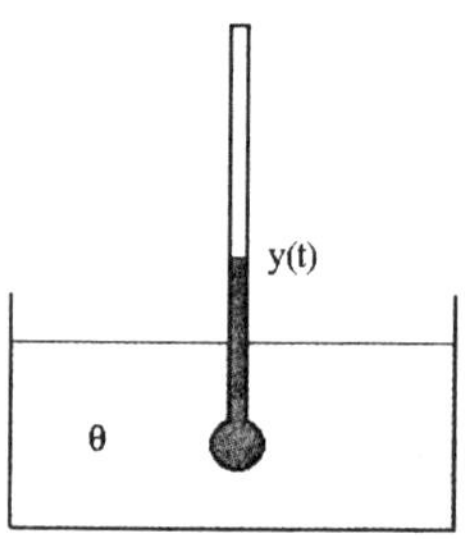

Figure 1.9. *Mercury thermometer*

PROBLEM 1.7. A mercury thermometer is used to measure the temperature $\theta(t)$ of a liquid contained in a tank. Initially, the thermometer is at the ambient temperature. Derive a model describing the evolution of the temperature of the mercury contained in this thermometer. A schematic diagram of this system is depicted in Figure 1.9.

SOLUTION 1.7. The heat transfer between the liquid and the thermometer (mercury) occurs by conduction. Let us denote by $F(t)$ the flow rate of the heat $Q(t)$ transferred to the mercury. We obtain:

$$F(t) = \frac{dQ(t)}{dt},$$

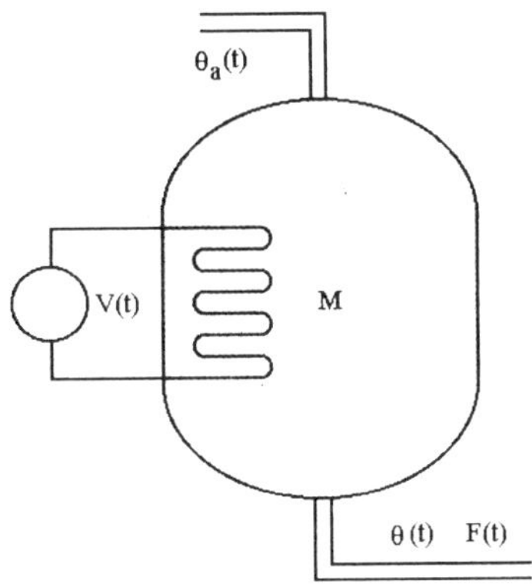

Figure 1.10. *Domestic water heating system*

which, from Newton's law, is proportional to the gradient of the temperature, i.e.:

$$F(t) = \alpha \left(\theta(t) - y(t) \right), \tag{1.8}$$

where $y(t)$ represents the temperature of the mercury. Notice also that the variation of $y(t)$ is a linear function of the heat flow rate $F(t)$:

$$F(t) = \beta \frac{dy(t)}{dt}, \tag{1.9}$$

where β represents the calorific capacity of the thermometer. From Equations (1.8) and (1.9), we derive:

$$\beta \frac{dy(t)}{dt} = \alpha \left(\theta(t) - y(t) \right),$$

$$y(t) + \frac{\beta}{\alpha} \frac{dy(t)}{dt} = \theta(t).$$

The static gain and the time constant of this first-order system are equal to $k = 1$ and $T = \beta/\alpha$, respectively.

PROBLEM 1.8. Consider the domestic water-heating system depicted in Figure 1.10. Derive its model under the following assumptions: (i) the thermal resistance of the water container and its insulation is equal to R; (ii) the electrical energy is totally converted into heat energy.

SOLUTION 1.8. We shall express the fact that the energy $V(t)$ provided by the electrical system is used (i) to heat the water (flow rate $F(t)$) from the ambient temperature

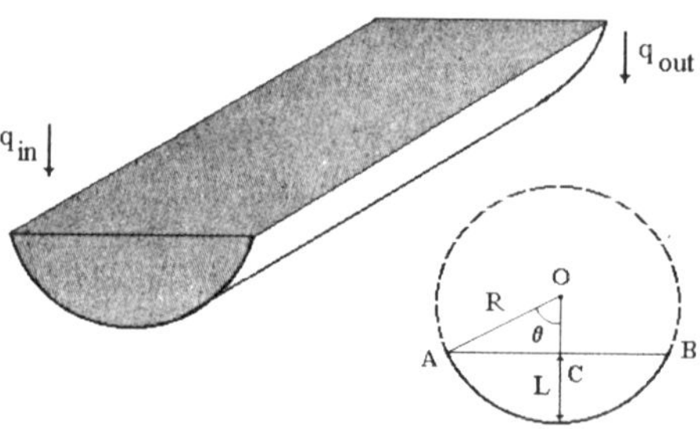

Figure 1.11. *Flow in a cylindrical tube*

θ_a (assumed to be constant) to the desired temperature $\theta(t)$, and (ii) to compensate the thermal loss (the energy exchanged by the water contained in the system with its environment). The remaining energy increases (or decreases) the temperature of the water contained in the heating system. We assume that the temperature of the inner wall of this heating system is equal to the temperature $\theta(t)$ of the heated water:

$$V(t) + F(t)(\theta(t) - \theta_a) + \frac{\theta(t) - \theta_a}{R} = M\frac{d\theta(t)}{dt}. \tag{1.10}$$

where M denotes the mass of water contained in the domestic water-heating system.

PROBLEM 1.9. Consider the flow of a water in a sloping irrigation channel of semi-cylindrical form and of length Y (see Figure 1.11). Determine a model which relates the liquid level L in the channel to the input and output flow rates. Assume that the slope is negligible.

SOLUTION 1.9. For an interval of time dt, the mass balance leads to:

$$dV = q_{in}\, dt - q_{out}\, dt, \tag{1.11}$$

where V represents the volume of the liquid contained in the tube. In order to calculate the variation of this volume, let us first calculate the area of the circle located under the chord AB (the area of the sector AOB minus the area of the triangle AOB). From Figure 1.11, we obtain:

$$OC = R\cos\theta, \quad AC = R\sin\theta.$$

The area of the triangle AOB is equal to:

$$A_T = R^2\cos\theta\sin\theta.$$

The area A_S of the sector AOB is given by:

$$A_S = R^2\theta,$$

and the difference A_{TMS} between these two areas is given by:

$$A_{TMS} = R^2\left(\theta - \cos\theta\sin\theta\right).$$

Therefore, the volume of water contained in the tube is equal to:

$$V = R^2\left(\theta - \cos\theta\sin\theta\right)Y.$$

Let us now express this volume as a function of the level L; we obtain:

$$OC = R\cos\theta = R - L,$$

$$\cos\theta = \frac{R-L}{R}$$

and

$$V = R^2Y\left(\arccos\left(\frac{R-L}{R}\right) - \left(\frac{R-L}{R}\right)\sqrt{1 - \left(\frac{R-L}{R}\right)^2}\right).$$

Let us introduce the following variable change:

$$x = \frac{R-L}{R}, \quad dx = -\frac{1}{R}dL.$$

The variation of the liquid volume[4] is given by:

$$dV = 2Y\sqrt{L\left(2R - L\right)}dL. \tag{1.12}$$

From equations (1.11) and (1.12), we obtain:

$$\frac{dL}{dt} = \frac{q_{in} - q_{out}}{2Y\sqrt{L\left(2R - L\right)}},$$

which corresponds to the desired result.

The next problem deals with the modeling of a conical tank which is characterized by a varying gain, according to the tank level.

PROBLEM 1.10. Consider the conical tank depicted in Figure 1.12. Derive a phenomenological model of this system.

---

1. Recall that $\dfrac{d}{dx}\arccos x = -\dfrac{1}{\sqrt{1 - x^2}}$.

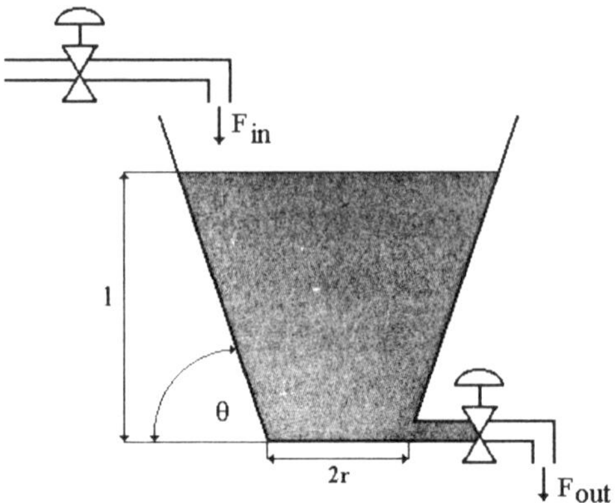

Figure 1.12. *Conical tank*

SOLUTION 1.10. During an interval of time dt, a mass balance analysis leads to the following equation:

$$F_{in}\,dt - F_{out}\,dt = \text{variation of the volume } V, \tag{1.13}$$

where V represents the volume of the liquid contained in the conical tank. The output flow rate F_{out} will be modeled on the basis of Bernoulli's law:

$$F_{out}(t) = k\sqrt{l\,(t)}.$$

Recall that the volume of a cone is equal to:

$$V_c = \frac{\pi}{3}R^2 h,$$

where R and h represent the radius of the base, and the height, respectively. From Figure 1.13, we obtain:

$$V = V_{cone\,1} - V_{cone\,2} = \frac{\pi}{3}R^2\,(l + l_1) - \frac{\pi}{3}r^2\,(l_1),$$

where:

$$R = r + \frac{l}{\tan\theta}, \quad l_1 = r\tan\theta.$$

It follows that:

$$V = \pi l\left[r^2 + \frac{lr}{\tan\theta} + \frac{l^2}{3\,(\tan\theta)^2}\right].$$

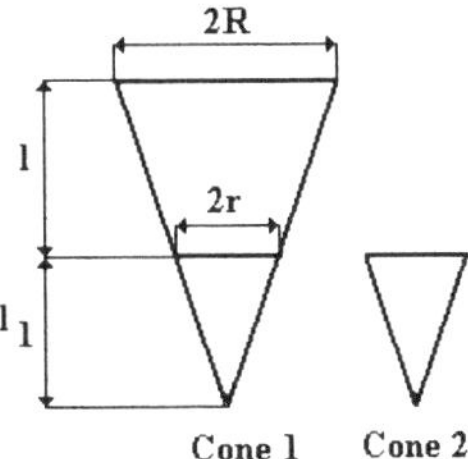

Figure 1.13. *Volume of water contained in the conical tank*

The final dynamic model is given by:

$$\frac{dl\,(t)}{dt} = \frac{F_{in}\,(t) - k\sqrt{l\,(t)}}{\pi\left[r^2 + 2lr/\tan\theta + l^2/(\tan\theta)^2\right]}. \tag{1.14}$$

This model is non-linear. It can be modelled by a Hammerstein structure (see Chapter 5).

The next problems concern the modeling process for chemical and biotechnological reactors. Notice that the reaction kinetics play an important role in this process [NAJ 89]. In a chemical reactor where two reactants A and B are involved, the reaction rate is expressed by:

$$r = kC_A^n C_B^n,$$

where m and n are the orders of reaction with respect to the components A and B, respectively. C_A and C_B represent the concentrations of the reactants A and B, respectively. The rate constant k is a function of the reactants and the temperature and depends on the presence of catalysts. The total order of the reaction considered is $m + n$. For both chemical and biotechnological bioreactors, the mass and energy balance equations are similar. The only difference concerns the kinetics; there exist specific kinetics for enzymatic reactions and the growth of microorganisms.

Control problems in biotechnological processes have gained increasing interest because of the great number of applications, mainly in the pharmaceutical industry and in biological depollution [NAJ 89]. The next problems concern the modeling of chemical reactors, of a fermentation process and of a distillation column which is used to separate the components contained in a mixture. Notice that batch processes are frequently utilized because of the inherent flexibility they possess in meeting market demand.

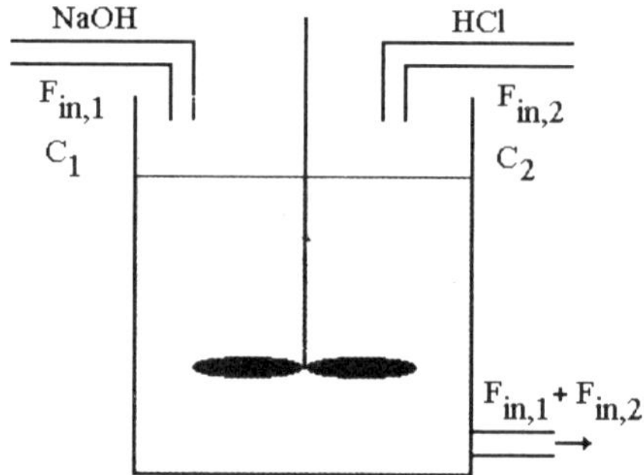

Figure 1.14. *Schematic diagram of a stirred tank reactor*

PROBLEM 1.11. Chemical reactors are used to manufacture a wide variety of materials. Consider a stirred tank reactor where neutralization takes place. In this chemical reaction, H^+ and OH^- combine to form H_2O (water) molecules, and the remaining components lead to a salt. A schematic diagram of a stirred chemical reactor used for a neutralization is depicted in Figure 1.14. The two input streams are sodium hydroxide ($NaOH$) and hydrochloric acid (HCl). The concentrations of $NaOH$ and HCl in these inputs are C_1 and C_2, respectively. The volume of the reactor is assumed to be constant and equal to V.

SOLUTION 1.11. Neutralization is a chemical reaction between acids and bases which produces a neutral solution ($pH = 7$) consisting of water and a salt. The neutralization reaction is accompanied by the production of heat, called the heat of neutralization. Some examples are given below:

$$HCl + KOH \rightarrow H_2O + KCl,$$
$$HNO_3 + KOH \rightarrow H_2O + KNO_3,$$
$$HCl + NaOH \rightarrow NaCl + H_2O,$$
$$Ca(OH)_2 + H_2CO_3 \rightarrow CaCO_3 + 2H_2O.$$

where HCl is hydrochloric acid, KOH is potassium hydroxide, KCl is potassium chloride, HNO_3 is nitric acid, KNO_3 is potassium nitrate, $NaOH$ is sodium hydroxide (caustic soda), $NaCl$ is sodium chloride (rock salt), $Ca(OH)_2$ is calcium hydroxide, $CaCO_3$ is calcium carbonate, and H_2CO_3 is carbonic acid.

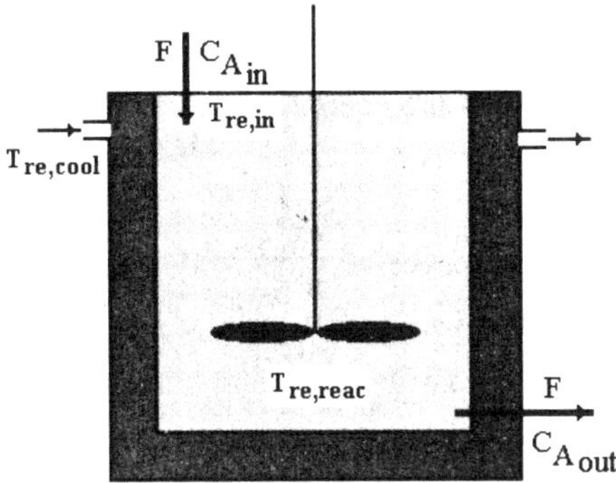

Figure 1.15. *Continuous stirred tank reactor*

Let us denote by x_1 and x_2 the concentrations of sodium ions Na^+ and chloride ions Cl^-, respectively. From material balance, we obtain:

$$V\frac{dx_1}{dt} = F_{in,1}C_1 - (F_{in,2} + F_{in,1})\,x_2,$$

$$V\frac{dx_2}{dt} = F_{in,2}C_2 - (F_{in,2} + F_{in,1})\,x_1.$$

The condition of electrical neutrality is expressed by:

$$[H^+] + [Na^+] = [Cl^-] + [OH]^-.$$

The heat of neutralization is equal to $56\ kJ/mole$.

PROBLEM 1.12. Consider a continuous stirred tank reactor (CSTR) where an exothermic chemical reaction takes place. A schematic diagram of this CSTR is shown in Figure 1.15. A component A is fed into the reactor and reacts with a component contained in the reactor. Derive a dynamic model for the process.

SOLUTION 1.12. Let us denote by V the volume of the reactor [CAL 88].

Mass balance. We express the fact that the variation of the quantity of material related to the component A is equal to the quantity fed into the reactor minus the quantity carried away by the output flow rate F, and minus the quantity which reacts with the reactant contained in the reactor:

$$V\frac{dC_A}{dt} = F\left(C_{A_{in}} - C_{A_{reac}}\right) - kC_{A_{out}}\exp\left(-\frac{E}{RT_{re,reac}}\right), \quad C_{A_{reac}} = C_{A_{out}}.$$

Energy balance. We express the fact that the rate of change of the energy stored by the reactor (a volume V of chemical products) is equal to the energy supplied by the quantity of chemical products associated with the flow rate F plus the energy produced by the chemical reaction minus the energy carried away by the coolant, minus the energy in the outflow:

$$\frac{dT_{re,reac}}{dt} = \frac{F}{V}\left(T_{re,in} - T_{re,reac}\right) + \frac{kC_{A_{reac}}}{\rho C_{p,reac}}\left(-\Delta H\right)\exp\left(-\frac{E}{RT_{re,reac}}\right)$$
$$-\frac{aC_h}{V\rho C_{p,reac}}\left(T_{re,out} - T_{re,cool}\right),$$

where the notation is given in Table 1.1.

The feedback linearization of this non-linear system is presented in [CAL 88]. Different techniques are used to transform a non-linear system into a linear system. The main commonly used approach is to linearize the non-linear system by transforming the input co-ordinate with a state feedback. The feedback linearization deals with the transformation of non-linear systems with control inputs but no input disturbances. However, in chemical industry, the processes (reactors, distillation columns, etc.) are affected by external disturbances. The methodology presented in [CAL 88] is based on a combined utilization of the mathematical machinery of feedforward/feedback linearization and internal model control approach.

$C_{A_{in}}$	Concentration of A in the inlet flow
$C_{A_{reac}}$	Concentration of A in the reactor
k	Frequency factor
E	Activation energy
R	Gas constant
$T_{re,in}$	Temperature of the inlet flow
$T_{re,reac}$	Temperature of the reactor content
ρ	Density of the reactor content
$C_{p,reac}$	Heat capacity of the reactor content
$T_{re,cool}$	Temperature of the cooling fluid
ΔH	Molar heat of the reaction
a	Overall heat transfer area
C_h	Overall heat transfer coefficient

Table 1.1. *Notation used in Solution 12*

PROBLEM 1.13. In biotechnological processes, fermentation, oxidation and/or reduction of a substrate (feedstuff) by micro-organisms such as yeasts and bacteria occur. Let us consider a continuous-flow fermentation process (see Figure 1.16). Derive its model.

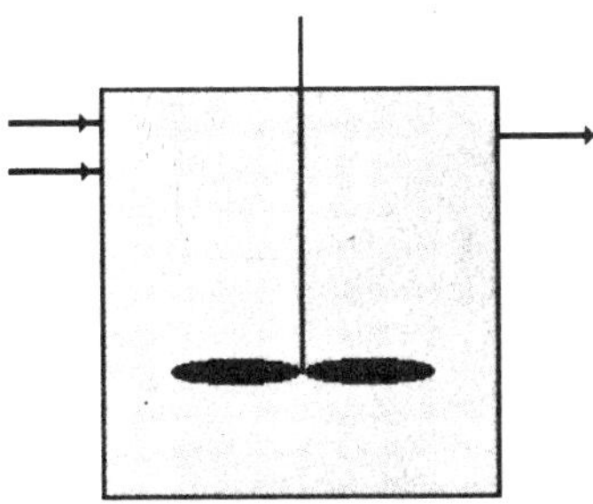

Figure 1.16. *Schematic diagram of a bioreactor*

SOLUTION 1.13. Let us denote by x, s, u, s_{in}, R and $\mu(x, s)$ the biomass concentration, the substrate concentration, the dilution rate, the substrate concentration in the inflow, the yield coefficient and the specific growth rate, respectively. From balance considerations we derive:

$$\frac{dx}{dt} = (\mu - u)\,x,$$
$$\frac{ds}{dt} = -\frac{1}{R}\mu x + u\,(s_{in} - s). \tag{1.15}$$

The specific growth rate is known to be a complex function of several parameters (biomass and substrate concentrations, pH, etc.). Many analytical formulae for the specific growth rate have been proposed in the literature [AND 68, MON 42, NAJ 89]. The Monod equation is frequently used as a kinetic description for the growth of micro-organisms and the formation of metabolic products:

$$\mu = \mu_{\max}\frac{s}{K_M + s}, \tag{1.16}$$

where $\mu_{\max}$ is the maximum growth rate and K_M is the Michaelis–Menten constant [NAJ 89]. Equations (1.15) and (1.16) show clearly that this model is non-linear, but many applications of control theory have been carried out on the basis of linear models of the input-output form and have led to good control performance.

PROBLEM 1.14. Consider a lake where some chemical product is being poured in. Derive a model governing the pollution concentration in this lake (see Figure 1.17).

SOLUTION 1.14. We assume that the volume V of the lake remains constant. A simple model for the pollution of a lake was given in [AGU 99]. On the basis of mass

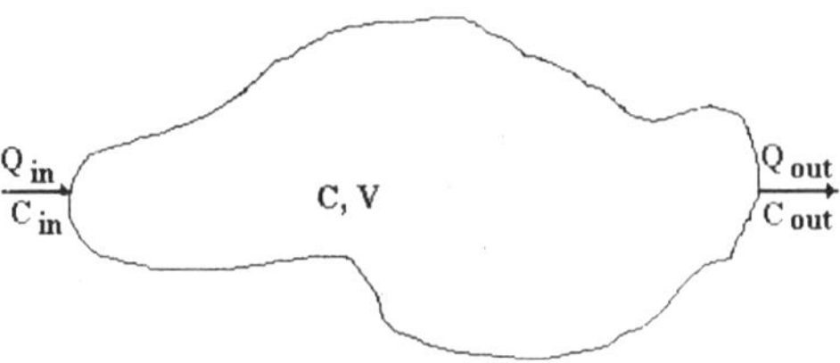

Figure 1.17. *Schematic diagram of a lake*

balance, we obtain:

$$\frac{dM}{dt} = F_{in}C_{in} - F_{out}C_{out} - kCV,$$

where M, k, F_{in}, F_{out}, C_{in} and C_{out} represent the mass of the contaminant, the rate of reaction of the chemical pollutant, the input flow rate, the output flow rate, the concentration of the contaminant at the input and the concentration of the contaminant at the output, respectively. C denotes the concentration of the contaminant in the lake. k represents a rate constant associated with a first-order chemical reaction.

PROBLEM 1.15. Distillation columns are commonly used in the petroleum, chemical and pharmaceutical industries as separation or purification units. Let us consider the binary (two-component) distillation column depicted in Figure 1.18. This column consists of N trays. Notice that a binary distillation column works as a still. Derive its dynamic model.

SOLUTION 1.15. The behavior of a distillation column is based on the fact that when a mixture, say $A + B$, is heated in the reboiler, then the component which is the most volatile is transformed first into vapor [RAD 75]. Therefore, to recover this component, it is sufficient to condense it.

A distillation column consists of a reboiler located at the bottom of the column, a column which contains a number of trays, and a condenser located at the top of the distillation column. The mixture enters the column near the center of the column and flows down. The vapors are condensed, and may be removed as overhead distillate or partially returned to the column as a reflux flow. In order to derive a dynamic model from mass and energy balances, we shall decompose the column into N stages, numbered from top to bottom ($i = 1, \ldots N$), including the condenser and the reboiler. The feed tray is numbered N_{in}. In order to make the modeling task easy, we shall assume that the vapor flow rate from tray to tray is constant over the column (no

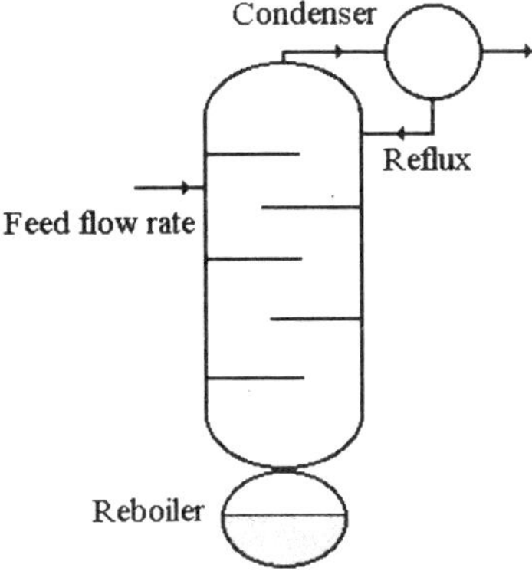

Figure 1.18. *Schematic diagram of a distillation column*

vapor hold-up on the trays), and the heat losses to the surroundings of the column are negligible. In view of these considerations we have only to derive mass conservation equations.

Mass balance. We consider four parts of the column separately, as follows.

1. Reboiler (bottom of column, stage 1). A schematic diagram of the bottom of the column is shown in Figure 1.19. Let us denote by M_b and x_b the mass of the liquid contained in the reboiler and the molar fraction of the component A in this liquid, respectively. The conservation of matter yields:

$$\frac{dM_b}{dt} = L_1 - V - F_b$$

$$\frac{dM_b x_b}{dt} = L_1 x_1 - V y_b - F_b x_b \quad \text{or} \quad \frac{dx_b}{dt} = \frac{1}{M_b}\left(L_1 x_1 - V y_b - F_b x_b\right). \tag{1.17}$$

The first equation expresses the fact that the variation of the mass M_b is equal to the mass carried by the flow rate L_1 minus the mass carried away by the vapor flow rate V minus the mass carried away by the flow rate F_b. The second equation is related to the variation of the molar fraction of the component A. In this equation, y_b and x_1 represent the molar fraction of the component A contained in the vapor flow rate V, and in the flows rate L_1 and F_b, respectively.

2. Stage i ($i = 1, ..., N - 2$). Figure 1.21 shows a schematic diagram of two consecutive stages (trays). From this figure, we obtain:

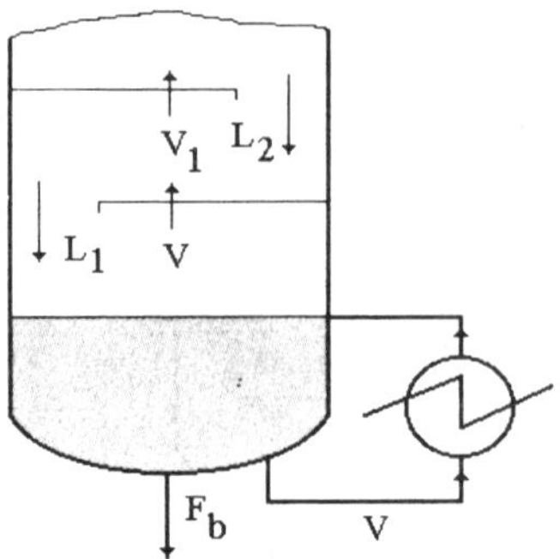

Figure 1.19. *Schematic diagram of the bottom of the distillation column*

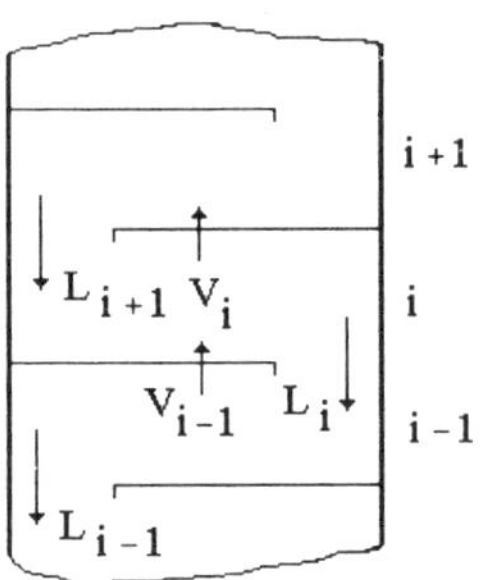

Figure 1.20. *Schematic diagram of two consecutive trays*

$$\frac{dM_i}{dt} = L_{i+1} - L_i + V_{i-1} - V_i = L_{i+1} - L_i,$$

$$\frac{dM_i x_i}{dt} = L_{i+1} x_{i+1} - L_i x_i + V_{i-1} y_{i-1} - V_i y_i \quad \text{or} \qquad (1.18)$$

$$\frac{dx_i}{dt} = \frac{1}{M_i}\left(L_{i+1} x_{i+1} - L_i x_i + V_{i-1} y_{i-1} - V_i y_i\right),$$

where L_i, V_i, x_i and y_i denote the flow rate of the liquid in each tray, the vapor flow rate, the molar fraction of the component A in the flow of rate L_i and the molar fraction of the component A contained in the vapor flow of rate V_i, respectively. Observe that,

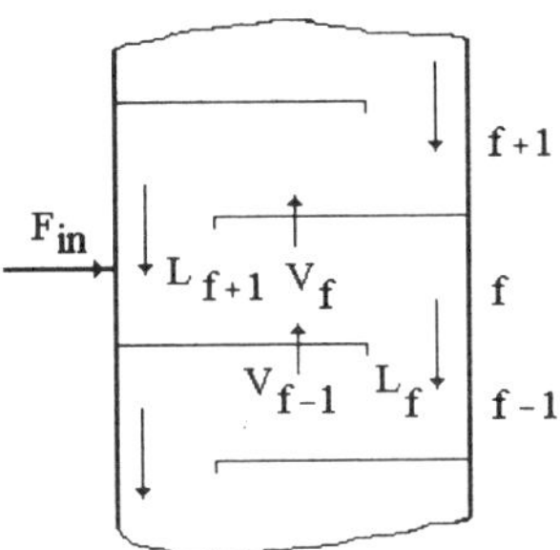

Figure 1.21. *Feeding plate*

in view of the assumption made above, $V_{i-1} - V_i = 0\ \forall i$. The last equation is obtained under the assumption that $M_i = const.$

3. Feed tray. A schematic diagram of the column, representing the neighborhood of feed tray, is depicted in Figure 1.21. From this figure, we derive the following differential equations:

$$\frac{dM_f}{dt} = F_{in} + L_{f+1} - L_f + V_{f-1} - V_f = F_{in} + L_{f+1} - L_f,$$
$$\frac{dM_f x_f}{dt} = F_{in} c_{in} + L_{f+1} x_{f+1} - L_f x_f + V_{f-1} y_{f-1} - V_f y_f \quad \text{or}$$
$$\frac{dx_f}{dt} = \frac{1}{M_f}\left(F_{in} c_{in} + L_{f+1} x_{f+1} - L_f x_f + V_{f-1} y_{f-1} - V_f y_f\right),$$

(1.19)

where c_{in} represents the molar fraction of the component A in the feed flow rate F_{in}. Now, it remains to model the top of the column.

4. Condenser (top of the column, stage N). The top of the column is depicted in Figure 1.22. From this schematic diagram, we obtain:

$$\frac{dM_c}{dt} = R - L_{N+1} + V_{N-1} - V_N = R - L_{N+1},$$
$$\frac{dM_c x_N}{dt} = R x_d - L_{N+1} x_N + V_{N-1} y_{N-1} - V_N y_N \quad \text{or}$$
$$\frac{dx_N}{dt} = \frac{1}{M_c}\left(R c_d - L_{N+1} x_N + V_{N-1} y_{N-1} - V_N y_N\right),$$

(1.20)

where R represents the reflux flow rate, and c_d is the composition with respect to the component A of both the reflux and the distillate stream.

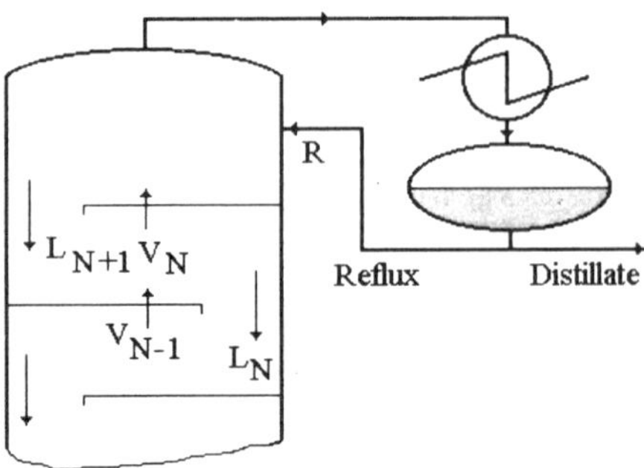

Figure 1.22. *Schematic diagram of the top of the distillation column*

To complete the derivation of the model of the binary distillation column, it remains to give the vapor composition at the various trays:

$$y_n = \frac{\alpha x_i}{1 + (\alpha - 1) x_i}, \quad i = 1, \cdots, N$$

and

$$y_b = \frac{\alpha x_b}{1 + (\alpha - 1) x_b},$$

where α represents the relative volatility, which is a function of temperature.

PROBLEM 1.16. Rapid thermal processing (RTP) is commonly used in semiconductor manufacturing processes such as annealing, oxidation and chemical vapor deposition. A schematic diagram of an RTP system is shown in Figure 1.23. The heat is provided by incandescent lamps. The heat intensity (control action), the furnace temperature, the wafer temperature (the controlled variable) and the ambient temperature are denoted by u, T_F, T_w and T_a, respectively. Derive a simplified dynamic model of this RTP system.

SOLUTION 1.16. We shall present a simplified model (see [GOR 02]). More detailed and complex models can be found in [BOR 90, NOR 92]. Observe that radiation is the main way in which heat is transferred. Heat transfer by radiation is governed by the Stefan–Boltzmann law:

$$Q = \alpha \left(T_r^4 - T^4 \right),$$

where Q, T_r and T represent the radiated heat (energy), the temperature of the radiator and the temperature of the receiver (the surroundings), respectively. From energy

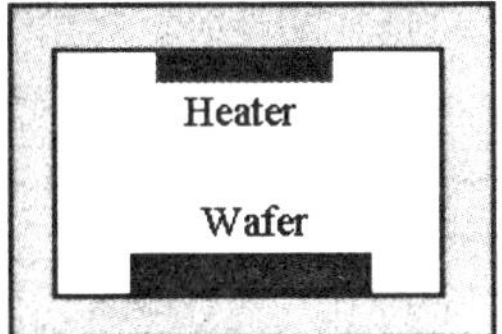

Figure 1.23. *Rapid thermal processing system*

balance, we obtain:

$$u\left(t\right) - \alpha_{FW}\left(T_F^4 - T_W^4\right) - \alpha_{FA}\left(T_F - T_a\right) \quad = \quad \beta_1\frac{dT_F}{dt}, \tag{1.21}$$

$$\beta_2\left(T_F^4 - T_W^4\right) \quad = \quad \beta_2\frac{dT_w}{dt}, \tag{1.22}$$

where $\alpha_{FW}\left(T_F^4 - T_W^4\right)$ represents the heat transmitted by radiation to the wafer, and $\alpha_{FA}\left(T_F - T_a\right)$ represents the heat transmitted by conduction to the surroundings of the system (losses).

If the right-hand side of Equation (1.21) is positive, the wafer temperature increases, and if this right-hand side is negative, the temperature decreases. Equation (1.22) expresses the fact that the variation of the wafer temperature is proportional to the radiated heat received by the wafer.

Observe that the term:

$$\left(T + 273\right)^4$$

can be approximated as follows:

$$\left(T + 273\right)^4 = \left(T + \alpha\right)^4 = \alpha^4\left(1 + \frac{T}{\alpha}\right)^4 \simeq \alpha^4\left(1 + 4\frac{T}{\alpha} + \cdots\right) \simeq \alpha^4 + 4\alpha^3 T.$$

Observe that the main non-linearity encountered in chemical engineering is of product form. If we are dealing with a flow rate $F\left(t\right)$ of a mixture of, for example, three components A, B and C, and if we are interested in the component B, we have to consider the quantity:

$$F\left(t\right)C_B\left(t\right),$$

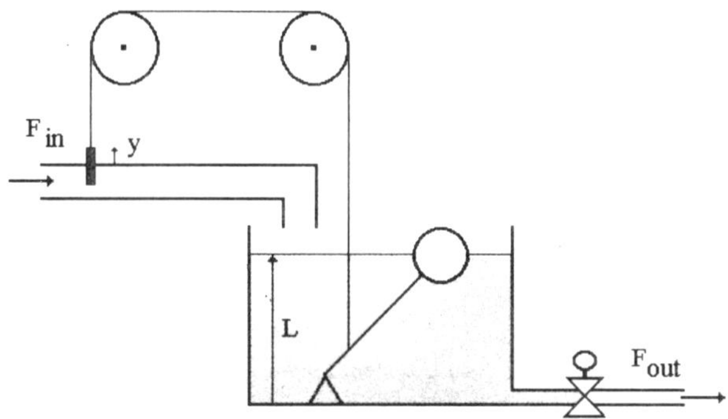

Figure 1.24. *Level regulation for a toilet*

where $C_B(t)$ represents the concentration of the component B. Observe also that the energy carried away by a flow rate $F(t)$ is proportional to $F(t)\,\theta(t)$, where $\theta(t)$ represents the temperature of the flow considered.

PROBLEM 1.17. Consider the level-regulation system depicted in Figure 1.24. This system represents a simplified scheme for a toilet. The ballcock (a spherical float at the end of a lever) is connected to a system for closing off the feed via two pulleys and a cable. The variable y represents the opening ratio of the feeding system. The cross-section of the tank is assumed to be constant and equal to S. We assume that the length of the cable is such that when the feed valve is closed ($y = 0$), the level L is equal to the desired level L_d.

1. Derive a model for this regulation system.

2. Draw its block diagram.

SOLUTION 1.17. 1. From mass balance considerations, we obtain:

$$F_{in}\,dt - F_{out}\,dt = S\,dL,$$
$$S\frac{dL}{dt} = F_{in} - F_{out},$$
$$L = \frac{1}{S}\left(F_{in} - F_{out}\right)dt.$$

The valve can be modeled on the basis of Bernoulli's law.

2. The block diagram is shown in Figure 1.25. Observe that the tank of the toilet is modelled by an integrator. Integration can be physically interpreted as an accumulation or as a model of the experience accumulated by an operator. This interpretation

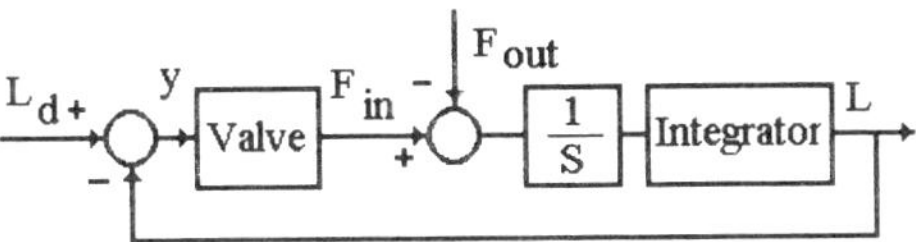

Figure 1.25. *Level regulation*

can be used to explain, partly, the introduction of the integral effect in the PID regulator, in the sense that the PID mimics the behavior of an operator.

1.2.2. *Mechanical systems*

In what follows, we shall consider some mechanical systems. On the basis of the Euler–Lagrange equations, models related to the dynamics of conservative mechanical systems (robots, etc.) can easily be derived. In Lagrangian mechanics, the Lagrangian is defined as the difference between the kinetic energy and the potential energy:

$$L\left(\mathbf{q}, \dot{\mathbf{q}}, t\right) = T\left(\mathbf{q}, \dot{\mathbf{q}}, t\right) - P\left(\mathbf{q}, \dot{\mathbf{q}}, t\right),$$

$$\mathbf{q}^T = [q_1, \cdots, q_n],$$

$$\dot{\mathbf{q}}^T = \left[\dot{q}_1, \cdots, \dot{q}_n\right],$$

$$\dot{q}_i = \frac{dq_i}{dt}$$

where q_i, $i = \overline{1, n}$, are generalized co-ordinates, and the $2n$-dimensional space with co-ordinates $\left(q_1, \cdots, q_n, \dot{q}_1, \cdots, \dot{q}_n\right)$ represents the position–velocity space. In Cartesian space, we have $q_1 = x$, $q_2 = y$, $q_3 = z$, $\dot{q}_1 = \dot{x}$, $\dot{q}_2 = \dot{y}$ and $\dot{q}_3 = \dot{z}$. These co-ordinates represent the components of the position and of the velocity.

The equations of motion are:

$$\frac{d}{dt}\left(\frac{\partial(T - P)}{\partial \dot{q}_i}\right) - \frac{\partial(T - P)}{\partial q_i} = F_i, \quad i = 1, \cdots, n, \tag{1.23}$$

where the F_i are the forces acting on the system. Observe that the Euler–Lagrange equations of motion are equivalent to Newton's laws of motion. If we consider a force F exerted on a particle of mass m, we obtain:

$$T = \frac{1}{2}mv^2,$$

and from (1.23) we derive:

$$\frac{d}{dt}\left(\frac{\partial(T-P)}{\partial\,\dot{q}_i}\right) - \frac{\partial(T-P)}{\partial q_i} = m\gamma = F, \quad \gamma = \frac{d}{dt}v,$$

which corresponds to Newton's law.

From the Euler–Lagrange equations, we may derive the Hamiltonian description by introducing the generalized momenta:

$$p_i\left(\mathbf{q},\dot{\mathbf{q}}\right) = \frac{\partial L}{\partial\,\dot{\mathbf{q}}}, \quad i = \overline{1,n}, \quad \mathbf{p}^T = [p_1,\cdots,p_n]. \tag{1.24}$$

The Hamiltonian is expressed as follows:

$$H(\mathbf{q},\mathbf{p},t) = \sum_{i=1}^{n} p_i\,\dot{q}_1 - L\left(\mathbf{q},\dot{\mathbf{q}},t\right). \tag{1.25}$$

The total derivative leads to:

$$dH = \sum_{i}^{n}\left(\frac{\partial H}{\partial q_i}dq_i + \frac{\partial H}{\partial p_i}dp_i\right) + \frac{\partial H}{\partial t}dt.$$

From (1.25), we derive:

$$dH = \sum_{i}\left(\dot{q}\,dp_i + p_i\,d\,\dot{q}_i - \frac{\partial L}{\partial q_i}dq_i - \frac{\partial L}{\partial\,\dot{q}_i}d\,\dot{q}_i\right) - \frac{\partial L}{\partial t}dt.$$

This, together with (1.25) and (1.24), yields[5]:

$$\frac{\partial H}{\partial t} = -\frac{\partial L}{\partial t},$$

and

$$\begin{cases} \dot{q}_i = \dfrac{\partial H}{\partial p_i}, i = \overline{1,n}, \\[2mm] \dot{p}_i = -\dfrac{\partial H}{\partial q_i}, i = \overline{1,n}. \end{cases}$$

---

5. The term $p_i\,d\,\dot{q}_i - \partial L/\partial\,\dot{q}_i\,d\,\dot{q}_i$ is equal to zero. Indeed, $p_i = \partial L/\partial\,\dot{q}_i$.

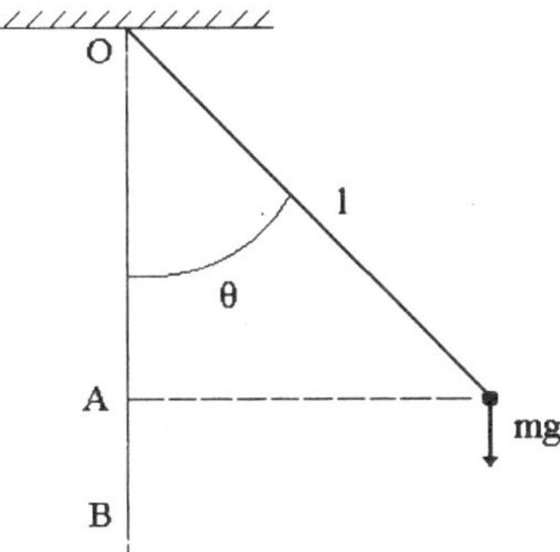

Figure 1.26. *Simple pendulum*

These $2n$ first-order differential equations correspond to the equations of motion (Hamilton's equations).

The next problems illustrate the application of Lagrangian mechanics.

PROBLEM 1.18. Consider the pendulum depicted in Figure 1.26. Derive its equation of motion.

SOLUTION 1.18. Let us assume that friction is negligible and consider the co-ordinate defined by θ, the angle of deflection of the rod. The kinetic energy and potential energy are given by:

$$T = \frac{1}{2}m\left(l\frac{d\theta}{dt}\right)^2, \quad P = mg\,|AB|,$$

$$|AB| = |OB| - |OA| = l - l\cos\theta \implies P = mgl\left(1 - \cos\theta\right),$$

where $ld\theta/dt$ represents the angular velocity. The initial value (0) of the potential energy corresponds to the vertical position of the pendulum (B). The Euler–Lagrange

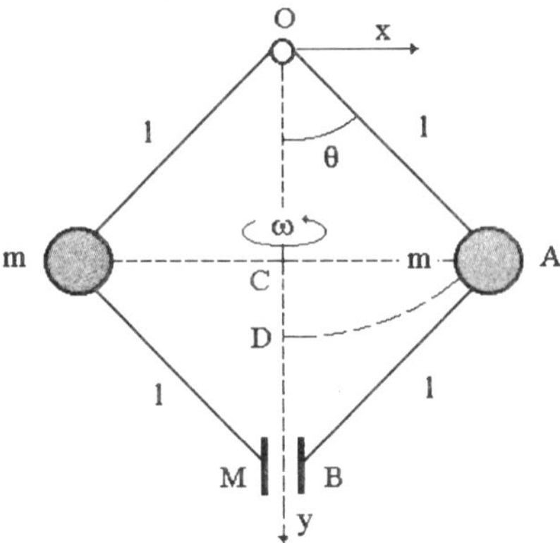

Figure 1.27. *Schematic diagram of a governor*

equation leads to the following equation of motion:

$$\frac{d}{dt}\left(\frac{\partial\,(T-P)}{\partial\,\dot{\theta}}\right)=\frac{\partial\,(T-P)}{\partial\theta}, \quad x=\theta \quad \text{and} \quad v=\frac{d\theta}{dt},$$

$$\frac{d}{dt}\left(\frac{1}{2}ml^2\left(\frac{d\theta}{dt}\right)^2\right)-(-mgl\sin\theta)=0, \tag{1.26}$$

$$l\frac{d^2\theta}{dt^2}+g\sin\theta=0.$$

For small values of θ, equation (1.26) can be linearized as follows:

$$\sin\theta\simeq\theta\implies l\frac{d^2\theta}{dt^2}+g\theta=0.$$

PROBLEM 1.19. A schematic diagram of a governor is shown in Figure 1.27. Governors were developed in the 18th century. They were used to operate machinery (steam engines) at constant speed. The governor represents an early controller designed for steam engines. It is connected to a throttle valve (a valve for regulating the supply of steam to an engine) and to the prime mover (not shown). The action of the governor depends on centrifugal force. The governor shown consists of two masses m rotating around the vertical (symmetry) axis at an angular velocity ω, four massless arms of length l, and a collar of mass M which slides along the vertical shaft. Derive a model for this system.

SOLUTION 1.19. Let us consider the co-ordinates of the masses m and M:

$$A\left(\begin{array}{l} x = l\sin\theta \\ y = l\cos\theta \end{array}\right), \quad B\left(\begin{array}{l} x = 0 \\ y = 2l\cos\theta \end{array}\right).$$

The potential energies associated with the two masses m and M are given by:

$$P_m = mgl\left(1 - \cos\theta\right), \quad P_M = 2Mgl\left(1 - \cos\theta\right).$$

The total potential energy is:

$$P = 2P_m + P_M = 2gl\left(m + M\right)\left(1 - \cos\theta\right).$$

The velocity of the mass m has three components:

$$v_A = \left(\begin{array}{c} \dot{x} \\ \dot{y} \\ \omega r \end{array}\right) = \left(\begin{array}{c} l\,\dot{\theta}\cos\theta \\ -l\,\dot{\theta}\sin\theta \\ \omega r \end{array}\right), \quad \dot{\theta} = \frac{d\theta}{dt},$$

where $r = AC = l\sin\theta$. The velocity of the mass M is given by:

$$v_B = \left(\begin{array}{c} 0 \\ \dfrac{d\left(2l\cos\theta\right)}{dt} = -2l\left(\sin\theta\right)\dfrac{d\theta}{dt} \end{array}\right).$$

The total kinetic energy is the sum of the kinetic energies of the two masses m $(T_m + T_m)$ and M (T_M):

$$T_m = \frac{1}{2}mv_A^2 = \frac{1}{2}m\left(\left(l\,\dot{\theta}\cos\theta\right)^2 + \left(-l\,\dot{\theta}\sin\theta\right)^2 + (\omega r)^2\right)$$

$$= \frac{1}{2}ml^2\left(\left(\dot{\theta}\right)^2 + \omega^2\sin\theta^2\right),$$

$$T_M = \frac{1}{2}Mv_B^2 = \frac{1}{2}M\left(-2l\left(\sin\theta\right)\dot{\theta}\right)^2 = 2Ml^2\left(\sin\theta\right)^2\left(\dot{\theta}\right)^2,$$

$$T = 2T_m + T_M = m\left(l^2\dot{\theta}^2 + \omega^2 l^2\left(\sin\theta\right)^2\right) + 2Ml^2\left(\sin\theta\right)^2\left(\dot{\theta}\right)^2$$

The Lagrangian function is given by:

$$L := T - P.$$

The generalized variables are θ, the angle between the vertical axis and the link OA, and φ, the angle of rotation around the axis y; $\dot{\varphi} = \omega$.

The equations of motion are given by:

$$\left\{\begin{array}{l} \dfrac{d}{dt}\dfrac{\partial}{\partial\dot{\theta}}L - \dfrac{\partial}{\partial\theta}L = 0, \\[2ex] \dfrac{d}{dt}\dfrac{\partial}{\partial\dot{\varphi}}L - \dfrac{\partial}{\partial\varphi}L = 0. \end{array}\right.$$

We have:

$$\begin{cases} \dfrac{\partial}{\partial \dot\theta} L = 2ml^2\,\dot\theta + 4Ml^2 (\sin\theta)^2\,\dot\theta, \\[2mm] \dfrac{\partial}{\partial \dot\varphi} L = \dfrac{\partial}{\partial \omega} L = 2ml^2\omega\,(\sin\theta)^2, \end{cases}$$

$$\begin{cases} \dfrac{\partial}{\partial \theta} L = \left(2m\omega^2 l^2 + 4Ml^2 \left(\dot\theta\right)^2\right)\sin\theta\cos\theta - 2gl\,(m+M)\sin\theta, \\[2mm] \dfrac{\partial}{\partial \varphi} L = 0. \end{cases}$$

So,

$$0 = \frac{d}{dt}\frac{\partial}{\partial \dot\theta}L - \frac{\partial}{\partial \theta}L$$

$$= 2ml^2\,\ddot\theta + 4Ml^2\left[(\sin\theta)^2\,\ddot\theta + 2\sin\theta\cos\theta\left(\dot\theta\right)^2\right] - \frac{\partial}{\partial\theta}L$$

$$= 2\left[ml^2 + 2Ml^2(\sin\theta)^2\right]\ddot\theta + 4Ml^2\left(\dot\theta\right)^2 \sin(2\theta)$$

$$- \left(m\omega^2 l^2 + 2Ml^2\left(\dot\theta\right)^2\right)\sin(2\theta) + 2gl\,(m+M)\sin\theta$$

and

$$0 = \frac{d}{dt}\frac{\partial}{\partial \dot\varphi}L - \frac{\partial}{\partial \varphi}L = 4m\omega l\,\dot\theta\sin\theta\cos\theta. \tag{1.27}$$

Equation (1.27) is no of interest. Finally, we obtain:

$$2\left[ml^2 + 2Ml^2(\sin\theta)^2\right]\ddot\theta + 2Ml^2\left(\dot\theta\right)^2\sin(2\theta) - \left(m\omega^2 l^2\right)\sin(2\theta)$$
$$+2gl\,(m+M)\sin\theta = 0.$$

PROBLEM 1.20. Consider the frictionless double pendulum (rigid two-link robot manipulator) [NIJ 90] depicted in Figure 1.28. We assume that the links are massless.

The control torques applied at the joints are denoted by u_1 and u_2. Derive a dynamic model for this system.

SOLUTION 1.20. We shall use the Lagrangian approach. The kinetic energy consists of two terms, associated with the masses m_1 and m_2. These terms will be denoted by $T_1(\theta_1)$ and $T_2(\theta_2)$, respectively. We have to find first the co-ordinates of the points A and B corresponding to the positions of the masses m_1 and m_2, respectively. Let us use the rectangular co-ordinates (x, y). From Figure 1.29, we derive:

$$\overrightarrow{OB} = \overrightarrow{OA} + \overrightarrow{AB}, \quad \overrightarrow{OA} = \begin{bmatrix} l_1\cos\theta_1 \\ l_1\sin\theta_1 \end{bmatrix}, \quad \overrightarrow{AB} = \begin{bmatrix} l_2\cos(\theta_1+\theta_2) \\ l_2\sin(\theta_1+\theta_2) \end{bmatrix},$$

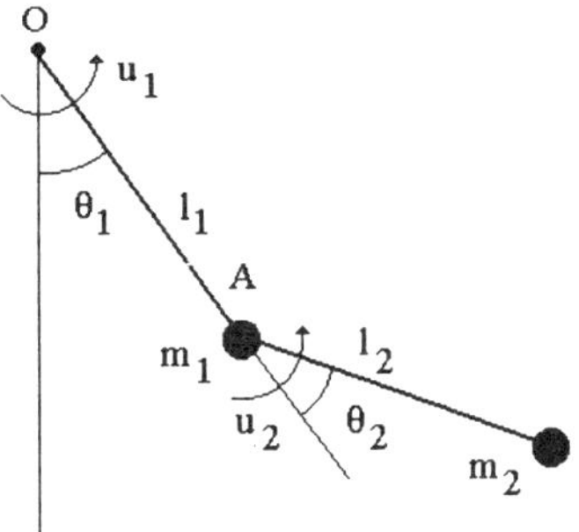

Figure 1.28. *Schematic diagram of a double pendulum*

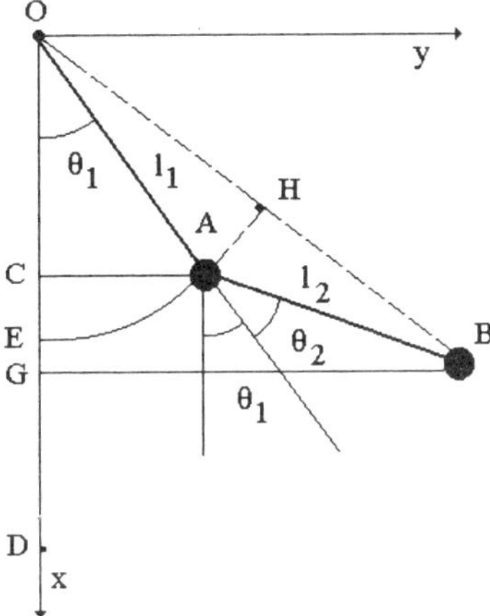

Figure 1.29. *Analysis of the vectors related to the double pendulum*

$$\overrightarrow{OB} = \begin{bmatrix} l_1 \cos\theta_1 + l_2 \cos(\theta_1 + \theta_2) \\ l_1 \sin\theta_1 + l_2 \sin(\theta_1 + \theta_2) \end{bmatrix}. \tag{1.28}$$

The above considerations are the main calculations involving the geometry and algebra of vectors that need to be done for the modeling of a robot. These calculations give the spatial positions from which the velocities are derived. The velocities of the masses m_1 and m_2 resulting from the action of forces acting over finite displacements are given by:

$$\vec{v_1} = \frac{d\overrightarrow{OA}}{dt} = \begin{bmatrix} -l_1 \sin \theta_1 \dfrac{d\theta_1}{dt} \\ l_1 \cos \theta_1 \dfrac{d\theta_1}{dt} \end{bmatrix},$$

$$|\vec{v_1}|^2 = \left(-l_1 \sin \theta_1 \frac{d\theta_1}{dt}\right)^2 + \left(l_1 \cos \theta_1 \frac{d\theta_1}{dt}\right)^2$$

$$|\vec{v_1}|^2 = v_1^2 = l_1^2 \left(\frac{d\theta_1}{dt}\right)^2,$$

$$\vec{v_2} = \frac{d\overrightarrow{OB}}{dt} = \begin{bmatrix} -l_1 \sin \theta_1 \dfrac{d\theta_1}{dt} - l_2 \sin(\theta_1 + \theta_2)\left(\dfrac{d\theta_1}{dt} + \dfrac{d\theta_2}{dt}\right) \\ l_1 \cos \theta_1 \dfrac{d\theta_1}{dt} + l_2 \cos(\theta_1 + \theta_2)\left(\dfrac{d\theta_1}{dt} + \dfrac{d\theta_2}{dt}\right) \end{bmatrix},$$

$$|\vec{v_2}|^2 = v_2^2 = \left[-l_1 \sin \theta_1 \frac{d\theta_1}{dt} - l_2 \sin(\theta_1 + \theta_2)\left(\frac{d\theta_1}{dt} + \frac{d\theta_2}{dt}\right)\right]^2$$

$$+ \left[l_1 \cos \theta_1 \frac{d\theta_1}{dt} + l_2 \cos(\theta_1 + \theta_2)\left(\frac{d\theta_1}{dt} + \frac{d\theta_2}{dt}\right)\right]^2$$

$$= l_1^2 \left(\frac{d\theta_1}{dt}\right)^2 + l_2^2 \left(\frac{d\theta_1}{dt} + \frac{d\theta_2}{dt}\right)^2 + 2l_1 l_2 \left(\frac{d\theta_1}{dt} + \frac{d\theta_2}{dt}\right) \frac{d\theta_1}{dt}$$

$$\times (\sin \theta_1 \sin(\theta_1 + \theta_2) + \cos \theta_1 \cos(\theta_1 + \theta_2)),$$

$$v_2^2 = l_1^2 \left(\frac{d\theta_1}{dt}\right)^2 + l_2^2 \left(\frac{d\theta_1}{dt} + \frac{d\theta_2}{dt}\right)^2 + 2l_1 l_2 \left(\frac{d\theta_1}{dt} + \frac{d\theta_2}{dt}\right) \frac{d\theta_1}{dt} \cos \theta_2.$$

Now, we are ready to calculate the kinetic energies:

$$T_1(\theta_1) = \frac{1}{2} m_1 v_1^2 = \tfrac{1}{2} m_1 l_1^2 \left(\frac{d\theta_1}{dt}\right)^2,$$

$$T_2(\theta_1, \theta_2) = \frac{1}{2} m_2 v_2^2 = \frac{1}{2} m_2 \left[l_1^2 \left(\frac{d\theta_1}{dt}\right)^2 + l_2^2 \left(\frac{d\theta_1}{dt} + \frac{d\theta_2}{dt}\right)^2 \right.$$

$$\left. + 2l_1 l_2 \left(\frac{d\theta_1}{dt} + \frac{d\theta_2}{dt}\right) \frac{d\theta_1}{dt} \cos \theta_2\right].$$

From Equation (1.28), we derive the gravitational potential energies of the double pendulum:

$$P_1(\theta_1) = m_1 g \left|\overrightarrow{CE}\right| = m_1 g l_1 (1 - \cos \theta_1),$$

$$P_2(\theta_1, \theta_2) = m_2 g \left|\overrightarrow{GD}\right|,$$

where $\left|\overrightarrow{ED}\right| = l_2$. Finally, we obtain:

$$P_2\left(\theta_1, \theta_2\right) = m_2 g\left[(l_1 + l_2) - l_1\cos\theta_1 - l_2\cos\left(\theta_1 + \theta_2\right)\right]$$

and

$$L\left(\theta, \dot{\theta}\right) = \frac{1}{2}m_1 l_1^2\left(\frac{d\theta_1}{dt}\right)^2 + \frac{1}{2}m_2\left[l_1^2\left(\frac{d\theta_1}{dt}\right)^2 + l_2^2\left(\frac{d\theta_1}{dt} + \frac{d\theta_2}{dt}\right)^2\right.$$

$$\left. + 2l_1 l_2\left(\frac{d\theta_1}{dt} + \frac{d\theta_2}{dt}\right)\frac{d\theta_1}{dt}\cos\theta_2\right] - m_1 g l_1\left(1 - \cos\theta_1\right) - m_2 g\left(l_1 + l_2\right)$$

$$+ m_2 g l_1\cos\theta_1 + m_2 g l_2\cos\left(\theta_1 + \theta_2\right), \quad \theta = \begin{bmatrix}\theta_1 \\ \theta_2\end{bmatrix}, \quad \dot{\theta} = \begin{bmatrix}\dot{\theta}_1 \\ \dot{\theta}_2\end{bmatrix}.$$

The Euler–Lagrange equations lead to:

$$\frac{d}{dt}\left(\frac{\partial L}{\partial\dot{\theta}_1}\right) - \frac{\partial L}{\partial\theta_1} = u_1, \quad L = T - P,$$

$$\frac{d}{dt}\left(\frac{\partial L}{\partial\dot{\theta}_2}\right) - \frac{\partial L}{\partial\theta_2} = u_1, \quad \dot{\theta}_i = \frac{d\theta_i}{dt}, \quad i = 1, 2,$$

$$\frac{\partial L}{\partial\dot{\theta}_1} = m_1 l_1^2\,\dot{\theta}_1 + m_2 l_1^2\,\dot{\theta}_1 + m_2 l_2^2\left(\dot{\theta}_1 + \dot{\theta}_2\right) + m_2 l_1 l_2\left(2\,\dot{\theta}_1 + \dot{\theta}_2\right)\cos\theta_2,$$

$$\frac{d}{dt}\left(\frac{\partial L}{\partial\dot{\theta}_1}\right) = m_1 l_1^2\,\ddot{\theta}_1 + m_2 l_2^2\left(\ddot{\theta}_1 + \ddot{\theta}_2\right) + m_2 l_1 l_2\left(2\,\ddot{\theta}_1 + \ddot{\theta}_2\right)\cos\theta_2$$

$$- m_2 l_1 l_2\left(2\,\dot{\theta}_1 + \dot{\theta}_2\right)\dot{\theta}_2\sin\theta_2 + m_2 l_1^2\,\ddot{\theta}_1,$$

$$\frac{\partial L}{\partial\theta_1} = -m_1 g l_1\sin\theta_1 - m_2 g l_1\sin\theta_1 - m_2 g l_2\sin\left(\theta_1 + \theta_2\right),$$

$$\frac{\partial L}{\partial\dot{\theta}_2} = m_2 l_2^2\left(\dot{\theta}_1 + \dot{\theta}_2\right) + m_2 l_1 l_2\left(\cos\theta_2\right)\dot{\theta}_1,$$

$$\frac{d}{dt}\left(\frac{\partial L}{\partial\dot{\theta}_2}\right) = m_2 l_2^2\left(\ddot{\theta}_1 + \ddot{\theta}_2\right) + m_2 l_1 l_2\left(\cos\theta_2\right)\ddot{\theta}_1 - m_2 l_1 l_2\sin\theta_2\,\dot{\theta}_1\dot{\theta}_2,$$

$$\frac{\partial L}{\partial\theta_2} = -m_2 l_1 l_2\,\dot{\theta}_1\left(\dot{\theta}_1 + \dot{\theta}_2\right)\sin\theta_2 - m_2 g l_2\sin\left(\theta_1 + \theta_2\right).$$

These equations may be written in a vector form as follows:

$$A\left(\theta\right)\ddot{\theta} + B\left(\theta, \dot{\theta}\right) + C\left(\theta\right) - U = 0, \tag{1.29}$$

where:

$$\ddot{\theta}=\begin{bmatrix}\dfrac{d^2\theta_1}{dt^2}\\[2mm]\dfrac{d^2\theta_2}{dt^2}\end{bmatrix},\quad U=\begin{bmatrix}u_1\\u_2\end{bmatrix},$$

$$A(\theta)=\begin{bmatrix}m_1l_1^2+\sum_{i=1}^{2}m_2l_i^2+2m_2l_1l_2\cos\theta_2 & m_2l_2^2+m_2l_1l_2\cos\theta_2\\[2mm]m_2l_2^2+m_2l_1l_2\cos\theta_2 & m_2l_2^2\end{bmatrix},$$

$$B\left(\theta,\dot{\theta}\right)=\begin{bmatrix}-m_2l_1l_2\left(\sin\theta_2\right)\dot{\theta}_2\left(2\,\dot{\theta}_1+\dot{\theta}_2\right)\\[2mm]m_2l_1l_2\left(\sin\theta_2\right)\left(\dot{\theta}_1\right)^2\end{bmatrix},$$

$$C(\theta)=\begin{bmatrix}m_1gl_1\sin\theta_1+m_2gl_1\sin\theta_1+m_2gl_2\sin\left(\theta_1+\theta_2\right)\\[2mm]m_2gl_2\sin\left(\theta_1+\theta_2\right)\end{bmatrix}.$$

Observe that if we consider the control strategy:

$$U=B\left(\theta,\dot{\theta}\right)+C(\theta)+A(\theta)V$$

which, with Equation (1.29), yields:

$$A(\theta)\,\ddot{\theta}+B\left(\theta,\dot{\theta}\right)+C(\theta)-B\left(\theta,\dot{\theta}\right)-C(\theta)-A(\theta)V=0,$$
$$A(\theta)\,\ddot{\theta}=A(\theta)V,$$

the determinant of the matrix $A(\theta)$ is given by:

$$\det A(\theta)=m_2l_2^2m_1l_1^2+m_2^2l_2^2l_1^2-m_2^2l_1^2l_2^2\cos^2\theta_2.$$

It follows that:

$$\ddot{\theta}=V,$$

which corresponds to linear dynamics (see, for instance, [NIJ 90]).

PROBLEM 1.21. Derive a dynamic model for and the transfer function of the mass–spring–damper system depicted in Figure 1.30. The stiffness, the damping and the displacement beyond equilibrium are denoted by k_0, c and y, respectively. Determine the damping factor, the natural frequency and the Q factor.

SOLUTION 1.21. The system considered is similar to a seismic transducer. The latter comprises two springs supporting a seismic mass m. Let us recall Newton's law

$$\sum\vec{F}=m\vec{\gamma}\tag{1.30}$$

where $\sum\vec{F}$ and $\vec{\gamma}$ represent the resultant of the forces acting upon the mechanical system, and the acceleration, respectively. The equation of motion follows quite

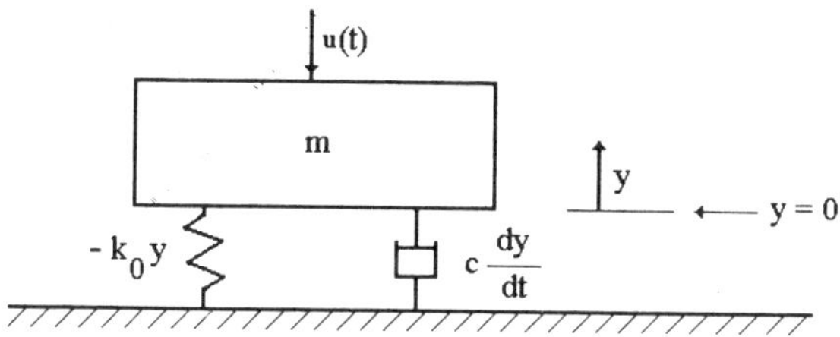

Figure 1.30. *Mechanical system*

straightforwardly from equation (1.30):

$$m\frac{d^2y}{dt^2} = -k_0 y - c\frac{dy}{dt} + u\left(t\right),$$

where $\overrightarrow{cdy/dt}$ and $k_0\,\overrightarrow{y}$ represent the frictional damping force, which is proportional to the velocity, and the force following the displacement of the spring from equilibrium, respectively. The parameter k_0 represents the stiffness, or spring constant, or modulus. The transfer function of this system is given by:

$$F\left(s\right) = \frac{Y\left(s\right)}{U\left(s\right)} = \frac{1}{ms^2 + cs + k_0}.$$

Recall that the transfer function of a second-order system is:

$$F\left(s\right) = \frac{k\omega_n^2}{s^2 + 2\zeta\omega_n s + \omega_n^2} = \frac{1/m}{s^2 + \left(c/m\right)s + k_0/m}.$$

By identification, we obtain:

$$k\omega_n^2 = \frac{1}{m}, \quad 2\zeta\omega_n = \frac{c}{m}, \quad \omega_n^2 = \frac{k_0}{m},$$

and

$$\omega_n = \sqrt{\frac{k_0}{m}}, \quad k = \frac{1}{k_0}, \quad \zeta = \frac{c}{2m\omega_n} = \frac{c}{2k_0}\sqrt{\frac{k_0}{m}}.$$

The Q factor which is a measure of the dissipation in a system, is given by:

$$Q = \frac{1}{2\zeta} = \frac{m}{c}\sqrt{\frac{k_0}{m}}.$$

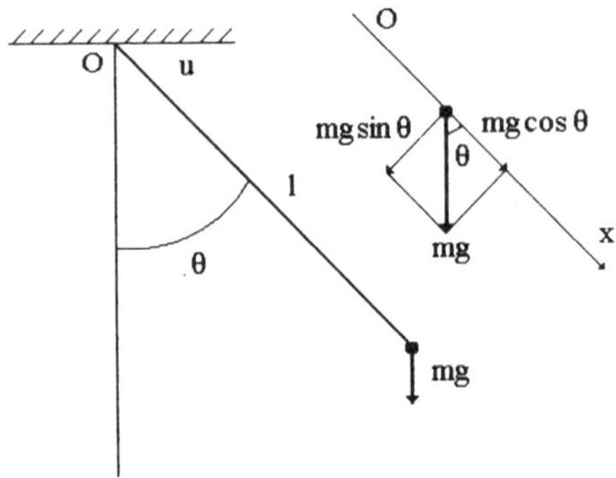

Figure 1.31. *Single-link manipulator*

Observe that in electronics, the damping factor is defined as the ratio between the nominal load impedance and the source impedance (the amplifier).

PROBLEM 1.22. Consider the single-link manipulator depicted in Figure 1.31, where θ is the angular position, m is a mass of the end of the rod , l is the length of the rod, v is the friction coefficient at the pivot point and u is the applied torque at the pivot point. Derive its model.

SOLUTION 1.22. From Figure 1.31 and Newton's law (momentum), we obtain:

$$\frac{d^2\theta}{dt^2} = -\frac{g}{l}\sin\theta - \frac{v}{ml^2}\frac{d\theta}{dt} + \frac{1}{ml^2}u.$$

For the control strategy:

$$u = ml^2\left(v + \frac{g}{l}\sin\theta\right),$$

we obtain:

$$\frac{d^2\theta}{dt^2} = -\frac{v}{ml^2}\frac{d\theta}{dt} + v,$$

which is a linearized system.

1.2.3. *Electrical systems*

PROBLEM 1.23. Determine models for the systems depicted in Figures 1.32 and 1.33. Show that these systems can be used as low- and high-pass filters. Answer the same question when the inductor is replaced by a capacitor.

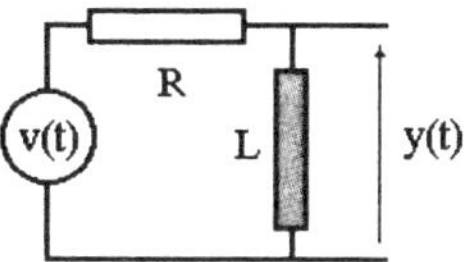

Figure 1.32. *Analogue filter (High-pass)*

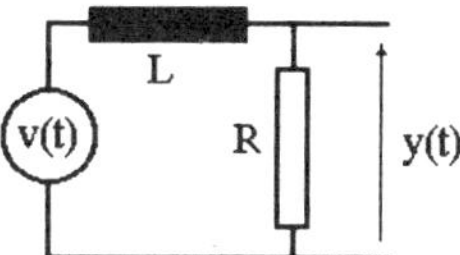

Figure 1.33. *Analogue filter (Low-pass)*

SOLUTION 1.23. The impedance of the inductor is $L\omega$. For low frequencies, $L\omega$ is close to zero. Therefore, the inductor behaves like a closed switch. For high frequencies, $L\omega \to \infty$, and the inductor behaves like an open switch, or, in other words, the impedance becomes infinite. Therefore, the system depicted in Figure 1.32 constitutes a high-pass filter.

From Figure 1.33, and following the comments made above, we derive the result that for low frequencies, the inductor acts as a closed switch, and for high frequency the inductor behaves like an open switch (infinite impedance). Consequently, this system represents a low-pass filter. Observe that this kind of reasoning and this kind of the behavior of an electronic circuit is commonly used by electronic engineers.

If the inductor is replaced by a capacitor, its impedance is given by:

$$z = \frac{1}{C\omega},$$

and tends to zero and infinity for $\omega \to \infty$ and 0, respectively. Therefore, for the system analogous to the system depicted in Figure 1.32, for high frequencies the capacitor behaves like a closed switch and for low frequencies it behaves like an open switch. This system corresponds to a low-pass filter. It readily follows that the second system is a high-pass filter.

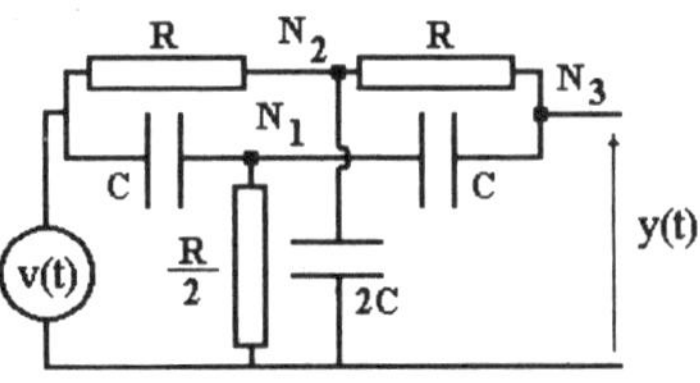

Figure 1.34. *Notch filter*

PROBLEM 1.24. Determine the equations governing the behavior of the network depicted in Figure 1.34.

SOLUTION 1.24. The electrical circuit shown in Figure 1.34 corresponds to a parallel-T notch filter. It is mainly used for rejecting a specific frequency (60 Hz or 50 Hz). We use the Kirchhoff's current law (node current law) which states that the sum of the currents flowing into any node is equal to the sum of all the currents flowing out of that node. For the various nodes, we obtain the following:

Node N_1:

$$[V_{N_1}(s) - V(s)] Cs + [V_{N_1}(s) - Y(s)] Cs + \frac{V_{N_1}(s)}{R/2} = 0. \qquad (1.31)$$

Node N_2:

$$[V_{N_2}(s) - V(s)] \frac{1}{R} + [V_{N_2}(s) - Y(s)] \frac{1}{R} + V_{N_2}(s) 2Cs = 0. \qquad (1.32)$$

Node N_3:

$$[Y(s) - V_{N_1}(s)] Cs + [Y(s) - V_{N_2}(s)] \frac{1}{R} = 0. \qquad (1.33)$$

Where $V_{N_1}(s)$, $V_{N_2}(s)$ and $V_{N_3}(s)$ are the Laplace transforms of the potentials associated with the nodes N_1, N_2 and N_3, respectively.

PROBLEM 1.25. A schematic diagram of a high-voltage line of length l is depicted in Figure 1.35.

High-voltage lines transport electrical power from generators at power plants to substations and, ultimately, consumers. Derive the equations governing the behavior of this system.

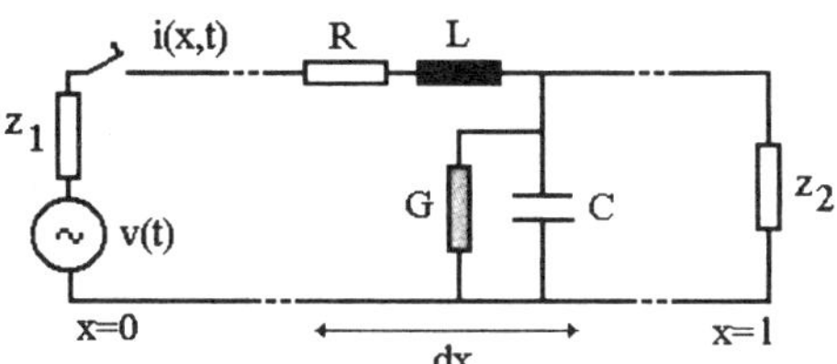

Figure 1.35. *Schematic diagram of a high-voltage line*

SOLUTION 1.25. Let us denote by R, L, C and G the resistance, inductance, capacitance and cross-conductance, respectively, per unit length of line, and consider an infinitesimal portion Δx of the line. From Ohm's law, we obtain:

$$v\left(x,t\right) - v\left(x + \Delta x, t\right) \;=\; L\,\Delta x \frac{di\left(x,t\right)}{dt} + R\,\Delta x\, i\left(x,t\right),$$

$$i\left(x,t\right) - i\left(x + \Delta x, t\right) \;=\; C\,\Delta x \frac{dv\left(x,t\right)}{dt} + G\,\Delta x\, v\left(x,t\right),$$

where $v\left(x,t\right)$ and $i\left(x,t\right)$ represent the voltage and the current along the line. For $\Delta x \to 0$, we obtain:

$$-\frac{dv\left(x,t\right)}{dx} = L\frac{di\left(x,t\right)}{dt} + Ri\left(x,t\right),$$
$$-\frac{di\left(x,t\right)}{dx} = C\frac{dv\left(x,t\right)}{dt} + Gv\left(x,t\right). \tag{1.34}$$

By taking the derivative with respect to x, we derive the following from the current equation:

$$-\frac{\partial^2 i\left(x,t\right)}{\partial x^2} = C\frac{\partial^2 v\left(x,t\right)}{\partial t\,\partial x} + G\frac{\partial v\left(x,t\right)}{\partial x}.$$

Taking the voltage equation (1.34) into account, we obtain:

$$-\frac{\partial^2 i\left(x,t\right)}{\partial x^2} = -C\frac{\partial}{\partial t}\left[L\frac{\partial i\left(x,t\right)}{\partial t} + Ri\left(x,t\right)\right] - G\left[L\frac{\partial i\left(x,t\right)}{\partial t} + Ri\left(x,t\right)\right],$$

$$LC\frac{\partial^2 i\left(x,t\right)}{\partial t^2} + \left(RC + LG\right)\frac{\partial i\left(x,t\right)}{\partial t} + RGi\left(x,t\right) = \frac{\partial^2 i\left(x,t\right)}{\partial x^2}.$$

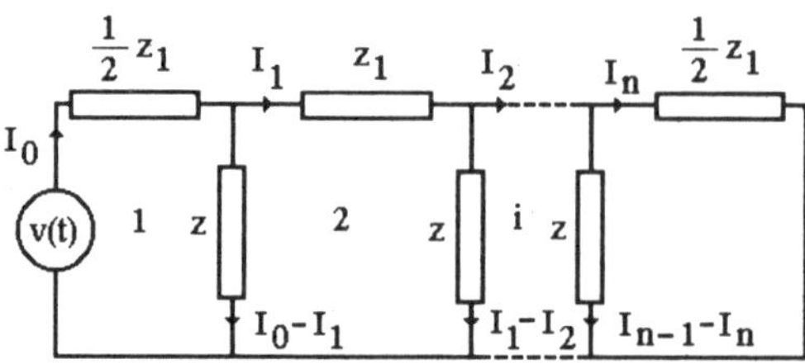

Figure 1.36. *Electrical circuit*

In the same manner (by taking the derivative of the voltage equation with respect to x), we obtain:

$$
\begin{aligned}
\frac{\partial^2 v\left(x,t\right)}{\partial x^2} &= L\frac{\partial}{\partial t}\left[C\frac{\partial v\left(x,t\right)}{\partial t} + Gv\left(x,t\right)\right] + R\left[C\frac{\partial v\left(x,t\right)}{\partial t} + Gv\left(x,t\right)\right] \\
&= LC\frac{\partial^2 v\left(x,t\right)}{\partial t^2} + [LG + RC]\frac{\partial v\left(x,t\right)}{\partial t} + RGv\left(x,t\right).
\end{aligned}
$$

PROBLEM 1.26. Determine the equations governing the behavior of the electrical system depicted in Figure 1.36.

For:

$$
z = \frac{1}{Cs}, \quad z_1 = R + Ls \quad \text{and } v\left(t\right) = v_0,
$$

calculate the Laplace transform of the current I_k.

SOLUTION 1.26. From Ohm's law, we obtain:

$$
V = \frac{1}{2}z_1 I_0 + z\left(I_0 - I_1\right), \tag{1.35}
$$

$$
z\left(I_0 - I_1\right) - z_1 I_1 - z\left(I_1 - I_2\right) = 0, \tag{1.36}
$$

$$
\cdots,
$$

$$
z\left(I_{n-1} - I_n\right) - \frac{1}{2}z_1 I_n = 0. \tag{1.37}
$$

For the $(k+1)$th loop of this circuit, we have:

$$(2z + z_1)\, I_k - z\,(I_{k+1} + I_{k-1}) = 0. \tag{1.38}$$

For the sake of simplicity of presentation, we have omitted the (s) from these equations. Observe that Equation (1.36) is a particular case of Equation (1.38) for $k = 1$. Let us look for solutions of the form[6]:

$$I_k = \alpha \exp(\lambda k) + \beta \exp(-\lambda k). \tag{1.39}$$

We have to calculate the parameters α, β and λ. If we insert Equation (1.39) into Equation (1.38), we obtain:

$$(2z + z_1)\,[\alpha \exp(\lambda k) + \beta \exp(-\lambda k)] - z\alpha \exp(\lambda(k+1))$$
$$-z\beta \exp(-\lambda(k+1)) - z\,[\alpha \exp(\lambda(k-1)) + \beta \exp(-\lambda(k-1))] = 0,$$
$$[\alpha \exp(\lambda k) + \beta \exp(-\lambda k)]\,[2z + z_1 - z \exp\lambda - z \exp(-\lambda)] = 0.$$

It follows readily that:

$$2z + z_1 - z \exp\lambda - z \exp(-\lambda) = 0. \tag{1.40}$$

Recall the expressions for the hyperbolic cosine and sine functions,

$$\cosh x = \frac{\exp x + \exp(-x)}{2} \quad \text{and} \quad \sinh x = \frac{\exp x - \exp(-x)}{2}.$$

---

6. Let us introduce the notation:

$$\Delta I_k = I_{k+1} - I_k, \quad \Delta^2 I_k = \Delta I_{k+1} - \Delta I_k,$$

which leads to:

$$\Delta^2 I_{k-1} = \Delta I_k - \Delta I_{k-1} = I_{k+1} - 2I_k + I_{k-1}.$$

The recurrence equation governing I_k may be written as follows:

$$(2z + z_1)\, I_k - z\,(I_{k+1} + I_{k-1})$$
$$= z_1 I_k - z\,(I_{k+1} + I_{k-1} - 2I_k) = 0,$$
$$z\,\Delta^2 I_{k-1} - z_1 I_k = 0.$$

We may associate a characteristic equation with this recurrence equation as for a differential equation. Therefore, the solution of this recurrence equation is the same as the solution of the differential equation:

$$\frac{d^2 y}{dt^2} = \alpha^2 y$$

the solution of which is of the form:

$$a \exp(\alpha t) + b \exp(-\alpha t).$$

Equation (1.40) may be written as follows:

$$1 + \frac{z_1}{2z} - \frac{1}{2}\exp\lambda - \frac{1}{2}\exp(-\lambda) = 1 + \frac{z_1}{2z} - \cosh\lambda = 0,$$

$$\cosh\lambda = 1 + \frac{z_1}{2z}. \tag{1.41}$$

The coefficients α and β are determined from the boundary conditions (1.35) and (1.37). Inserting expression (1.39) for I_0, I_1, I_n and I_{n-1} into Equations (1.35) and (1.37) gives:

$$V = \frac{1}{2}z_1(\alpha + \beta) + z(\alpha + \beta - \alpha\exp\lambda - \beta\exp(-\lambda)) \tag{1.42}$$

$$z(\alpha\exp(\lambda(n-1)) + \beta\exp(-\lambda(n-1)) - \alpha\exp(\lambda n) - \beta\exp(-\lambda n))$$
$$-\frac{1}{2}z_1(\alpha\exp(\lambda n) + \beta\exp(-\lambda n)) = 0$$

$$\tag{1.43}$$

Simple calculations lead to:

$$\frac{z_1}{2z}(\alpha + \beta) + (\alpha + \beta - \alpha\exp\lambda - \beta\exp(-\lambda)) = \frac{V}{z},$$

$$\alpha\left(\frac{z_1}{2z} + 1 - \exp\lambda\right) + \beta\left(\frac{z_1}{2z} + 1 - \exp(-\lambda)\right) = \frac{V}{z}. \tag{1.44}$$

We insert Equation (1.41) into this equation to obtain:

$$\alpha(\cosh\lambda - \exp\lambda) + \beta(\cosh\lambda - \exp(-\lambda)) = \frac{V}{z},$$

$$\alpha\left(\frac{\exp\lambda + \exp(-\lambda)}{2} - \exp\lambda\right) + \beta\left(\frac{\exp\lambda + \exp(-\lambda)}{2} - \exp(-\lambda)\right) = \frac{V}{z},$$

$$-\alpha\sinh\lambda + \beta\sinh\lambda = \frac{V}{z}$$

$$\tag{1.45}$$

and

$$z(I_{n-1} - I_n) - \frac{1}{2}z_1 I_n = 0$$

$$(\alpha\exp(\lambda(n-1)) + \beta\exp(-\lambda(n-1)) - \alpha\exp(\lambda n) - \beta\exp(-\lambda n))$$
$$-\frac{z_1}{2z}(\alpha\exp(\lambda n) + \beta\exp(-\lambda n)) = 0$$

$$\alpha\exp\lambda n\left[\exp(-\lambda) - 1 - \frac{z_1}{2z}\right] + \beta\exp(-\lambda n)\left[\exp\lambda - 1 - \frac{z_1}{2z}\right] = 0.$$

We obtain from Equation (1.41):

$$\alpha\exp\lambda n\left[\exp(-\lambda) - \cosh\lambda\right] + \beta\exp(-\lambda n)\left[\exp\lambda - \cosh\lambda\right] = 0,$$

$$\alpha\exp\lambda n\left[\frac{\exp(-\lambda) - \exp\lambda}{2}\right] + \beta\exp(-\lambda n)\left[\frac{\exp\lambda - \exp(-\lambda)}{2}\right] = 0,$$

$$-\alpha\exp\lambda n\sinh\lambda + \beta\exp(-\lambda n)\sinh\lambda = 0,$$

$$\alpha = \beta\exp(-2\lambda n). $$

$$\tag{1.46}$$

Finally, Equations (1.45) and (1.46) yield:

$$-\beta \exp\left(-2\lambda n\right) + \beta = \frac{V}{z \sinh \lambda},$$

$$\beta \left(1 - \exp\left(-2\lambda n\right)\right) = \beta \left(\exp\left(-\lambda n\right)\exp\left(\lambda n\right) - \exp\left(-2\lambda n\right)\right) = \frac{V}{z \sinh \lambda},$$

$$\beta = \frac{V \exp\left(\lambda n\right)}{2z \sinh \lambda \sinh \lambda n},$$

$$\alpha = \frac{V \exp\left(-\lambda n\right)}{2z \sinh \lambda \sinh \lambda n}.$$

In the above calculations, we have used the simple fact that $\exp x \exp\left(-x\right) = 1$. The expression for I_k is:

$$I_k = \frac{V \exp\left(-\lambda n\right)}{2z \sinh \lambda \sinh \lambda n} \exp \lambda k + \frac{V \exp\left(\lambda n\right)}{2z \sinh \lambda \sinh \lambda n} \exp\left(-\lambda k\right)$$

$$= \frac{V}{z} \frac{1}{2 \sinh \lambda \sinh \lambda n} \left(\exp\left(-\lambda n\right)\exp \lambda k + \exp\left(\lambda n\right)\exp\left(-\lambda k\right)\right), \qquad (1.47)$$

$$I_k = \frac{V}{z} \frac{\cosh\left(n - k\right)\lambda}{\sinh \lambda \sinh \lambda n}.$$

Now we shall determine the expression for I_k for a constant voltage v_0 we have:

$$z = \frac{1}{Cs} \quad z_1 = R + Ls \quad \text{and} \quad V\left(s\right) = \frac{v_0}{s}.$$

From Equations (1.41) and (1.47), we obtain:

$$s^2 + \frac{R}{L}s + \frac{2}{CL}\left(1 - \cosh \lambda\right) = 0$$

and

$$I_k(s) = Cv_0 \frac{\cosh\left(n - k\right)\lambda}{\sinh \lambda \sinh \lambda n} s.$$

Before we end this section, we shall present a simple model of the profits of two shopkeepers.

PROBLEM 1.27. Let us consider two shopkeepers (A and B) selling the same merchandise. We assume that their shops are very close together (i.e., in the same block of houses). Develop a model describing the evolution of their respective profits.

SOLUTION 1.27. Owing to the fact that the shops are very close together, the profits of each shopkeeper depend on the amount of merchandise sold by the other shopkeeper [CES 97]. It follows that:

$$\frac{dP_A}{dt} = \alpha_A P_A\left(t\right) + \beta_A P_B\left(t\right),$$

$$\frac{dP_B}{dt} = \alpha_B P_A\left(t\right) + \beta_B P_B\left(t\right),$$

where P_A and P_B represent the profits made by shopkeeper A and B, respectively.

The discussion below provides some insight into the linearization of non-linear models.

1.3. Linearization

In practice, processes are characterized by non-linear dynamics. Linear control theory, which is now well developed, does not permit the derivation of control strategies for this kind of system. Indeed, the analysis of non-linear systems is not nearly complete. This explains why linear models are commonly used for the control of processes. An approach which consists of considering the behavior of a process as linear around a given operating point is, fundamentally, an approximation of first order. From a mathematical point of view, this approximation corresponds to the linearization of the model. In what follows, we shall present some problems related to this question. In mechanics, and especially in robotics, a non-linearity of the form $\sin\theta$ appears frequently.

Let us consider a characteristic of a valve (see Figure 1.37). This characteristic is non-linear. Consider the operating point represented by the point A, which corresponds to a flow rate equal to y_a. If the variation of the opening ratio of the valve is small, we can, as an approximation of the first order, approximate the portion CD of this characteristic by the segment CD of the tangent to this characteristic at the point A. Consequently, a linear model of the form $y = a_A x + b_A$ can be used as a model of the valve. Observe that we have indexed the coefficient of x and the constant with A. This means that the approximation is valid only in the neighborhood of A, but the model structure $(y = ax + b)$ can be used to approximate the behavior of the valve around any operating point.

The characteristic of a valve can be modelled on the basis of Bernoulli's law. The derived non-linear model is of the form:

$$y = k\sqrt{x}.$$

Observe that:

$$y_a = k\sqrt{x_a},$$

and around this operating point we have:

$$y = y_a + \Delta y, \; x = x_a + \Delta x, \; y_a + \Delta y = k\sqrt{x_a + \Delta x} = k\sqrt{x_a\left(1 + \frac{\Delta x}{x_a}\right)} \simeq$$

$$k\sqrt{x_a}\left(1 + \frac{\Delta x}{2x_a}\right), \; \Delta y = k\sqrt{x_a}\,\frac{\Delta x}{2xa}, \; y = y_a + \Delta y = \frac{k}{2\sqrt{x_a}}\Delta x + k\sqrt{x_a},$$

$$y = \frac{k}{2\sqrt{x_a}}\Delta x + k\sqrt{x_a}.$$

$$(1.48)$$

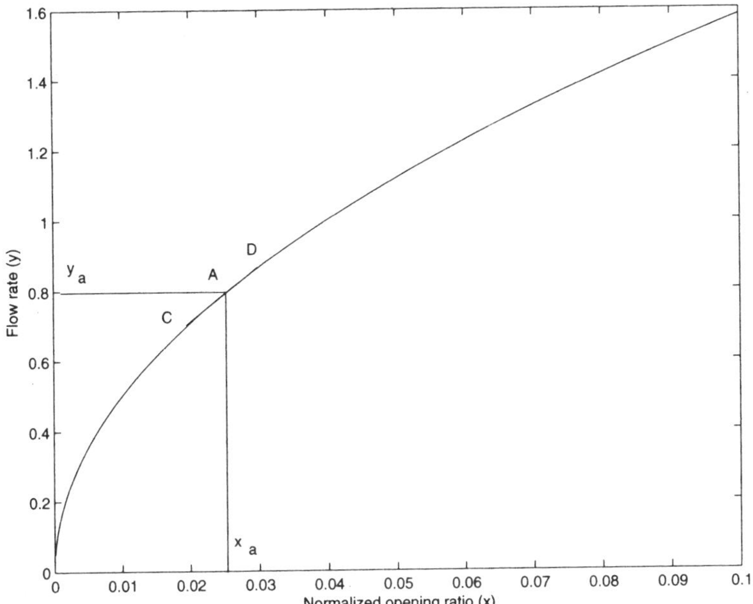

Figure 1.37. *Characteristics of a valve*

To obtain these expressions, we have used $\sqrt{1+\varepsilon} \simeq 1 + \varepsilon/2$. This linear expression corresponds to the equation of a straight line with a slope equal to $k/2\sqrt{x_a}$ and a y-intercept $k\sqrt{x_a}$. This slope is exactly the value of the derivative of $k\sqrt{x}$ at $x = x_a$, and we can easily verify that the co-ordinates of the point A satisfy this equation. Finally, we can show that Equation (1.48) represents the equation of the tangent to the characteristic of the valve at the point A.

PROBLEM 1.28. Linearize the model of a system consisting of two tanks (Problem 1.4) around an operating point defined by F_{in0}, L_{10} and L_{20}. We assume that the variables conditioning the behavior of this system fluctuate in the neighborhood of these values, namely $F_{in} = F_{in0} + f$, $L_1 = L_{10} + l_1$ and $L_2 = L_{20} + l_2$, where:

$$f \ll F_{in0}, \quad l_1 \ll L_{10} \quad \text{and } l_2 \ll L_2.$$

SOLUTION 1.28. In the Equations (1.5), the non-linearity is due to the presence of a square root. We can adopt one of two related solutions: a Taylor series expansion or the usual approximation [NAJ 99]

$$\sqrt[n]{1+\varepsilon} \simeq 1 + \frac{\varepsilon}{n}. \tag{1.49}$$

Let us consider again the model (1.5). For the steady-state operating point, we have:

$$\begin{cases} F_{in0} - k_1\sqrt{L_{10}} = A_1\dfrac{L_{10}}{dt} = 0, \\[2mm] k_1\sqrt{L_{10}} - k_2\sqrt{L_{20}} = A_2\dfrac{L_{20}}{dt} = 0. \end{cases}$$

In the previous derivation, we used the fact that L_{10} and L_{20} are constant. In general, we obtain:

$$\begin{cases} F_{in0} + f - k_1\sqrt{L_{10} + l_1} = A_1\dfrac{d\left(L_{10} + l_1\right)}{dt} = A_1\dfrac{dl_1}{dt}, \\[2mm] k_1\sqrt{L_{10} + l_1} - k_2\sqrt{L_{20} + l_2} = A_2\dfrac{d\left(L_{20} + l_2\right)}{dt} = A_2\dfrac{dl_2}{dt}, \end{cases}$$

$$\begin{cases} F_{in0} + f - k_1\sqrt{L_{10}\left(1 + \dfrac{l_1}{L_{10}}\right)} = A_1\dfrac{dl_1}{dt}, \\[2mm] k_1\sqrt{L_{10}\left(1 + \dfrac{l_1}{L_{10}}\right)} - k_2\sqrt{L_{20}\left(1 + \dfrac{l_2}{L_{20}}\right)} = A_2\dfrac{dl_2}{dt}. \end{cases}$$

Using the approximation (1.49), we derive the desired result:

$$\begin{aligned} f - k_1\sqrt{L_{10}}\dfrac{l_1}{2L_{10}} &= A_1\dfrac{dl_1}{dt}, \\[2mm] k_1\sqrt{L_{10}}\dfrac{l_1}{2L_{10}} - k_2\sqrt{L_{20}}\dfrac{l_2}{2L_{20}} &= A_2\dfrac{dl_2}{dt}. \end{aligned} \tag{1.50}$$

We shall now use the Taylor expansion to linearize this model. Let us rewrite the model equations in the following form:

$$\begin{aligned} g_1\left(f(t), l_1(t), \dfrac{dl_1(t)}{dt}\right) &= 0, \\[2mm] g_2\left(l_1(t), l_2(t), \dfrac{dl_2(t)}{dt}\right) &= 0, \end{aligned}$$

where:

$$\begin{aligned} g_1(\cdot) &= F_{in}(t) - k_1\sqrt{L_1(t)} - A_1\dfrac{dL_1(t)}{dt}, \\[2mm] g_2(\cdot) &= k_1\sqrt{L_1(t)} - k_2\sqrt{L_2(t)} - A_2\dfrac{dL_2(t)}{dt}. \end{aligned}$$

It follows that:

$$g_1(\cdot) \simeq \frac{\partial g_1(\cdot)}{\partial f(t)} f(t) + \frac{\partial g_1(\cdot)}{\partial l_1(t)} l_1(t) + \frac{\partial g_1(\cdot)}{\partial (dl_1(t)/dt)} \frac{dl_1(t)}{dt} = 0,$$

$$g_2(\cdot) \simeq \frac{\partial g_2(\cdot)}{\partial l_1(t)} l_1(t) + \frac{\partial g_2(\cdot)}{\partial l_2(t)} l_2(t) + \frac{\partial g_2(\cdot)}{\partial (dl_2(t)/dt)} \frac{dl_2(t)}{dt} = 0.$$

The derivatives involved in these equations are given by:

$$\frac{\partial g_1(\cdot)}{\partial f(t)} = 1, \quad \frac{\partial g_1(\cdot)}{\partial l_1(t)} = -\frac{k_1}{2\sqrt{L_{10} + l_1(t)}} \simeq -\frac{k_1}{2\sqrt{L_{10}}},$$

$$\frac{\partial g_1(\cdot)}{\partial (dl_1(t)/dt)} = -A_1, \quad \frac{\partial g_2(\cdot)}{\partial l_1(t)} \simeq \frac{k_1}{2\sqrt{L_{10}}}, \quad \frac{\partial g_2(\cdot)}{\partial l_2(t)} \simeq -\frac{k_2}{2\sqrt{L_{20}}}$$

and

$$\frac{\partial g_2(\cdot)}{\partial (dl_2(t)/dt)} = -A_2.$$

Finally, we obtain the same linearized model as in (1.50).

PROBLEM 1.29. Linearize the model for two communicating tanks developed in Problem 1.5.

SOLUTION 1.29. We consider the following case: $L_1(t) - L_2(t) > 0$ and $L_2(t) - L_3(t) > 0$. Let us consider the following operating point:

$$F_{in1} = F_{10}, F_{in2} = F_{20}, L_1 = L_{10}, L_2 = L_{20}, L_3 = L_{30}$$

and assume that the amplitudes of the fluctuations of the input flow rates and of the water levels around the values defined by this operating point are small, i.e.:

$$F_{in1} = F_{10} + f_1, \quad F_{in2} = F_{20} + f_2, \quad L_1 = L_{10} + l_1,$$

$$L_2 = L_{20} + l_2, L_3 = L_{30} + l_3.$$

From the model derived in Problem 1.5, we obtain the following for the steady-state operating point:

$$0 = F_{10} - k_1\sqrt{L_{10} - L_{20}},$$
$$0 = F_{20} + k_1\sqrt{L_{10} - L_{20}} - k_2\sqrt{L_{20} - L_{30}}. \tag{1.51}$$

Around the operating point, we obtain:

$$A\frac{dl_1}{dt} = F_{10} + f_1 - k_1 \left(L_{10} - L_{20}\right)^{\frac{1}{2}} \sqrt{\left(1 + \frac{l_1 - l_2}{L_{10} - L_{20}}\right)},$$

$$A\frac{dl_2}{dt} = F_{20} + f_2 + k_1 \left(L_{10} - L_{20}\right)^{\frac{1}{2}} \sqrt{\left(1 + \frac{l_1 - l_2}{L_{10} - L_{20}}\right)}$$

$$-k_2 \left(L_{20} - L_{30}\right)^{1/2} \sqrt{\left(1 + \frac{l_2 - l_3}{L_{20} - L_{30}}\right)}.$$

Taking into account the fact that $(1 + \varepsilon)^{1/2} \simeq 1 + \frac{1}{2}\varepsilon$, we obtain:

$$A\frac{dl_1}{dt} = F_{10} + f_1 - k_1 \left(L_{10} - L_{20}\right)^{1/2} \left(1 + \frac{l_1 - l_2}{2\left(L_{10} - L_{20}\right)}\right),$$

$$A\frac{dl_2}{dt} = F_{20} + f_2 + k_1 \left(L_{10} - L_{20}\right)^{1/2} \left(1 + \frac{l_1 - l_2}{2\left(L_{10} - L_{20}\right)}\right)$$

$$-k_2 \left(L_{20} - L_{30}\right)^{1/2} \left(1 + \frac{l_2 - l_3}{2\left(L_{20} - L_{30}\right)}\right).$$

Equations (1.51) imply that:

$$A\frac{dl_1}{dt} = f_1 - k_1 \left(L_{10} - L_{20}\right)^{1/2} \frac{l_1 - l_2}{2\left(L_{10} - L_{20}\right)},$$

$$A\frac{dl_2}{dt} = f_2 + k_1 \left(L_{10} - L_{20}\right)^{1/2} \frac{l_1 - l_2}{2\left(L_{10} - L_{20}\right)}$$

$$-k_2 \left(L_{20} - L_{30}\right)^{1/2} \frac{l_2 - l_3}{2\left(L_{20} - L_{30}\right)}.$$

For the other cases, the linearization approach presented above remains valid.

PROBLEM 1.30. Linearize the model of the water heater derived in Problem 1.8 (see Figure 1.10).

SOLUTION 1.30. This model is non-linear. Let us consider the operating point: V_0, F_0 and θ_0, and assume that $V(t)$, $F(t)$ and $\theta(t)$ fluctuate by a small amount[7] around this operating point, i.e., $V(t) = V_0 + v(t)$, $F(t) = F_0 + f(t)$ and $\theta(t) = \theta_0 + \theta(t)$. From (1.10), we derive the following:

Steady-state operating point:

$$V_0 + F_0 \left(\theta_0 - \theta_a\right) + \frac{\theta_0 - \theta_a}{R} = M\frac{d\theta_0}{dt} = 0. \tag{1.52}$$

---

7. Here $v(t)$, $f(t)$ and $\theta(t)$ represent approximations of the first order.

Around the operating point:

$$V_0 + v\left(t\right) + \left(F_0 + f\left(t\right)\right)\left(\theta_0 + \theta\left(t\right) - \theta_a\right) + \frac{\theta_0 + \theta\left(t\right) - \theta_a}{R} = M\frac{d\theta\left(t\right)}{dt}$$

$$= V_0 + v\left(t\right) + F_0\left(\theta_0 - \theta_a\right) + \frac{\theta_0 - \theta_a}{R} + F_0\theta\left(t\right) + f\left(t\right)\theta\left(t\right)$$

$$+ \left(\theta_0 - \theta_a\right)f\left(t\right) + \frac{\theta\left(t\right)}{R} = M\frac{d\theta\left(t\right)}{dt}.$$

$$(1.53)$$

Now we use Equation (1.52) in (1.53) to obtain:

$$v\left(t\right) + f\left(t\right)\theta\left(t\right) + \left(\theta_0 - \theta_a\right)f\left(t\right) + F_0\theta\left(t\right) + \frac{\theta\left(t\right)}{R} = M\frac{d\theta\left(t\right)}{dt}. \qquad (1.54)$$

By neglecting the term $f\left(t\right)\theta\left(t\right)$, which is an approximation of the second order, we obtain the linearized model:

$$v\left(t\right) + \underbrace{\left(\theta_0 - \theta_a\right)f\left(t\right)}_{const.} + F_0\theta\left(t\right) + \frac{\theta\left(t\right)}{R} = M\frac{d\theta\left(t\right)}{dt}.$$

For such a water-heating system, the voltage which provides the heat, via a resistor, is usually considered as the control variable. Observe that the model remains non-linear. Indeed:

$$v\left(t\right) = Ri^2\left(t\right).$$

It is advisable to adopt $v\left(t\right)$ as the control action in a linear model for describing the dynamics of this heating system. In this case, the resulting model is of gray-box type and is "less" non-linear.

PROBLEM 1.31. Derive a model relating the temperature in a conductor to the current.

SOLUTION 1.31. Let us denote by T and i the temperature of the conductor and the current, respectively. Observe first that a change in the direction of the current has no effect on the temperature of the conductor. Therefore, the temperature is a function of i^2. We may adopt the following model structure:

$$T = ai^2 + b. \qquad (1.55)$$

A high current leads to a high temperature. In this situation, the resistance does not remain constant but increases with the temperature. Energy balance yields:

$$W = Ri^2 = \alpha A\left(T - T_0\right),$$

where α, A and T_0 are the heat transfer coefficient, the cross-section of the conductor and the initial temperature of the conductor, respectively. The coefficient α varies at high temperature. The observation (a change in the direction of the current has no effect on the temperature of the conductor) made before remains valid. The relation (1.55) becomes inaccurate. We have to introduce an extra term that is a function of the current. On the basis of the observation that the temperature does not depend on the direction of the current, this term can be proportional to i^4. Therefore, we shall add ci^4. Then, expression (1.55) becomes:

$$T = ci^4 + ai^2 + b. \tag{1.56}$$

This model is non-linear. For $i = 0$, the temperature can be considered constant and equal to the ambient temperature.

Let us introduce the variables:

$$x = i^2 \quad \text{and} \quad y = \frac{T - b}{i^2}.$$

It follows that:

$$y = cx + a,$$

which is a linear model. This problem illustrates approach based on grey-box models.

We shall end this chapter with some comments concerning the time delay. When we start from a static behavior, the effect of a change in the control variable is not observed instantaneously but after a period of time, called the time delay. When a process involves mass (or energy) transport, a time delay (transportation lag) is associated with the motion of the material. This time delay is equal to the ratio L/V, where L represents the length of the process (a furnace for example), and V is the velocity of the material. In these cases, the phenomenological model comprises parabolic partial differential equations. All real systems exhibit a time delay. For biotechnological systems, the time delay is relatively high even if there is no material transport. In contrast, the time delay associated with a computer hard disk is very small.

PROBLEM 1.32. Let us consider again the drying furnace of Problem 1. Derive the equations governing the variation of the flow rates of the prunes and gas.

SOLUTION 1.32. Let us denote by V_p and V_a the velocities of the prunes and gas (air), respectively. The diffusion of water through the prunes and through the air, which can be expressed by a partial differential equation of second order, may be approximated by a term that is a function of the vaporization rate and of the latent heat [NAJ 89]. We assume that all the flow rates are constant in the z and y directions and that they depend only on the space variable x and on the time variable t. From the mass balances, we derive:

$$\frac{\partial P_{dr}(x,t)}{\partial x} + \frac{\partial P_{dr}(x,t)}{\partial t}\frac{dt}{\partial x} = -\lambda R_{H_2 0},$$
$$\frac{\partial A(x,t)}{\partial x} + \frac{\partial A(x,t)}{\partial t}\frac{dt}{\partial x} = \lambda R_{H_2 0},$$

where λ and R_{H_2O} represent the latent heat of vaporization and the vaporization rate, respectively. The right-hand sides of these equations represent the total derivative. Using the notation adopted for the velocities, we obtain:

$$\frac{\partial P_{dr}\left(x,t\right)}{\partial x} + \frac{1}{V_p}\frac{\partial P_{dr}\left(x,t\right)}{\partial t} = -\lambda R_{H_2O},$$

$$\frac{\partial A\left(x,t\right)}{\partial x} + \frac{1}{V_a}\frac{\partial A\left(x,t\right)}{\partial t} = \lambda R_{H_2O}.$$

These equations are of hyperbolic type. The first equation gives the total variation of the flow rate of the prunes, which is equal to the negative of the quantity of evaporated water. Of course, the total variation of the air flow rate is equal to the evaporated water. It can be shown that hyperbolic equations represent a time delay. In order to simulate this kind of model, it is advisable to use the following variable change:

$$z_1 = x + V_p t, \quad z_2 = x + V_a t.$$

Chapter 2

Laplace Transforms and Block Diagrams

2.1. The Laplace transform

The Laplace transform represents a powerful tool for solving a wide variety of problems (differential equations, etc.). The Laplace transform permits us to replace the derivative operator d/dt in differential equations by s, which leads to the use of algebraic methods. This is as if we had to carry out multiplication of roman numerals, and we converted them first to arabic numerals. A function $f(t)$ of a variable t is called *original* if the following conditions are fulfilled:

1) $f(t) = 0$ for $t < 0$;

2) $f(t)$ is piecewise[1] continuous on $[0, \infty[$;

3) $f(t)$ is of exponential order (it has bounded exponential growth), i.e., $|f(t)| < M \exp(s_0 t)$ or $\lim_{t \to \infty} |f(t) M \exp(-s_0 t)| = 0$. s_0 is called the *indicator*.

---

1. A function $f(x)$ defined on the interval $[a, b]$ is said to be piecewise continuous if and only if there exists a partition $\{\alpha_1, \alpha_2, \cdots, \alpha_x\}$ of the interval $[a, b]$ such that:

    a) $f(x)$ is continuous on $[a, b]$ except maybe for some points α_i,

    b) the right limit and the left limit of $f(x)$ exist for the points α_i.

The one-sided Laplace transform[2] (image) of the function $f(t)$ is denoted by $F(s)$ or $\mathcal{L}[f(t)]$ and is given by:

$$F(s) = \int_0^\infty f(t) \exp(-st)\, dt, \tag{2.1}$$

which corresponds to an integral transformation. This integral is convergent. On the basis of the third condition above, we obtain:

$$\left| \int_0^\infty f(t) \exp(-st)\, dt \right| \leq \int_0^\infty M \exp\left(-(s-s_0)t\right) dt = \frac{M}{s-s_0}. \tag{2.2}$$

In other words, the integral (2.1) does not converge for functions that increase faster than exponentially. We can easily verify that the Laplace transform of $f(t) = \exp(t^2)$ does not exist. Indeed, $\exp(t^2) > M \exp(t)$. From the inequality (2.2), we derive:

$$\lim_{s \to \infty} F(s) = 0. \tag{2.3}$$

Observe that for $s = jw$, the Laplace transform leads to the Fourier transform, which is often used in signal processing. The inverse Laplace transform is given by the following inversion integral:

$$f(t)\, 1(t) = \frac{1}{2\pi j} \int_{\sigma - j\infty}^{\sigma + j\infty} F(s) \exp(st)\, ds, \tag{2.4}$$

where the contour of integration is a straight line parallel to the imaginary axis and located to the right-hand side of any singularity of $F(s)$ (a singularity is a point where $F(s)$ blows up). In other words, σ must be selected greater than the real parts of the singularities of $F(s)$ (i.e., greater than the abscissa of convergence of $F(s)$). Expression (2.4) is known as Mellin's inverse formula. $1(t)$ represents the unit step, and $f(t)$ is multiplied by it in order to fulfil the first condition above. From Cauchy's residue theorem, we obtain:

$$f(t) = \sum \text{residues of } F(s).$$

---

2. The two-sided Laplace transform is defined for functions which do not need to be equal to zero for negative argument as:

$$F(s) = \int_{-\infty}^{+\infty} f(t) \exp(-st)\, dt.$$

$\mathcal{L}\left[\alpha f\left(t\right)+\beta g\left(t\right)\right]=\alpha F\left(s\right)+\beta G\left(s\right)$
$\mathcal{L}\left[f\left(\alpha t\right)\right]=\dfrac{1}{\alpha}F\left(\dfrac{s}{\alpha}\right)$
$\mathcal{L}\left[f^{(n)}\left(t\right)\right]=s^{n}F\left(s\right)-s^{n-1}f\left(0\right)-s^{n-2}f'\left(0\right)-\cdots-f^{(n-1)}\left(0\right)$
$\mathcal{L}\left[\left(-1\right)^{n}t^{n}f\left(t\right)\right]=F^{n}\left(s\right)$
$\mathcal{L}\left[\displaystyle\int_{0}^{t}f\left(\tau\right)d\tau\right]=\dfrac{1}{s}F\left(s\right)$
$\mathcal{L}\left[f\left(t-\tau\right)\right]=\exp\left(-\tau s\right)F\left(s\right)$
$\mathcal{L}\left[\exp\left(\lambda t\right)f\left(t\right)\right]=F\left(s-\lambda\right)$
$\mathcal{L}\left[f\left(t\right)*g\left(t\right)\right]=\mathcal{L}\left[g\left(t\right)*f\left(t\right)\right]=F\left(s\right)G\left(s\right)$
$\mathcal{L}\left[\dfrac{f\left(t\right)}{t}\right]=\displaystyle\int_{s}^{\infty}F\left(s\right)ds$
$\displaystyle\int_{0}^{\infty}f\left(y\right)\dfrac{dy}{y}=\int_{0}^{\infty}F\left(z\right)dz$
$f\left(0\right)=\displaystyle\lim_{s\to\infty}sF\left(s\right),\quad f\left(\infty\right)=\lim_{s\to0}sF\left(s\right)$

Table 2.1. *Properties of the Laplace transform*

The function $F\left(s\right)=P\left(s\right)/Q\left(s\right)$ has a residue at each pole, i.e., at each root of the polynomial $Q\left(s\right)$. The residue of $F\left(s\right)$ at a simple pole p is equal to:

$$\lim_{s\to p}\left(s-p\right)F\left(s\right)\exp\left(st\right).$$

For a pole of order n, the residue of $F\left(s\right)$ is given by:

$$\lim_{s\to p}\frac{1}{\left(n-1\right)!}\frac{d^{(n-1)}}{ds^{n-1}}\left[\left(s-p\right)^{n}F\left(s\right)\exp\left(st\right)\right].$$

The main properties of the Laplace transform are summarized in Table 2.1.

2.2. Transfer functions

Let us consider a system governed by the following differential equation:

$$a_{0}\frac{d^{n}}{dt^{n}}y\left(t\right)+a_{1}\frac{d^{n-1}}{dt^{n-1}}y\left(t\right)+\cdots+y\left(t\right)=b_{0}\frac{d^{m}}{dt^{m}}u\left(t\right)+\cdots+b_{m}u\left(t\right). \quad (2.5)$$

Assume that the initial conditions are equal to zero, and denote $\mathcal{L}\left[y\left(t\right)\right]$ and $\mathcal{L}\left[u\left(t\right)\right]$ by $Y\left(s\right)$ and $U\left(s\right)$, respectively. The Laplace transform of (2.5) yields:

$$F\left(s\right)=\frac{Y\left(s\right)}{U\left(s\right)}=\frac{b_{0}s^{m}+b_{1}s^{m-1}+\cdots+b_{m}}{a_{0}s^{n}+a_{1}s^{n-1}+\cdots+1}. \quad (2.6)$$

This Laplace transform represents the transfer function of the system considered. The roots of the numerator of the transfer function represent the zeros (transmission zeros) of the system, while the roots of the denominator represent the poles (natural modes). Equation (2.6) may be written as follows:

$$Y(s) = F(s)U(s),$$

$$y(t) = \int_{\tau=0}^{\infty} g(t-\tau)u(\tau)d\tau = u(t) * g(t) = g(t) * u(t),$$

where:

$$\mathcal{L}\left[g(t)\right] = F(s)$$

and "$*$" represents the convolution.[3] The order[4] of the system corresponds to the highest power of s in the denominator of the transfer function. The transfer function is a useful tool for representing the behavior of a system or of a set of systems in series, in parallel or in feedback. Let us consider the following cases:

1) N systems in series (see Figure 2.1). We obtain:

---

3. The convolution theorem:

$$F(s) * G(s) = \mathcal{L}\left[f * g\right] = \mathcal{L}\left[g * f\right],$$

is a particular case of the theorem:

$$\mathcal{L}\left[\int_0^{\infty} f(\tau)g(t,\tau)d\tau\right] = F(q(s))G(s),$$

where $G(s)$ and $q(s)$ are analytic functions such that:

$$\mathcal{L}\left[g(t,\tau)\right] = G(s)\exp\left(-\tau q(s)\right) \quad \text{and } \mathcal{L}\left[f(t)\right] = F(s).$$

In particular, by choosing $q(s) = s$, we obtain:

$$\mathcal{L}\left[g(t,\tau)\right] = G(s)\exp\left(-\tau s\right) = \mathcal{L}\left[g(t-\tau)\right]$$

and

$$F(s) * G(s) = \mathcal{L}\left[f * g\right].$$

In order to prove this theorem, we may use the method used in Solution 2.37.

4. For analogue electrical circuits, the order of the system corresponds to the number of capacitor or inductors. There is an analogy between mechanical and electrical systems: capacitor $\rightarrow$ spring; inductor $\rightarrow$ mass; voltage across an inductor $\rightarrow$ force used to accelerate a mass m; current $i(t) \rightarrow$ velocity of the mass ($Ldi/dt \rightarrow mdv/dt = m\gamma$, where γ represents the acceleration).

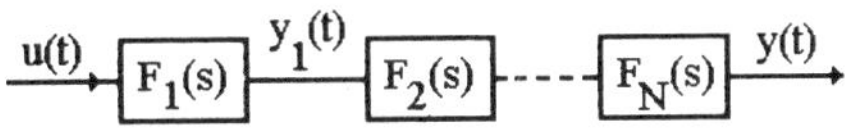

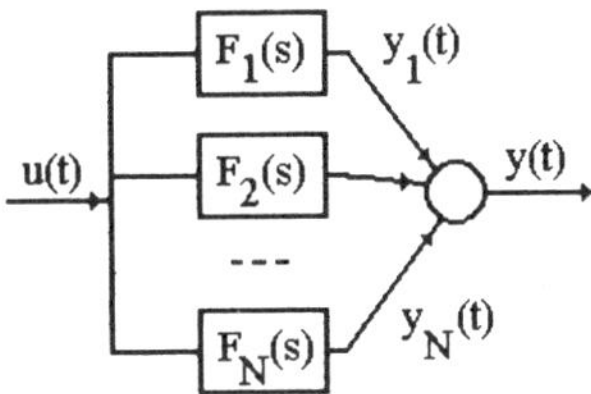

Figure 2.1. *Systems in series*

Figure 2.2. *Systems in parallel*

$$F_1\left(s\right) = \frac{Y_1\left(s\right)}{U\left(s\right)}, \quad F_2\left(s\right) = \frac{Y_2\left(s\right)}{Y_1\left(s\right)}, \cdots, \quad F_N\left(s\right) = \frac{Y\left(s\right)}{Y_{N-1}\left(s\right)}.$$

The Laplace transform of the global output $Y\left(s\right)$ is equal to:

$$\begin{aligned} Y\left(s\right) \;=\;& F_N\left(s\right) Y_{N-1}\left(s\right) = F_N\left(s\right) F_{N-1}\left(s\right) Y_{N-2}\left(s\right) = \cdots \\ =\;& \prod_{i=1}^{N} F_i\left(s\right) U\left(s\right) \end{aligned}$$

and the equivalent transfer function is:

$$F\left(s\right) = \frac{Y\left(s\right)}{U\left(s\right)} = \prod_{i=1}^{N} F_i\left(s\right).$$

2) N systems in parallel (see Figure 2.2). We obtain:

$$Y_1\left(s\right) = F_1\left(s\right) U\left(s\right), \cdots, \; Y_i\left(s\right) = F_i\left(s\right) U\left(s\right), \cdots, \; Y_N\left(s\right) = F_N\left(s\right) U\left(s\right)$$

and

$$Y\left(s\right) = \sum_{i=1}^{N} Y_i\left(s\right) = \sum_{i=1}^{N} F_i\left(s\right) U\left(s\right),$$

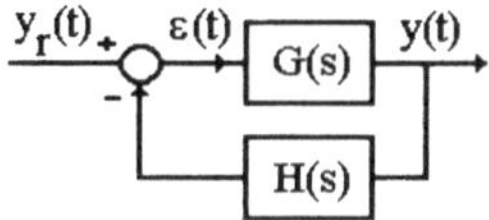

Figure 2.3. *Feedback system*

which implies:

$$F\left(s\right) = \frac{Y\left(s\right)}{U\left(s\right)} = \sum_{i=1}^{N} F_i\left(s\right).$$

3) A feedback system (see Figure 2.3). The equations governing the behavior of this system are:

$$\Xi\left(s\right) = Y_r\left(s\right) - H\left(s\right)Y\left(s\right), \quad Y\left(s\right) = G\left(s\right)\Xi\left(s\right),$$

which yields the closed-loop transfer function:

$$F\left(s\right) = \frac{Y\left(s\right)}{Y_r\left(s\right)} = \frac{G\left(s\right)}{1 + G\left(s\right)H\left(s\right)} = \frac{G\left(s\right)}{1 + T\left(s\right)},$$

where $T\left(s\right) = G\left(s\right)H\left(s\right)$ represents the open-loop transfer function. For positive feedback, we obtain:

$$F\left(s\right) = \frac{Y\left(s\right)}{Y_r\left(s\right)} = \frac{G\left(s\right)}{1 - G\left(s\right)H\left(s\right)}.$$

2.3. Laplace transform calculations

PROBLEM 2.1. Calculate the inverse Laplace transform of the following function:

$$Y\left(s\right) = \frac{1}{s^2\left(s + a\right)}.$$

SOLUTION 2.1. Let us consider the following Laplace transforms:

$$Y_1\left(s\right) = \frac{1}{\left(s + a\right)} \quad \text{and } Y_2\left(s\right) = \frac{1}{s}Y_1\left(s\right).$$

Their inverse Laplace transforms will be denoted by $y_1(t)$ and $y_2(t)$, respectively. It follows that:

$$y_1(t) = \mathcal{L}^{-1}\left[Y_1\left(s\right)\right] = \exp\left(-at\right)$$

$$y_2(t) = \mathcal{L}^{-1}\left[Y_2(s)\right] = \int_0^t y_1(t)dt = \frac{1}{a}\left[1 - \exp(-at)\right].$$

Taking into account the fact that:

$$Y(s) = \frac{1}{s}Y_2(s),$$

we derive:

$$y(t) = \int_0^t y_2(t)dt = \frac{1}{a^2}\left[\exp(-at) + at - 1\right].$$

PROBLEM 2.2. Calculate the inverse Laplace transform of the following function:

$$Y(s) = \frac{2s + 1}{s\left(s^2 + 1\right)}.$$

SOLUTION 2.2. The poles (natural modes) of this function are:

$$p_1 = 0, p_2 = j \text{ and } p_3 = -j.$$

Let us now calculate the residues associated with these singularities. The residue related to the pole $p_1 = 0$ is given by:

$$r_0 = \lim_{s \to 0}\left[\frac{s\left(2s + 1\right)}{s\left(s^2 + 1\right)}\exp(st)\right] = 1.$$

The residue associated with the pole $p_2 = j$ is:

$$r_j = \lim_{s \to j}\left[\frac{(s - j)\left(2s + 1\right)}{s\left(s^2 + 1\right)}\exp(st)\right] = -\frac{1 + 2j}{2}\exp(jt).$$

Finally, the residue associated with the pole $p_3 = -j$ is:

$$r_{-j} = \lim_{s \to -j}\left[\frac{(s + j)\left(2s + 1\right)}{s\left(s^2 + 1\right)}\exp(st)\right] = -\frac{1 - 2j}{2}\exp(-jt).$$

Therefore:

$$y(t) = r_0 + r_j + r_{-j} = 1 - \cos t + 2\sin t.$$

PROBLEM 2.3. Calculate the inverse Laplace transform of the following function:

$$Y(s) = \frac{s+3}{s^3 + 8s^2 + 17s + 10}.$$

SOLUTION 2.3. Let us calculate the roots of the denominator of this function. Observe that $s = -1$ is a root. Therefore, we obtain:

$$s^3 + 8s^2 + 17s + 10 = (s+1)\left(s^2 + 7s + 10\right) = (s+1)(s+2)(s+5).$$

Let us now determine the partial fraction expansion[5] of $Y(s)$:

$$Y(s) = \frac{\alpha}{(s+1)} + \frac{\beta}{(s+2)} + \frac{\delta}{(s+5)}.$$

Multiplying both sides of this equation by $(s+1)$ and setting $s = -1$, we obtain:

$$\alpha = \frac{1}{2}.$$

Multiplying both sides of the equation by $(s+2)$ and setting $s = -2$, we obtain:

$$\beta = -\frac{1}{3}.$$

Now, proceeding as before for the pole $s = -5$, we derive:

$$\delta = -\frac{1}{6}.$$

Finally, the partial fraction expansion of $Y(s)$ is:

$$Y(s) = -\frac{1}{6(s+5)} - \frac{1}{3(s+2)} + \frac{1}{2(s+1)}$$

from which we derive:

$$y(t) = \frac{1}{2}\exp(-t) - \frac{1}{3}\exp(-2t) - \frac{1}{6}\exp(-5t).$$

---

5. The partial fraction expansion provides a way to integrate all rational functions and to convert a continuous transfer function to discrete time by use of a z-transform table.

Observe that the residues associated with the poles -1, -3, and -4 are:

$$r_{-1} = \lim_{s \to -1} \left[\frac{(s+1)(s+3)}{(s+1)(s+2)(s+5)} \exp(st) \right] = \frac{1}{2} \exp(-t),$$

$$r_{-2} = \lim_{s \to -1} \left[\frac{(s+2)(s+3)}{(s+1)(s+2)(s+5)} \exp(st) \right] = -\frac{1}{3} \exp(-2t),$$

$$r_{-5} = \lim_{s \to -1} \left[\frac{(s+5)(s+3)}{(s+1)(s+2)(s+5)} \exp(st) \right] = -\frac{1}{6} \exp(-5t),$$

$$y(t) = r_{-1} + r_{-2} + r_{-5} = \frac{1}{2} \exp(-t) - \frac{1}{3} \exp(-2t) - \frac{1}{6} \exp(-5t).$$

Of course, we obtain the same result.

PROBLEM 2.4. Prove the following equalities.

1.

$$\int_0^x (x-t)^n \, t^m dt = \frac{n! m!}{(n+m+1)!} t^{n+m+1}. \tag{2.7}$$

2.

$$\mathcal{L}\left((-1)^n t^n y(t)\right) = Y^{(n)}(s) = \frac{d^n}{ds^n} Y(s).$$

SOLUTION 2.4. 1. The Laplace transform of the convolution defined as:

$$\int_0^t f(t-\tau) g(\tau) \, d\tau = \int_0^t f(\tau) g(t-\tau) \, d\tau$$

is given by:

$$\mathcal{L}[f(t)] \, \mathcal{L}[g(t)].$$

Let us substitute $(t-\tau)^n$ and t^m for $f(t-\tau)$ and $g(t)$, respectively. From Equation (2.7), we derive:

$$\mathcal{L}\left[\int_0^t f(t-\tau) g(\tau) \, d\tau \right] = \mathcal{L}[f(t)] \, \mathcal{L}[g(t)] = \frac{n!}{s^{n+1}} \frac{m!}{s^{m+1}}.$$

Taking the inverse Laplace transform, we obtain:

$$\mathcal{L}^{-1}\left[\frac{n! m!}{p^{n+m+2}} \right] = \frac{n! m!}{(n+m+1)!} t^{n+m+1}.$$

2. Let us calculate the derivative of $Y(s)$:

$$Y(s) = \int_0^\infty y(t) \exp(-st)\, dt, \quad Y'(s) = -\int_0^\infty ty(t) \exp(-st)\, dt,$$

$$\frac{d^2}{ds^2} Y(s) = (-1)^2 \int_0^\infty t^2 y(t) \exp(-st)\, dt, \quad \cdots,$$

$$\frac{d^n}{ds^n} Y(s) = (-1)^n \int_0^\infty t^n y(t) \exp(-st)\, dt,$$

which yields the desired result:

$$\frac{d^n}{ds^n} Y(s) = \mathcal{L}\left[(-1)^n t^n y(t)\right].$$

PROBLEM 2.5. Express the integral:

$$J_2 = \int_0^\infty [t\varepsilon(t)]^2\, dt$$

as an integral involving the Laplace transform $G(s) = \mathcal{L}[t\varepsilon(t)]$.

SOLUTION 2.5. If $\varepsilon(t)$ represents the error $(y_r(t) - y(t))$, the integral J_2 corresponds to the integral square time error (ISTE) criterion associated with PID synthesis (see Chapter 5). In this case:

$$G(s) = -\frac{d}{ds}\mathcal{L}[\varepsilon(t)].$$

On the basis of the Laplace transform, we obtain:

$$t\varepsilon(t) = \frac{1}{2\pi j} \int_{\sigma-j\infty}^{\sigma+j\infty} G(s) \exp(st)\, ds,$$

which inserted in the integral J_2 yields[6]:

$$J_2 = \int_0^\infty [t\varepsilon(t)]^2\, dt = \int_0^\infty t\varepsilon(t) \left[\frac{1}{2\pi j} \int_{\sigma-j\infty}^{\sigma+j\infty} G(s) \exp(st)\, ds \right] dt.$$

---

6. This technique may be used in the proof of Efros's theorem, which can be stated as follows:

$$F(q(s)) G(s) = \mathcal{L}\left[\int_0^\infty f(\tau) g(t;\tau)\, d\tau \right],$$

where $F(s) = \mathcal{L}[f(t)]$, $G(s)$ and $q(s)$ are analytic functions such that:

$$\mathcal{L}[g(t;\tau)] = G(s) \exp(-\tau q(s)).$$

As the integrals are convergent (Fubini's theorem), reversing the order of the integrations yields:

$$J_2 = \int_0^\infty [t\varepsilon\,(t)]^2\,dt = \frac{1}{2\pi j} \int_{\sigma-j\infty}^{\sigma+j\infty} G\,(s) \underbrace{\left[\int_0^\infty [t\varepsilon\,(t)] \exp\,(st)\,dt \right]}_{G(-s)} ds,$$

$$J_2 = \frac{1}{2\pi j} \int_{\sigma-j\infty}^{\sigma+j\infty} G\,(s)\,G\,(-s)\,ds.$$

This integral appears also in the evaluation of the loss function (performance index) for a dynamical system whose input is stochastic. A recursive algorithm and a Fortran program for evaluating this kind of integral are given in [AST 70]. This recursive algorithm is presented in Appendix A.

PROBLEM 2.6. The Laguerre polynomial of order n *is defined by:*

$$L_n\,(x) = \exp\,(x)\,\frac{d^n}{dx^n}\,[x^n \exp\,(-x)] \quad (n = 0, 1, 2, \cdots).$$

Show that $L_n\,(x)$ *is a polynomial of order* n.

SOLUTION 2.6. Let us first calculate the Laplace transform of the function $x^n \exp\,(x)$ (see Problem 2.4):

$$\mathcal{L}\,[x^n \exp\,(-x)] = \frac{n!}{(s+1)^{n+1}} \quad (n \text{ integer and } \geq 0).$$

For $x = 0$, this function is equal to zero, and so is its first $(n-1)$ derivatives. Therefore:

$$\mathcal{L}\left[\frac{d^n}{dx^n}\,[x^n \exp\,(-x)] \right] = \frac{n!s^n}{(s+1)^{n+1}} = F\,(s).$$

For $q\,(s) = s$, we obtain $g\,(t; \tau) = g\,(t - \tau)$ and:

$$F\,(q\,(s))\,G\,(s) = \mathcal{L}\left[\int_0^\infty f\,(\tau)\,g\,(t - \tau)\,d\tau \right].$$

As $g\,(t - \tau) = 0$ for $\tau < t$ (the first condition related to the existence of the Laplace transform of a given function), we obtain the convolution theorem.

Using the translation theorem,[7] we derive:

$$\mathcal{L}\left[L_n\left(x\right)\right] = \mathcal{L}\left[\exp\left(x\right)\frac{d^n}{dx^n}\left[x^n\exp\left(-x\right)\right]\right] = F\left(s-1\right) = \frac{n!}{s}\left(1-\frac{1}{s}\right)^n. \quad (2.8)$$

The Laplace transform of $L_n\left(x\right)$ contains a term $\left(-1\right)^n n!/s^{n+1}$ which is the inverse Laplace transform of $\left(-1\right)^n x^n$. Therefore, the degree of the *Laguerre polynomial* $L_n\left(x\right)$ *is equal to* n. From Equation (2.8), it is easy to derive the first few Laguerre polynomials:

$$\mathcal{L}\left[L_0\left(x\right)\right] = \frac{1}{s} \implies L_0\left(x\right) = 1,$$

$$\mathcal{L}\left[L_1\left(x\right)\right] = \frac{1}{s}\left(1-\frac{1}{s}\right) \implies L_1\left(x\right) = 1 - x,$$

$$\mathcal{L}\left[L_2\left(x\right)\right] = \frac{2!}{s}\left(1-\frac{1}{s}\right)^2 \implies L_2\left(x\right) = 2!\left(1 - 2x + \frac{x^2}{2!}\right),$$

and so on.

PROBLEM 2.7. Let the Laplace transform of the function $g\left(t\right)$ (see Figure 2.4) be $G\left(s\right)$. Calculate the Laplace transform of the periodic function $f\left(t\right)$ depicted in Figure 2.4.

SOLUTION 2.7. Let us express $f\left(t\right)$ as a function of $g\left(t\right)$:

$$f\left(t\right) = g\left(t\right) + g\left(t-T\right) + g\left(t-2T\right) + \cdots + g\left(t-nT\right) + \cdots.$$

---

7. The Laplace transform of $\exp\left(\lambda t\right) f\left(t\right)$ is given by:

$$\mathcal{L}\left[\exp\left(\lambda t\right) f\left(t\right)\right] = \int_0^\infty \exp\left(\lambda t\right) f\left(t\right)\exp\left(-st\right) dt =$$

$$\int_0^\infty f\left(t\right)\exp\left(-\left(s-\lambda\right)t\right) dt = F\left(s-\lambda\right)$$

where:

$$F\left(s\right) = \mathcal{L}\left[f\left(t\right)\right].$$

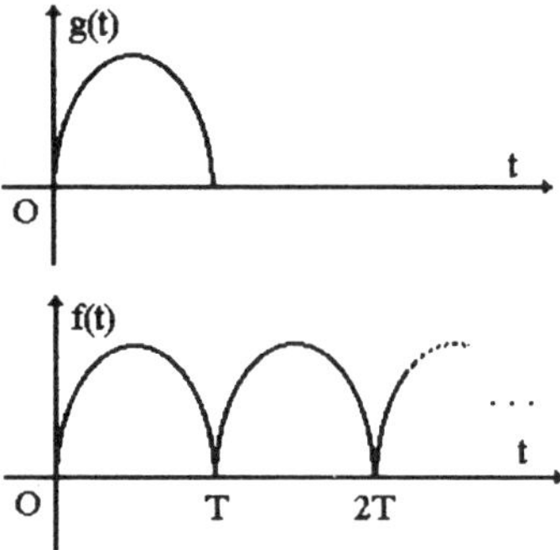

Figure 2.4. *Periodic function (Problem 2.7)*

Taking the Laplace transform,[8] we obtain:

$$[\mathcal{L}f(t)] = F(s) = \mathcal{L}[g(t)] + \mathcal{L}[g(t-T)] + \cdots,$$
$$F(s) = G(s) + \exp(-Ts)G(s) + \exp(-2Ts)G(s) + \cdots,$$
$$F(s) = [1 + \exp(-Ts) + \exp(-2Ts) + \cdots]G(s),$$
$$F(s) = \frac{1}{1-\exp(-Ts)}G(s).$$

PROBLEM 2.8. Calculate the Laplace transform of the function:

$$f(t) = \begin{cases} 0 \text{ for } 0 \le t \le T, \\ k \text{ for } T \le t \le 2T, \\ 0 \text{ for } t > 2T. \end{cases}$$

SOLUTION 2.8. Let us decompose the function $f(t)$ into elementary functions as follows:

$$f(t) = k1(t-T) - k1(t-2T),$$

where $1(t)$ represents the unit step function. Taking the Laplace transform, we obtain:

$$F(s) = \frac{k}{s}[\exp(-Ts) - \exp(-2Ts)].$$

---

8. Recall that:

$$\frac{1}{1-a} = 1 + a + a^2 + \ldots + a^n + \ldots.$$

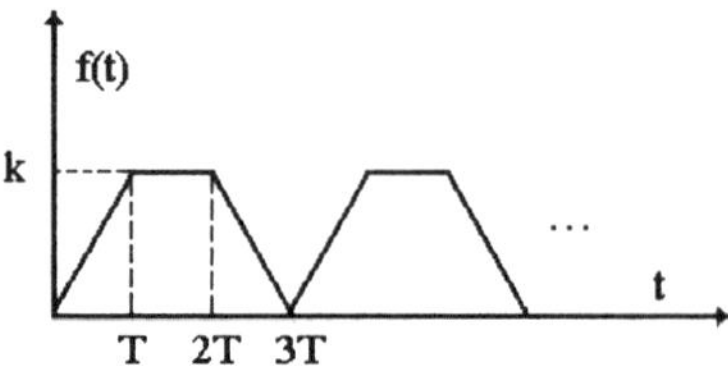

Figure 2.5. *Periodic function (Problem 2.9)*

PROBLEM 2.9. Calculate the Laplace transform of the periodic function depicted in Figure 2.5.

SOLUTION 2.9. Let us denote by $g(t)$ the function:

$$g(t) = \begin{cases} f(t) \text{ for } 0 \leq t \leq 3T, \\ 0 \text{ otherwise.} \end{cases}$$

The periodic function $f(t)$ is equal to:

$$f(t) = g(t) + g(t - 3T) + g(t - 6T) + \dots .$$

The decomposition of $g(t)$ into elementary functions leads to (see Figure 2.6):

$$g(t) = g_1(t) + g_2(t) + g_3(t) + g_4(t),$$

where:

$$g_1(t) = \frac{k}{T}t1(t),$$

$$g_{i+1}(t) = \begin{cases} 0 \text{ for } 0 < t \leq iT, \\ -\dfrac{k}{T}(t - iT) \text{ for } t \geq iT,\ i = 1, 2, \end{cases}$$

$$g_4(t) = \begin{cases} 0 \text{ for } 0 < t \leq 3T, \\ \dfrac{k}{T}(t - 3T) \text{ for } t \geq 3T. \end{cases}$$

Taking the Laplace transform, we obtain:

$$G(s) = \frac{k}{Ts^2}\left[1 - \exp(-Ts) - \exp(-2Ts) + \exp(-3Ts)\right].$$

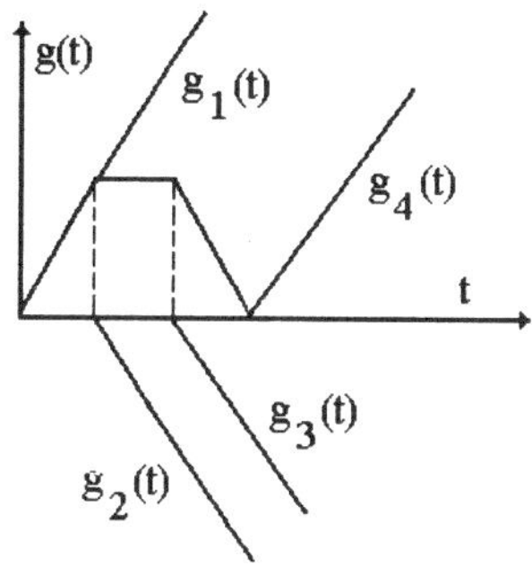

Figure 2.6. *Decomposition of $g(t)$ into elementary functions*

Finally, the Laplace transform of $f(t)$ is (see Problem 2.7):

$$F(s) = \frac{1}{1 - \exp(-3Ts)} G(s).$$

PROBLEM 2.10. Calculate the Laplace transforms of the functions:

$$f_1(t) = \frac{\sin t}{t} \quad \text{and} \quad f_2(t) = t^2 \exp(3t),$$

and show that:

$$\mathcal{L}[f(\alpha t)] = \frac{1}{\alpha} F\left(\frac{s}{\alpha}\right),$$

$$\mathcal{L}[\cosh wt] \neq \frac{1}{s^2 + w^2}.$$

SOLUTION 2.10. 1. Recall some properties of the Laplace transform:[9]

$$\mathcal{L}\left[\frac{f(t)}{t}\right] = \int_s^\infty F(s)\,ds,$$
$$\mathcal{L}\left[\exp(\alpha t)\,f(t)\right] = F(s-\alpha).$$

On the basis of these properties, we derive:

$$\mathcal{L}\left[f_1(t)\right] = \int_s^\infty \mathcal{L}\left[\sin t\right]ds = \int_s^\infty \frac{1}{s^2+1}\,ds = \left[ar\tan s\right]_s^\infty = \frac{\pi}{2} - ar\tan s,$$
$$F(s) = \mathcal{L}\left[t^2\right] = \frac{2}{s^3},$$
$$\mathcal{L}\left[f_2(t)\right] = \mathcal{L}\left[t^2\exp(3t)\right] = F(s-3) = \frac{2}{(s-3)^3}.$$

2. By definition, the Laplace transform of $f(\alpha t)$ is given by:

$$\mathcal{L}\left[f(\alpha t)\right] = \int_0^\infty f(\alpha t)\exp(-st)\,dt.$$

Let us introduce the variable change $z = \alpha t$. It follows that:

$$\int_0^\infty f(z)\exp\left(-s\frac{z}{\alpha}\right)\frac{dz}{\alpha} = \frac{1}{\alpha}\int_0^\infty f(z)\exp\left(-\frac{s}{\alpha}z\right)dz = \frac{1}{\alpha}F\left(\frac{s}{\alpha}\right).$$

3. Using the initial-value theorem, we obtain:

$$\lim_{s\to\infty} s\frac{1}{s^2+w^2} = 0.$$

----

9. Let us consider the integral:

$$\int_s^\infty F(s)\,ds = \int_s^\infty ds \int_0^\infty f(t)\exp(-st)\,dt.$$

The second integral is convergent. Reversing the order of the integrations yields:

$$\int_s^\infty F(s)\,ds = \int_0^\infty f(t)\,dt \int_s^\infty \exp(-st)\,ds = \int_0^\infty f(t)\,dt\left[-\frac{\exp(-st)}{t}\right]_0^\infty$$
$$= \int_0^\infty \left(\frac{f(t)}{t}\right)\exp(-st)\,dt = \mathcal{L}\left[\frac{f(t)}{t}\right]$$

However,

$$\cosh wt = \frac{\exp(wt) + \exp(-wt)}{2}$$

leads to:

$$\cosh(0) = 1.$$

Therefore, the expression associated with the Laplace transform of $\cosh wt$ is false. Observe that, in general, this theorem permits us to check quickly the validity of Laplace transform calculations.

PROBLEM 2.11. Prove the initial- and final-value (Tauber) theorem,

$$f(0) = \lim_{s \to \infty} sF(s), \quad f(\infty) = \lim_{t \to \infty} f(t) = \lim_{s \to 0} sF(s), \quad F(s) = \mathcal{L}[f(t)].$$

SOLUTION 2.11. Let us consider the Laplace transform of the derivative of $f(t)$,

$$\mathcal{L}\left[\frac{df(t)}{dt}\right] = G(s) = sF(s) - f(0).$$

Using equation (2.3), we derive:

$$\lim_{s \to \infty} G(s) = 0 \Rightarrow \lim_{s \to \infty} (sF(s) - f(0)) = 0,$$
$$f(0) = \lim_{s \to \infty} sF(s).$$

Taking into account the fact that:

$$G(s) = \int_0^\infty \frac{df(t)}{dt} \exp(-st)\, dt = sF(s) - f(0) \tag{2.9}$$

$$\text{and} \int_0^\infty \frac{df(t)}{dt} = f(\infty) - f(0),$$

we derive:

$$\lim_{s \to 0} \int_0^\infty \frac{df(t)}{dt} \exp(-st)\, dt = \int_0^\infty \frac{df(t)}{dt}\, dt = \lim_{s \to 0} [sF(s) - f(0)]. \tag{2.10}$$

Comparing Equations (2.9) and (2.10), we obtain the second part of the desired result:

$$\lim_{s \to 0} \left[sF(s) - f(0) \right] = f(\infty) - f(0) \implies \lim_{s \to 0} sF(s) = f(\infty).$$

PROBLEM 2.12. Calculate the Laplace transforms of the Fresnel integrals:

$$C(t) = \int_0^t \frac{\cos t}{\sqrt{2\pi t}}\, dt, \quad S(t) = \int_0^t \frac{\sin t}{\sqrt{2\pi t}}\, dt.$$

The Fresnel integrals play an important role in physics (theory of diffraction).

SOLUTION 2.12. Let us consider the following integral:

$$I(t) = \int_0^t \frac{\exp(jt)}{\sqrt{2\pi t}}\, dt = \int_0^t \frac{\cos t + j \sin t}{\sqrt{2\pi t}}\, dt.$$

From a table of Laplace transforms,[10] we obtain:

$$\mathcal{L}\left[\frac{1}{\sqrt{\pi t}} \right] = \frac{1}{\sqrt{s}}.$$

On the basis of the translation theorem, we derive:

$$\mathcal{L}\left[\frac{\exp(jt)}{\sqrt{2\pi t}} \right] = \frac{1}{\sqrt{2(s - j)}}.$$

From a theorem related to the integration of an original function (see section 2.1), we obtain:

$$\mathcal{L}\left[\int_0^t \frac{\exp(jt)}{\sqrt{2\pi t}} \right] = \frac{1}{s\sqrt{2(s - j)}}.$$

---

10. The Laplace transform of t^a is given by:

$$\mathcal{L}[t^a] = \frac{\Gamma(a + 1)}{s^{a+1}},$$

where $\Gamma(.)$ represents the gamma (Euler) function, and for integer values of $a = n$, $\Gamma(n + 1) = n!$ One property of the gamma function is:

$$\Gamma(a)\Gamma(1 - a) = \frac{\pi}{\sin a\pi}.$$

For $a = \frac{1}{2}$, we obtain:

$$\Gamma\left(\frac{1}{2}\right) = \sqrt{\pi}.$$

Similarly, we obtain:

$$\mathcal{L}\left[\int_0^t \frac{\exp(-jt)}{\sqrt{2\pi t}}\right] = \frac{1}{s\sqrt{2(s+j)}}.$$

The linearity property leads to:

$$\mathcal{L}\left[C(t)\right] = \mathcal{L}\left[\int_0^t \frac{\exp(jt) + \exp(-jt)}{2\sqrt{2\pi t}}\,dt\right]$$

$$= \frac{1}{2\sqrt{2}s}\left(\frac{1}{\sqrt{s+j}} + \frac{1}{\sqrt{s-j}}\right) = \frac{\sqrt{s+j} + \sqrt{s-j}}{2\sqrt{2}s\sqrt{s^2+1}}$$

$$= \frac{\sqrt{\left(\sqrt{s+j} + \sqrt{s-j}\right)^2}}{2\sqrt{2}s\sqrt{s^2+1}} = \frac{\sqrt{2s + 2\sqrt{s^2+1}}}{2\sqrt{2}s\sqrt{s^2+1}} = \frac{\sqrt{s + \sqrt{s^2+1}}}{2s\sqrt{s^2+1}}.$$

Similarly, we obtain:

$$\mathcal{L}\left[S(t)\right] = \mathcal{L}\left[\int_0^t \frac{\exp(jt) - \exp(-jt)}{2\sqrt{2\pi t}}\,dt\right]$$

$$= \frac{1}{2\sqrt{2}js}\left(\frac{1}{\sqrt{s-j}} - \frac{1}{\sqrt{s+j}}\right) = \frac{\sqrt{\sqrt{s^2+1} - s}}{2s\sqrt{s^2+1}}.$$

PROBLEM 2.13. Calculate the Laplace transform of the function:

$$\int_t^\infty \frac{\exp(-\tau)}{\tau}\,d\tau.$$

SOLUTION 2.13. The Laplace transform of the function $f(at)$, $a > 0$, is given by:

$$\mathcal{L}\left[f(at)\right] = \int_0^\infty f(at)\exp(-st)\,dt.$$

Let us make the following change $\tau = at$:

$$\mathcal{L}\left[f\left(at\right)\right] = \frac{1}{a}\int_0^\infty f\left(\tau\right)\exp\left(-\frac{s}{a}\tau\right)d\tau = \frac{1}{a}F\left(\frac{s}{a}\right),$$

where $F\left(s\right) = \mathcal{L}\left[f\left(t\right)\right]$. For $\alpha = 1/a$, we obtain:

$$\mathcal{L}\left[f\left(\frac{t}{\alpha}\right)\right] = \alpha F\left(\alpha s\right).$$

Integration over the interval $[0, 1]$ with respect to α yields:

$$\int_0^1 F\left(\alpha s\right)d\alpha = \int_0^1\int_0^\infty \frac{1}{\alpha}f\left(\frac{t}{\alpha}\right)\exp\left(-st\right)dt\,d\alpha.$$

These integrals are convergent. Reversing the order of integration, we obtain:

$$\int_0^1 F\left(\alpha s\right)d\alpha = \int_0^\infty \exp\left(-st\right)dt\int_0^1 \frac{1}{\alpha}f\left(\frac{t}{\alpha}\right)d\alpha = \mathcal{L}\left[\int_0^1 \frac{1}{\alpha}f\left(\frac{t}{\alpha}\right)d\alpha\right].$$

Let us now make the variable changes $\tau = t/\alpha$ and $p = \alpha s$; we obtain:

$$\frac{1}{s}\int_0^s F\left(p\right)dp = \mathcal{L}\left[\int_t^\infty \frac{f\left(\tau\right)}{\tau}d\tau\right]. \tag{2.11}$$

Recall that:

$$\mathcal{L}\left[\exp\left(-\tau\right)\right] = G\left(s\right) = \frac{1}{s+1}.$$

From Equation (2.11), we derive:

$$\mathcal{L}\left[\int_t^\infty \frac{g\left(\tau\right)}{\tau}d\tau\right] = \frac{1}{s}\int_0^s \frac{1}{p+1}dp = \frac{1}{s}\ln\left(s+1\right).$$

Observe that Equation (2.11) may be used to calculate the Laplace transforms of many functions, such as the cosine integral.

In what follows, we shall consider some differential, integral and partial differential equations.

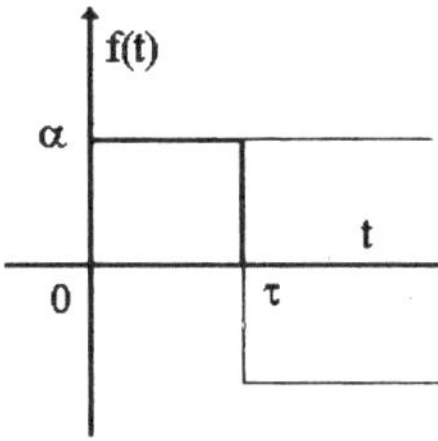

Figure 2.7. *The function $f(t)$*

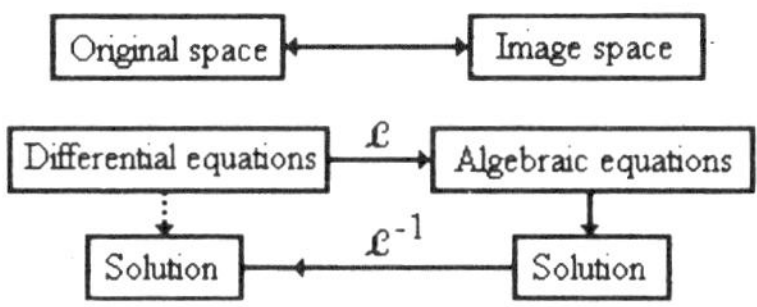

Figure 2.8. *The Laplace transform as a tool for solving differential equations*

2.4. Differential and integral equations

PROBLEM 2.14. Solve the following differential equation:

$$\frac{d^2y}{dt^2} + 4y = f(t), \quad y(0) = 0, \quad \left.\frac{dy}{dt}\right|_{t=0} = 0, \tag{2.12}$$

where $f(t)$ is plotted in Figure 2.7.

SOLUTION 2.14. The use of Laplace transforms for solving differential equations is illustrated by the diagram in Figure 2.8.

Let us decompose the function $f(t)$ into elementary functions:

$$f(t) = \alpha\left[1(t) - 1(t - \tau)\right],$$

where $1(t)$ represents the unit step. The Laplace transform of Equation (2.12) yields:

$$s^2Y(s) + 4Y(s) = \frac{\alpha}{s}\left(1 - \exp(-\tau s)\right),$$

$$Y(s) = \frac{\alpha\left(1 - \exp(-\tau s)\right)}{s(s^2 + 4)}.$$

Let us consider the following Laplace transform:

$$Y^{*}\left(s\right)=\frac{\alpha}{s\left(s^{2}+4\right)}=\frac{\alpha}{4s}-\frac{\alpha s}{4\left(s^{2}+4\right)},$$

which implies[11]:

$$y^{*}\left(t\right)=\frac{\alpha}{4}-\frac{\alpha}{4}\cos\left(2t\right)=\frac{\alpha}{2}\left(\sin^{2}t\right)1\left(t\right).$$

The inverse Laplace transform of:

$$-\frac{\alpha\exp\left(-\tau s\right)}{s\left(s^{2}+4\right)}$$

is

$$-\frac{\alpha}{2}\left[\sin^{2}\left(t-\tau\right)\right]1\left(t-\tau\right).$$

Finally, we obtain:

$$y\left(t\right)=\frac{\alpha}{2}\left[1\left(t\right)\sin^{2}t-1\left(t-\tau\right)\sin^{2}\left(t-\tau\right)\right].$$

PROBLEM 2.15. Solve the following differential equation:

$$\frac{d^{2}y}{dt^{2}}+a^{2}y=b\sin\left(at\right),\quad y\left(0\right)=y_{0},\quad\left.\frac{dy}{dt}\right|_{t=0}=y_{1}.$$

SOLUTION 2.15. The Laplace transform yields:

$$\left(s^{2}+a^{2}\right)Y\left(s\right)=\frac{ab}{s^{2}+a^{2}}+y_{0}s+y_{1},$$

$$Y\left(s\right)=\frac{ab}{\left(s^{2}+a^{2}\right)^{2}}+y_{0}\frac{s}{s^{2}+a^{2}}+\frac{y_{1}}{s^{2}+a^{2}}.$$

Recall that:

$$\mathcal{L}\left[\sin\left(\omega t\right)\right]=\frac{\omega}{s^{2}+\omega^{2}},\quad\mathcal{L}\left[\cos\left(\omega t\right)\right]=\frac{s}{s^{2}+\omega^{2}}$$

and

$$\mathcal{L}\left[t^{n}\sin\left(\omega t\right)\right]=n!\frac{Im\left(s+j\omega\right)^{n+1}}{\left(s^{2}+\omega^{2}\right)^{n+1}}.$$

---

11. Note that $\cos 2x=1-2\sin^{2}x$.

Therefore,

$$\mathcal{L}\left[t\sin\left(at\right)\right] = \frac{2as}{\left(s^2 + a^2\right)^2}$$

and

$$\mathcal{L}^{-1}\left[\frac{ab}{\left(s^2 + a^2\right)^2}\right] = \frac{b}{2}\int_0^t t\sin\left(at\right)dt = \frac{b}{2a^2}\left[\sin\left(at\right) - at\cos\left(at\right)\right].$$

Finally, we obtain:

$$y\left(t\right) = \frac{1}{a}\left(y_1 + \frac{b}{2a}\right)\sin\left(at\right) + \left(y_0 - \frac{b}{2a}t\right)\cos\left(at\right).$$

PROBLEM 2.16. For $t > 0$, solve the following system of differential equations:

$$\begin{cases} \dfrac{d}{dt}f\left(t\right) + \dfrac{d}{dt}g\left(t\right) = t1\left(t\right), \\ \dfrac{d^2}{dt^2}f\left(t\right) - g\left(t\right) = \exp\left(-t\right)1\left(t\right), \end{cases}$$

$$f\left(0\right) = 3, \quad \left.\frac{d}{dt}f\left(t\right)\right|_{t=0} = -2 \quad \text{and } g\left(0\right) = 0.$$

Observe that we have introduced the unit step $1\left(t\right)$ because the Laplace transform is defined for functions equal to zero for $t < 0$.

SOLUTION 2.16. We calculate the following Laplace transforms:

$$\mathcal{L}\left[\frac{d}{dt}f\left(t\right)\right] = sF\left(s\right) - 3, \quad \mathcal{L}\left[\frac{d^2}{dt^2}f\left(t\right)\right] = s^2F\left(s\right) - 3s + 2,$$

$$\mathcal{L}\left[\frac{d}{dt}g\left(t\right)\right] = sG\left(s\right), \quad \mathcal{L}\left[t\right] = \frac{1}{s^2}, \quad \mathcal{L}\left[\exp\left(-t\right)\right] = \frac{1}{s+1}.$$

Transforming the system of differential equations, we obtain a system of equations where the unknowns are the Laplace transforms $F\left(s\right)$ and $G\left(s\right)$:

$$\begin{cases} s\left[F\left(s\right) + G\left(s\right)\right] - 3 = \dfrac{1}{s^2}, \\ s^2F\left(s\right) - 3s + 2 - G\left(s\right) = \dfrac{1}{s+1}, \end{cases}$$

$$F(s) = \frac{s^4 + 2s^3 + 3s^2 + 3s^5 + s + 1}{(s+1)(1+s^2)s^3}, \quad G(s) = \frac{2s^2 + 2s + 1}{s(s+1)(1+s^2)}.$$

For $Re(s) > 0$, we derive the partial fraction expansions:

$$F(s) = \frac{2}{s} + \frac{1}{s^3} + \frac{1}{2(s+1)} + \frac{s-3}{2(s^2+1)},$$

$$G(s) = \frac{1}{s} - \frac{1}{2(s+1)} + \frac{3-s}{2(s^2+1)}.$$

Finally, the solution of the system of differential equations is obtained by taking the inverse Laplace transform:

$$f(t) = 2 + \frac{1}{2}\{t^2 + \cos t - 3\sin t + \exp(-t)\},$$

$$g(t) = 1 - \frac{1}{2}\{\cos t - 3\sin t + \exp(-t)\}.$$

PROBLEM 2.17. For $t > 0$, solve the following integral equation:

$$f(t) = t^2 + \int_0^t \sin(t-\tau)f(\tau)\,d\tau.$$

SOLUTION 2.17. The second term of the right-hand side of this equation corresponds to a convolution involving $\sin t$ and $f(t)$. Therefore, we obtain:

$$f(t) = t^2 + \sin t * f(t).$$

Taking the Laplace transform, we obtain:

$$\mathcal{L}[f(t)] = \mathcal{L}[t^2 + \sin t * f(t)] = \mathcal{L}[t^2] + \mathcal{L}[\sin t]\,\mathcal{L}[f(t)],$$

$$\mathcal{L}[f(t)] = F(s) = \frac{2}{s^3} + \frac{F(s)}{1+s^2}.$$

It follows that:

$$F(s) = \frac{2}{s^5} + \frac{2}{s^3}, \quad Re(s) > 0,$$

the inverse Laplace transform of which is:

$$f(t) = \frac{t^4}{12} + t^2.$$

PROBLEM 2.18. Solve the following differential equation:

$$\frac{d^2 f(t)}{dt^2} - 5\frac{df(t)}{dt} + 6f(t) = \exp(t)\,1(t),$$

$$f(0) = 0 \quad \text{and} \quad \left.\frac{df(t)}{dt}\right|_{t=0} = 0,$$

where $1(t)$ represents the unit step.

SOLUTION 2.18. A Laplace transform leads to:

$$\mathcal{L}[f(t)] = F(s) = \frac{1}{(s-1)(s^2 - 5s + 6)}.$$

A partial fraction expansion gives:

$$F(s) = \frac{1/2}{(s-1)} + \frac{-1}{(s-2)} + \frac{1/2}{(s-3)},$$

which corresponds to:

$$f(t) = \frac{1}{2}\exp(t) - \exp(2t) + \frac{1}{2}\exp(3t).$$

PROBLEM 2.19. Calculate the following integral:

$$f(x) = \int_0^\infty \frac{y\sin(xy)}{1+y^2}\,dy, \qquad (x > 0).$$

SOLUTION 2.19. The Laplace transform of $f(x)$ is:

$$F(s) = \int_0^\infty \exp(-sx)\,f(x)\,dx = \int_0^\infty \left[\int_0^\infty \frac{y\sin(xy)}{1+y^2}\,dy\right]\exp(-sx)\,dx$$

$$= \int_0^\infty \frac{y}{1+y^2}\,dy \int_0^\infty \exp(-sx)\sin(xy)\,dx.$$

Taking into account the fact that:

$$\mathcal{L}\left[\sin\left(xy\right)\right] = \frac{y}{s^2 + y^2},$$

we derive:

$$F\left(s\right) = \int_0^\infty \frac{y^2}{\left(1 + y^2\right)\left(s^2 + y^2\right)}\, dy.$$

The partial fraction expansion of $y^2 / \left(1 + y^2\right)\left(s^2 + y^2\right)$ gives:

$$\frac{y^2}{\left(1 + y^2\right)\left(s^2 + y^2\right)} = \frac{a}{\left(1 + y^2\right)} + \frac{b}{\left(s^2 + y^2\right)} = \frac{a\left(s^2 + y^2\right) + b\left(1 + y^2\right)}{\left(1 + y^2\right)\left(s^2 + y^2\right)},$$

$$a + b = 1 \quad \text{and} \quad as^2 + b = 0 \quad \Longrightarrow a = \frac{1}{1 - s^2}, \quad b = \frac{s^2}{s^2 - 1},$$

which yields:

$$F\left(s\right) = \frac{1}{1 - s^2} \int_0^\infty \frac{dy}{1 + y^2} + \frac{s^2}{s^2 - 1} \int_0^\infty \frac{dy}{s^2 + y^2}.$$

Let us make the variable change: $Y = y/s$. We obtain:

$$
\begin{aligned}
F\left(s\right) &= \frac{1}{1 - s^2} \int_0^\infty \frac{dy}{1 + y^2} + \frac{s^2}{s^2 - 1} \frac{1}{s} \int_0^\infty \frac{dY}{1 + Y^2} \\
&= \left(\frac{1}{s + 1}\right) \int_0^\infty \frac{dy}{1 + y^2}.
\end{aligned}
$$

Taking into account the fact that:

$$\int \frac{dy}{1 + y^2} = \arctan y,$$

we obtain:

$$F\left(s\right) = \frac{\pi}{2} \frac{1}{s + 1}.$$

Therefore,

$$f(x) = \mathcal{L}^{-1}[F(s)] = \frac{\pi}{2} \exp(-x) \quad (x > 0).$$

PROBLEM 2.20. Solve the following integral equation (Abel's integral equation):

$$f(t) = \int_0^t \frac{g(t)}{\sqrt{t-\tau}} \, d\tau, \quad \tau \ge 0,$$

where $f(t)$ is differentiable and $f(0) = 0$.

SOLUTION 2.20. The left-hand side of this equation corresponds to the convolution of $g(t)$ with $1/\sqrt{t}$. Taking the Laplace transform, we obtain:

$$F(s) = G(s) \, \mathcal{L}[t^{-\frac{1}{2}}], \tag{2.13}$$

where:

$$F(s) = \mathcal{L}[f(t)], \quad G(s) = \mathcal{L}[g(t)] \quad \text{and} \quad \mathcal{L}[t^\alpha] = \frac{\Gamma(\alpha+1)}{s^{\alpha+1}}.$$

$\Gamma(s) = \int_0^{+\infty} t^{s-1} \exp(-t) \, dt$ represents the gamma function and exhibits the following properties:

$$\begin{aligned}
\Gamma(n+1) &= n! \quad \text{for } n \text{ integer,} \\
\Gamma(s)\Gamma(1-s) &= \frac{\pi}{\sin \pi s},
\end{aligned}$$

which implies $\Gamma\left(\frac{1}{2}\right) = \sqrt{\pi}$. Equation (2.13) yields:

$$G(s) = \frac{1}{\sqrt{\pi}} \sqrt{s} F(s) = \frac{\sqrt{\pi}}{\pi \sqrt{s}} s F(s).$$

Observe that $sF(s)$ corresponds to the Laplace transform of $f'(t) = df(t)/dt$. Therefore,

$$g(t) = \frac{1}{\pi} \int_0^t \frac{f'(t)}{\sqrt{t-\tau}} \, d\tau.$$

PROBLEM 2.21. Solve the following partial differential equation (the diffusion equation):

$$\frac{dy\,(x,t)}{dt} = w^2 \frac{d^2 y\,(x,t)}{dx^2}, \qquad w = const, \quad 0 < x < \infty, \tag{2.14}$$

with

$$y\,(x,0) = 0,\, y\,(0,t) = f\,(t) = 1.$$

SOLUTION 2.21. We apply the Laplace transform to one selected variable. Let us denote by $Y\,(s)$ the Laplace transform of $y\,(x,t)$ with respect to the variable t. From Equation (2.14), we derive:

$$sY\,(s) = w^2 \frac{d^2}{dx^2} Y\,(s). \tag{2.15}$$

The solution of this differential equation is given by:

$$Y\,(s) = \alpha_1 \exp\left(-\frac{\sqrt{s}}{w}x\right) + \alpha_2 \exp\left(\frac{\sqrt{s}}{w}x\right).$$

For $\alpha_2 \neq 0$, $Y\,(s,x) \to \infty$. We set $\alpha_2 = 0$. From the boundary condition:

$$y\,(0,t) = f\,(t) \Rightarrow Y\,(s,0) = \mathcal{L}\,[f\,(t)] = F\,(s),$$

we derive:

$$Y = F\,(s) \exp\left(-x\frac{\sqrt{s}}{w}\right) = \frac{1}{s} \exp\left(-x\frac{\sqrt{s}}{w}\right). \tag{2.16}$$

From a table of Laplace transforms,[12] we obtain:

$$\mathcal{L}\left[erfc\left(\frac{\delta}{2\sqrt{t}}\right)\right] = \frac{1}{s} \exp\left(-\delta\sqrt{s}\right),$$

------

12. Recall that:

$$erfc\,(x) \;=\; 1 - erfx, \quad erfx = \frac{2}{\sqrt{\pi}} \int_0^x \exp\left(-\tau^2\right) d\tau,$$

$$erf\infty \;=\; 1 \quad \text{and } erfc\,(\infty) = 0.$$

The derivative of the error function is given by:

$$\frac{d}{dx} erfx = \frac{2}{\sqrt{\pi}} \exp\left(-x^2\right).$$

which leads to:

$$y\left(x,t\right) = erfc\left(\frac{x}{2w\sqrt{t}}\right) = 1 - \frac{2}{\sqrt{\pi}} \int\limits_{0}^{x/2w\sqrt{t}} \exp\left(-\tau^2\right) d\tau. \qquad (2.17)$$

For a different boundary condition ($f\left(t\right) \neq 1$), the Laplace transform of $Y\left(s\right)$ may be written in the following form:

$$Y\left(s\right) = F\left(s\right)\exp\left(-\frac{\sqrt{s}}{w}x\right) = sF\left(s\right)Y_1\left(s\right), \quad Y_1\left(s\right) = \frac{1}{s}\exp\left(-\frac{\sqrt{s}}{w}x\right).$$

Using Duhamel's integral (or theorem)[13] (which is related to a convolution), for:

$$g\left(t\right) = erfc\left(\frac{x}{2w\sqrt{t}}\right), \quad g\left(0\right) = erfc\left(\infty\right) = 0,$$

we derive[14]:

$$\frac{d}{dt}g\left(t\right) = \frac{x}{2w\sqrt{\pi}t^{3/2}}\exp\left(-\frac{x^2}{4w^2t}\right).$$

---

13. Recall that:

$$\mathcal{L}\left[f * g\right] = F\left(s\right)G\left(s\right), \quad F\left(s\right) = \mathcal{L}\left[f\right], \quad G\left(s\right) = \mathcal{L}\left[g\right],$$

and consider the function:

$$sF\left(s\right)G\left(s\right).$$

By adding and substracting $f\left(0\right)G\left(s\right)$, we obtain:

$$\begin{aligned} sF\left(s\right)G\left(s\right) &= sF\left(s\right)G\left(s\right) + f\left(0\right)G\left(s\right) - f\left(0\right)G\left(s\right) \\ &= f\left(0\right)G\left(s\right) + \left[sF\left(s\right) - f\left(0\right)\right]G\left(s\right). \end{aligned}$$

The term $sF\left(s\right) - f\left(0\right)$ represents the Laplace transform of $df\left(t\right)/dt$. It follows that:

$$\mathcal{L}^{-1}\left[f\left(0\right)G\left(s\right) + \left[sF\left(s\right) - f\left(0\right)\right]G\left(s\right)\right]$$
$$= f\left(0\right)g\left(t\right) + \int\limits_{0}^{t} g\left(\tau\right)\tfrac{d}{dt}f\left(t - \tau\right)d\tau.$$

This expression represents Duhamel's integral. This integral may also be written in the following form:

$$sF\left(s\right)G\left(s\right) = \mathcal{L}\left[g\left(0\right)f\left(t\right) + \int\limits_{0}^{t} \frac{d}{dt}g\left(t - \tau\right)f\left(\tau\right)d\tau\right].$$

14. Let $\varphi\left(\alpha\right)$ and $\psi\left(\alpha\right)$ be twice-differentiable functions in the interval $\left[c, d\right]$, and let $f\left(x, \alpha\right)$ be both integrable with respect to x over the interval $\left[\varphi\left(\alpha\right), \psi\left(\alpha\right)\right]$ and differentiable with

It follows that:

$$u(x,t) = \frac{x}{2w\sqrt{\pi}} \int_0^t \frac{f(\tau)}{(t-\tau)^{3/2}} \exp\left(-\frac{x^2}{4w^2(t-\tau)}\right) d\tau. \qquad (2.18)$$

Let us introduce the following variable change:

$$z = \frac{x}{2w\sqrt{t-\tau}}.$$

The integral (2.18) becomes:

$$u(x,t) = \frac{2}{\sqrt{\pi}} \int_{x/2w\sqrt{t}}^{\infty} f\left(t - \frac{x^2}{4w^2 z^2}\right) \exp\left(-z^2\right) dz.$$

PROBLEM 2.22. Let us consider Problem 25 of Chapter 1 again. Determine an expression of the current.

SOLUTION 2.22. Recall the equations governing the high-voltage line,

$$-\frac{\partial v(x,t)}{\partial x} = L\frac{\partial i(x,t)}{\partial t} + Ri(x,t), \qquad (2.19)$$

$$-\frac{\partial i(x,t)}{\partial x} = C\frac{\partial v(x,t)}{\partial t} + Gv(x,t).$$

Taking the Laplace transform, we obtain:

$$(Ls + R)I(s) = -\frac{dV(s)}{dx} + Li(0),$$

$$(Cs + G)V(s) = -\frac{dI(s)}{dx} + Cv(0),$$

---

respect to α. Then:

$$\frac{d}{d\alpha} \int_{\varphi(\alpha)}^{\psi(\alpha)} f(x,\alpha)\, dx = \frac{d\psi(\alpha)}{d\alpha} f(\psi(\alpha),\alpha) - \frac{d\varphi(\alpha)}{d\alpha} f(\varphi(\alpha),\alpha) + \int_{\varphi(\alpha)}^{\psi(\alpha)} \frac{\partial f(x,\alpha)}{\partial \alpha}\, dx.$$

where:
$$I(s) = \mathcal{L}[i(t)] \quad \text{and } V(s) = \mathcal{L}[v(t)].$$

By eliminating $I(s)$, we obtain:

$$\frac{d^2 V(s)}{dx^2} - (Ls + R)(Cs + G) V(s) = L\frac{di(0)}{dx} - C(Ls + R) v(0).$$

For initial conditions equal to zero, the solution of this equation is given by:

$$V(s) = \alpha_1 \exp(-wx) + \alpha_2 \exp(wx),$$

where:
$$w = \sqrt{(Ls + R)(Cs + G)}.$$

The coefficients α_1 and α_2 depend on the boundary conditions. For the current, we obtain:

$$I(s) = -\frac{1}{(Ls + R)}\left[\frac{dV(s)}{dx} - Li(0)\right].$$

For initial conditions equal to zero, we obtain:

$$I(s) = -\frac{1}{Ls + R}\frac{dV(s)}{dx} = -\frac{1}{Ls + R}\frac{d}{dx}[\alpha_1 \exp(-wx) + \alpha_2 \exp(wx)]$$
$$= \frac{w}{Ls + R}(\alpha_1 \exp(-wx) - \alpha_2 \exp(wx)) = Z(\alpha_1 \exp(-wx) - \alpha_2 \exp(wx)),$$

where:
$$Z = \sqrt{\frac{Cs + G}{Ls + R}}$$

corresponds to the characteristic impedance of the line.

PROBLEM 2.23. Let us consider the following 2×2 matrix:

$$\mathbf{A} = \begin{bmatrix} 0 & 1 \\ 6 & 1 \end{bmatrix}.$$

Calculate:

$$\exp(\mathbf{A}t).$$

SOLUTION 2.23. Let us consider the following system of differential equations:

$$\frac{d\mathbf{x}}{dt} = \mathbf{A}\mathbf{x}, \quad \mathbf{x}(0) = \mathbf{x}_0, \quad \mathbf{x} = \begin{bmatrix} x_1 \\ x_2 \end{bmatrix}. \tag{2.20}$$

The solution of this system is:

$$\mathbf{x}(t) = \exp(\mathbf{A}t)\mathbf{x}_0. \tag{2.21}$$

Transforming Equation (2.20) yields:

$$sX(s) - x_0 = AX(s),$$

$$X(s) = [sI - A]^{-1} x_0, \tag{2.22}$$

where $X(s) = \mathcal{L}[x(t)]$, and I represents the identity matrix. From Equations (2.21) and (2.22), we derive:

$$\exp(At) = \mathcal{L}^{-1}\left[[sI - A]^{-1}\right]$$

$$[sI - A] = \begin{bmatrix} s & -1 \\ -6 & s-1 \end{bmatrix}, \quad \det[sI - A] = (s-3)(s+2),$$

$$[sI - A]^{-1} = \frac{1}{(s-3)(s+2)} \begin{bmatrix} s-1 & 1 \\ 6 & s \end{bmatrix}.$$

The partial fraction expansion of the terms:

$$\frac{s-1}{(s-3)(s+2)}, \quad \frac{1}{(s-3)(s+2)} \quad \text{and} \quad \frac{s}{(s-3)(s+2)}$$

leads to:

$$\frac{s}{(s-3)(s+2)} = \frac{3}{5(s-3)} + \frac{2}{5(s+2)},$$

$$\frac{1}{(s-3)(s+2)} = \frac{1}{5(s-3)} - \frac{1}{5(s+2)},$$

$$\frac{s-1}{(s-3)(s+2)} = \frac{2}{5(s-3)} + \frac{3}{5(s+2)}.$$

Finally, the inverse Laplace transform yields the desired result:

$$\exp(At) = \frac{1}{5} \begin{bmatrix} 2\exp(3t) + 3\exp(-2t) & \exp(3t) - \exp(-2t) \\ 6\exp(3t) - 6\exp(-2t) & 3\exp(3t) + 2\exp(-2t) \end{bmatrix}.$$

PROBLEM 2.24. In reliability engineering, the renewal function $M(t)$, which is defined as the expected (mean) number of renewals, satisfies an equation called "the fundamental renewal equation". This equation is given by:

$$M(t) = F(t) + \int_0^t M(t - x)\, dF(x) \tag{2.23}$$

where $f(x) = dF(x)/dx$ and $F(x)$ represent the density function of the time to failure and its corresponding probability distribution, respectively. Calculate the renewal function for $f(t) = \lambda \exp(-\lambda t)$.

SOLUTION 2.24. Observe that the integral in Equation (2.23) corresponds to a convolution. A Laplace transform leads to:

$$
\begin{aligned}
\mathcal{L}[M(t)] &= \mathcal{L}\left[F(t) + \int_0^t M(t-x)\,dF(x)\right] \\
&= \mathcal{L}[F(t)] + \mathcal{L}[M(t)]\,\mathcal{L}[dF(t)].
\end{aligned}
$$

Recall that:

$$
f(t) = \frac{dF(t)}{dt} \Rightarrow \mathcal{L}[F(t)] = \frac{1}{s}\mathcal{L}[f],
$$

which implies that:

$$
\mathcal{L}[M(t)] = \frac{1}{s}\frac{\mathcal{L}[f]}{1 - \mathcal{L}[f]}.
$$

The inverse transform of $\mathcal{L}[M(t)]$ gives $M(t)$. For:

$$
f(t) = \lambda \exp(-\lambda t),
$$

we derive:

$$
\mathcal{L}[M(t)] = \frac{\lambda}{s^2}.
$$

Finally, we obtain:

$$
M(t) = \lambda t.
$$

2.5. Block diagrams

PROBLEM 2.25. Show that the systems depicted in Figure 2.9, and in Figure 2.10, are equivalent (we have left out the argument s in the transfer functions of the various blocks).

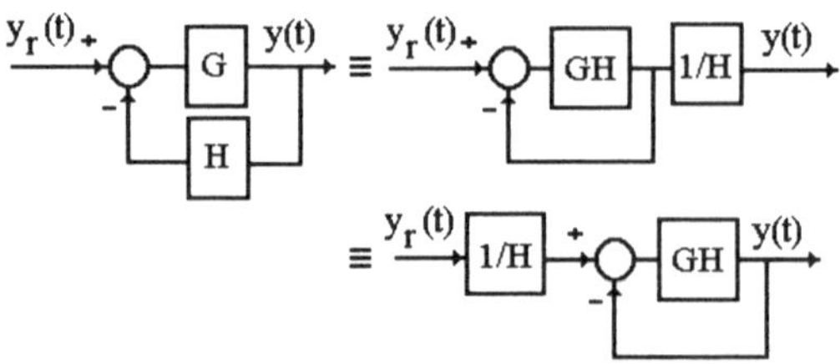

Figure 2.9. *Equivalent systems*

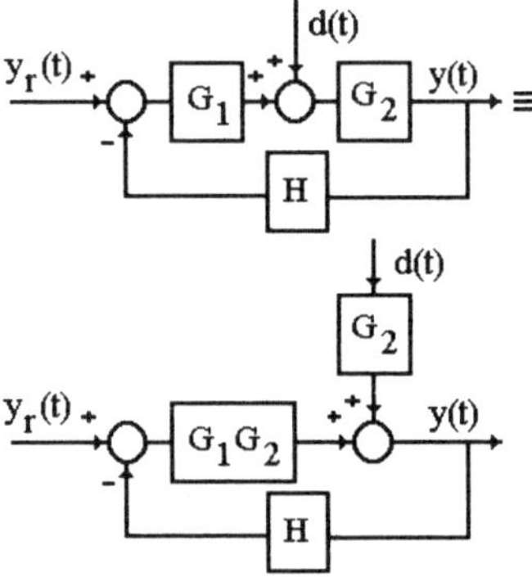

Figure 2.10. *Equivalent systems: Disturbed case*

SOLUTION 2.25. Recall that two systems are equivalent if their transfer functions are equal. The transfer functions of the systems depicted in Figure 2.9 are:

$$F_1(s) = \frac{G}{1+GH},$$

$$F_2(s) = \frac{GH}{1+GH}\frac{1}{H} = \frac{G}{1+GH},$$

$$F_3(s) = \frac{1}{H}\frac{GH}{1+GH} = \frac{G}{1+GH}.$$

These three systems are equivalent.

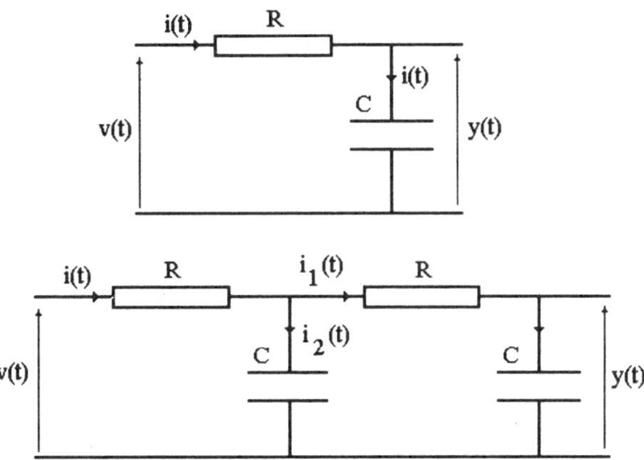

Figure 2.11. *RC systems*

For the first system depicted in Figure 2.10, we have:

$$[(Y_r\,(s) - HY\,(s))\,G_1 + D\,(s)]\,G_2 = Y\,(s)\,,$$

$$Y\,(s) = \frac{G_1 G_2}{1 + G_1 G_2 H}Y_r\,(s) + \frac{G_2}{1 + G_1 G_2 H}D\,(s)\,.$$

For the second system, we obtain:

$$(Y_r\,(s) - HY\,(s))\,G_1 G_2 + D\,(s)\,G_2 = Y\,(s)\,,$$

which leads to the same expression as before. Consequently, these two systems are equivalent.

PROBLEM 2.26.
1. Calculate the transfer function $G\,(s)$ of the RC system depicted at the top of Figure 2.11.
2. Explain why the transfer function of the two RC systems depicted at the bottom of Figure 2.11 is not equal to the product $G\,(s)\,G\,(s)$.
3. Show that the RC system is equivalent to a closed-loop system with:

$$G\,(s) = \frac{\beta}{s}\,, \quad H\,(s) = 1.$$

SOLUTION 2.26. 1. From Ohm's law, we derive:

$$v\left(t\right) \;=\; Ri(t) + \frac{1}{C}\int i(t)\,dt, \tag{2.24}$$

$$y\left(t\right) \;=\; \frac{1}{C}\int i(t)\,dt,$$

where $Ri(t)$ and $1/C\int i(t)\,dt$ represent the voltage drops across the resistor and capacitor, respectively. Equation (2.24) expresses the fact that the input voltage $v\left(t\right)$ is the sum of the voltage drops across the resistor and capacitor. Taking the Laplace transform,

$$V\left(s\right) = RI\left(s\right) + \frac{I\left(s\right)}{sC}, \quad Y\left(s\right) = \frac{I\left(s\right)}{sC},$$

we derive:

$$G\left(s\right) = \frac{Y\left(s\right)}{V\left(s\right)} = \frac{1}{1+RCs}.$$

This simple system is commonly used as a low-pass filter.[15]

2. Now consider the equations related to the second system:

$$v\left(t\right) \;=\; Ri(t) + \frac{1}{C}\int i_2(t)\,dt,$$

$$i(t) \;=\; i_1(t) + i_2(t),$$

$$\frac{1}{C}\int i_2(t)dt \;=\; Ri_1(t) + \frac{1}{C}\int i_1(t)\,dt,$$

$$y\left(t\right) \;=\; \frac{1}{C}\int i_1(t)\,dt.$$

Taking the Laplace transform, we obtain:

$$V\left(s\right) \;=\; RI(s) + \frac{1}{sC}I_2(s),$$

$$I\left(s\right) \;=\; I_1(s) + I_2(s),$$

$$\frac{1}{sC}I_2(s) \;=\; RI_1(s) + \frac{1}{sC}I_1(s)$$

$$Y\left(s\right) \;=\; \frac{1}{sC}I_1(s),$$

---

15. Observe that for $\omega = 1/RC$, $|G\left(j\omega\right)|_{dB} = \left(\sqrt{2}/2\right)_{dB} = -3\,dB$. This means that at this frequency, the amplitude of a sine wave input to an RC filter is attenuated by $3\,dB$ (about 30%). The frequency response will be considered in Chapter 3.

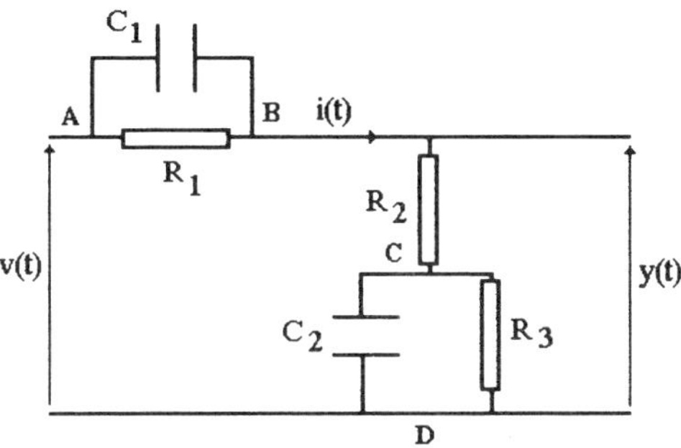

Figure 2.12. *Electrical circuit (Problem 2.27)*

which leads to:

$$\frac{Y\left(s\right)}{V\left(s\right)} = \frac{1}{1 + 3RCs + R^2C^2s^2}.$$

Therefore:

$$\frac{Y\left(s\right)}{V\left(s\right)} \neq G\left(s\right)G\left(s\right).$$

This result was expected because the current $i_1\left(t\right)$ is not equal to zero, unlike the case for one RC system (open circuit, i.e., with an impedance equal to infinity).

3. The closed-loop transfer function is given by:

$$F\left(s\right) = \frac{G\left(s\right)}{1 + G\left(s\right)H\left(s\right)} = \frac{\beta/s}{1 + \beta/s} = \frac{\beta}{s + \beta} = \frac{1}{1 + \left(1/\beta\right)s}.$$

For $\beta = 1/RC$, this closed-loop system is equivalent to an RC system.

PROBLEM 2.27. Calculate the transfer function of the system shown in Figure 2.12.

SOLUTION 2.27. The transfer function of the system between A and B is:

$$\frac{1}{z_1} = \frac{1}{R_1} + C_1s \Rightarrow z_1 = \frac{R_1}{1 + R_1C_1s}.$$

This system is similar to the system located between C and D, the transfer function of which is:

$$\frac{1}{z_2} = \frac{1}{R_3} + C_2s \Rightarrow z_2 = \frac{R_3}{1 + R_3C_2s}.$$

We obtain:

$$
\begin{aligned}
V(s) &= z_1 I(s) + [R_2 + z_2] I(s), \\
Y(s) &= [R_2 + z_2] I(s), \\
I(s) &= \frac{Y(s)}{R_2 + z_2},
\end{aligned}
$$

which leads to:

$$
\begin{aligned}
V(s) &= [R_2 + z_1 + z_2] \frac{Y(s)}{R_2 + z_2}, \\
F(s) &= \frac{Y(s)}{V(s)} = \frac{R_2 + z_2}{R_2 + z_1 + z_2}, \\
F(s) &= \frac{num}{den},
\end{aligned}
$$

where:

$$
\begin{aligned}
num &= R_2 + R_3 + (R_2 R_3 C_2 + R_2 R_1 C_1 + R_3 R_1 C_1) s + R_1 R_2 R_3 C_1 C_2 s^2 \\
den &= R_1 + R_2 + R_3 + (R_2 R_3 C_2 + R_2 R_1 C_1 + R_1 R_3 C_2 + R_1 R_3 C_1) s \\
&\quad + R_1 R_2 R_3 C_1 C_2 s^2.
\end{aligned}
$$

PROBLEM 2.28. Determine the transfer function $F(s) = Y(s)/V(s)$ of the system depicted in Figure 2.13.

SOLUTION 2.28. Using Ohm's law, we obtain:

$$
\begin{aligned}
v(t) &= R i_1(t) + 1/C_2 [i_1(t) - i_3(t)] \, dt, \\
y(t) &= 1/C_2 [i_1(t) - i_3(t)] \, dt - L \frac{di_3}{dt}, \\
1/C_1 i_2 dt &= R i_1(t) + L \frac{di_3}{dt}, \\
i_2(t) &= -i_3(t), \quad i_1(t) = i_3(t) + i_4(t),
\end{aligned}
\tag{2.25}
$$

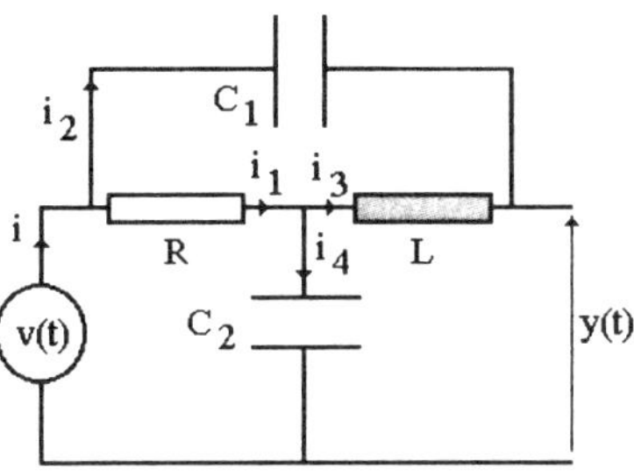

Figure 2.13. *Electrical circuit (Problem 2.28)*

where the term $L di_3/dt$ represents the voltage drop across the inductor L. Taking the Laplace transform, we obtain:

$$\begin{cases} V(s) = RI_1(s) + \dfrac{I_1(s) - I_3(s)}{C_2 s}, \\[2mm] Y(s) = \dfrac{I_1(s) - I_3(s)}{C_2 s} - LsI_3(s), \\[2mm] \dfrac{1}{C_1 s} I_2(s) = RI_1(s) + LsI_3(s), \\[2mm] I_2(s) = -I_3(s), \quad I_1(s) = I_3(s) + I_4(s). \end{cases}$$

Let us rewrite these equations in the following form:

$$V(s) = z_1 I_1(s) + z_2 \left[I_1(s) - I_3(s) \right], \tag{2.26}$$

$$Y(s) = \left[I_1(s) - I_3(s) \right] z_2 - z_3 I_3(s), \tag{2.27}$$

$$z_4 I_2(s) = z_1 I_1(s) + z_3 I_3(s), \tag{2.28}$$

$$I_1(s) = I_3(s) + I_4(s), \tag{2.29}$$

$$I_2(s) = -I_3(s), \tag{2.30}$$

where the z_i, $i = 1, \ldots, 4$, represent the impedances. Combining Equations (2.29) and (2.28), we obtain:

$$z_4 I_2(s) = z_1 \left(I_3(s) + I_4(s) \right) + z_3 I_3(s). \tag{2.31}$$

From Equation (2.30), it follows that:

$$I_4(s) = -\frac{1}{z_1} \left[z_4 + z_1 + z_3 \right] I_3(s).$$

Equation (2.29) leads to:

$$I_1\left(s\right) = -\frac{1}{z_1}\left[z_4 + z_3\right]I_3\left(s\right).$$

On the basis of Equation (2.27), we derive:

$$I_3\left(s\right) = -\frac{Y\left(s\right)}{z_2/z_1\left[z_4 + z_3\right] + z_2 + z_3}.$$

Finally, equation (2.26) yields:

$$V\left(s\right) = -\left[\frac{1}{z_1}\left[z_1 + z_2\right]\left[z_4 + z_3\right] + z_2\right]I_3\left(s\right) \tag{2.32}$$

and

$$V\left(s\right) = \frac{1/z_1\left[z_1 + z_2\right]\left[z_4 + z_3\right] + z_2}{z_2/z_1\left[z_4 + z_3\right] + z_2 + z_3}Y\left(s\right) \tag{2.33}$$

Taking into account the expressions for z_i, $i = \overline{1,4}$,

$$z_1 = R, \quad z_2 = \frac{1}{C_2 s}, \quad z_3 = Ls, \quad z_4 = \frac{1}{C_1 s},$$

we obtain:

$$\frac{Y\left(s\right)}{V\left(s\right)} = \frac{\dfrac{1}{RC_2 s}\left[1/C_1 s + Ls\right] + 1/C_2 s + Ls}{\dfrac{1}{R}\left[R + 1/C_2 s\right]\left[1/C_1 s + Ls\right] + 1/C_2 s}.$$

PROBLEM 2.29. An equivalent schematic diagram of an operational amplifier is given in Figure 2.14. Show that this system is equivalent to a feedback control system.

SOLUTION 2.29. The function of an operational amplifier is to multiply a voltage level by the gain of the amplifier. It represents a basic component of active filters. Let us derive the model equations of this system. From Ohm's law, we derive:

$$\frac{v - \varepsilon}{z_{in}} = \frac{\varepsilon - y}{z_{out}}, \quad y = A\varepsilon,$$

which leads to:

$$\frac{z_{out}}{z_{in} + z_{out}}v + \frac{z_{in}}{z_{in} + z_{out}}y = \varepsilon, \quad y = A\varepsilon.$$

This expression leads to the block diagram depicted in Figure 2.15.

The operational amplifier is a basc component of analogue computers.

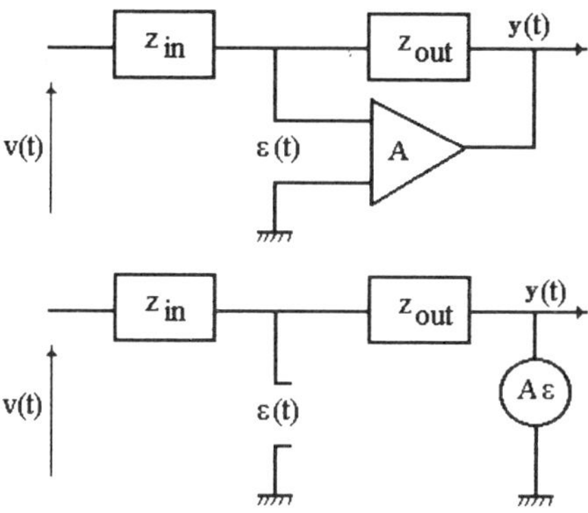

Figure 2.14. *Schematic diagram of an operational amplifier*

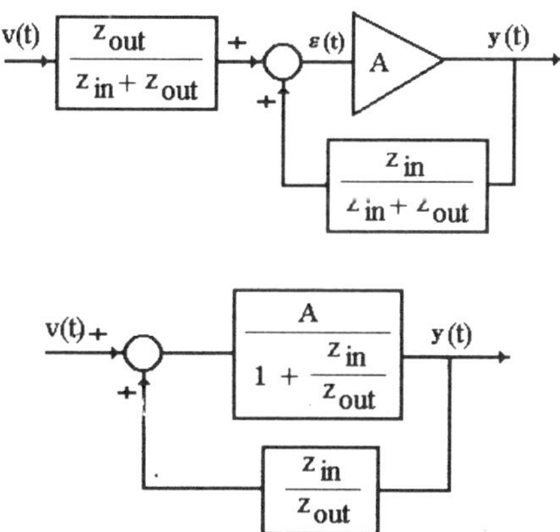

Figure 2.15. *Closed-loop representation of an operational amplifier*

PROBLEM 2.30. For a relative variation of the gain dA/A equal to 5% and for $A = 500$, determine the maximal value of the ratio z_{in}/z_{out} such that the relative variation of the transfer function of an operational amplifier is less than $1/1,000$, i.e., $dF/F < 1/1,000$.

SOLUTION 2.30. The transfer function of an operational amplifier is given by (see Problem 2.29):

$$\frac{z_{out}}{z_{in} + z_{out}} V(s) + \frac{z_{in}}{z_{in} + z_{out}} Y(s) = \frac{1}{A} Y(s), \qquad (2.34)$$

$$F(s) = \frac{Y(s)}{V(s)} = \frac{z_{out}}{(z_{in} + z_{out})/A - z_{in}}.$$

The relative variation of this transfer function is:

$$\frac{dF}{F} = \frac{z_{in}/z_{out} + 1}{z_{in}/z_{out}(1 - A) + 1} \frac{dA}{A} < \frac{1}{1000},$$

which leads to:

$$\frac{z_{out}}{z_{in}} \simeq -11.$$

For high gain, $A \to \infty$, Equation (2.34) leads to:

$$F(s) = \frac{S(s)}{E(s)} = -\frac{z_{out}}{z_{in}}.$$

If we choose $z_{out} = z_{in} = R$ (an electrical resistance), we obtain an inverter. For $z_{out} = 1/Cs$ (capacitance) and $z_{in} = R$ (resistance) with $RC = 1$, we obtain the opposite of an integrator ($F(s) = -1/RCs = -1/s$).

PROBLEM 2.31.
1. Determine the transfer function relating the input $v(t)$ to the electrical charge $q(t)$ of the capacitor C for the system shown in Figure 2.16.
2. Determine the resonant frequency.
3. Determine the transfer function relating the input $v(t)$ to the voltage drop across the resistor R. Determine also the resonant frequency and the bandwidth.

SOLUTION 2.31. 1. From Ohm's law, we derive:

$$v(t) = Ri(t) + L\frac{di(t)}{dt} + \frac{1}{C}\int i(t)\, dt, \qquad (2.35)$$

$$i(t) = \frac{dq(t)}{dt}.$$

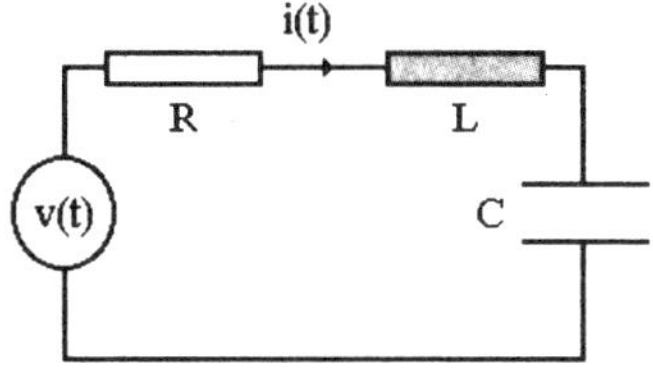

Figure 2.16. *Electrical system (RLC)*

Taking the Laplace transform of equation (2.35), we obtain:

$$V\left(s\right) = RI\left(s\right) + sLI\left(s\right) + \frac{I\left(s\right)}{sC},$$
$$I\left(s\right) = sQ\left(s\right),$$

which leads to:

$$F\left(s\right) = \frac{Q\left(s\right)}{V\left(s\right)} = \frac{C}{1 + RCs + LCs^2}.$$

2. The magnitude of the transfer function is given by:

$$|F\left(j\omega\right)| = \left|\frac{C}{1 - LC\omega^2 + jRC\omega}\right| = \frac{C\left|1 - LC\omega^2 - jRC\omega\right|}{\left(1 - LC\omega^2\right)^2 + \left(RC\omega\right)^2},$$

and $|F\left(j\omega\right)|$ attains its maximal value for:

$$1 - LC\omega^2 = 0,$$

which implies:

$$\omega_r = \frac{\sqrt{LC}}{LC} \quad \text{and} \quad |F\left(j\omega_r\right)| = \frac{\sqrt{LC}}{R}.$$

3. The equations governing the behavior of this system are:

$$V\left(s\right) = RI\left(s\right) + sLI\left(s\right) + \frac{I\left(s\right)}{sC},$$
$$Y\left(s\right) = RI\left(s\right),$$

and the transfer function is given by:

$$F\left(s\right) = \frac{RCs}{1 + RCs + LCs^2}.$$

The resonant frequency corresponds to:

$$\max_{\omega} |F(j\omega)| = \max_{\omega} \left| \frac{jRC\omega}{1 - LC\omega^2 + jRC\omega} \right| = \max_{\omega} \left| \frac{RC\omega}{RC\omega + j(LC\omega^2 - 1)} \right|$$

which leads to:

$$LC\omega^2 - 1 = 0 \Longrightarrow \omega_r = \frac{\sqrt{LC}}{LC} \quad \text{and} \quad \max_{\omega} |F(j\omega)| = 1.$$

Let us now compute the frequency for which $|F(j\omega)|_{dB} = -3\ dB$ ($|F(j\omega)| \simeq \sqrt{2}/2$):

$$\left| \frac{RC\omega}{RC\omega + j(LC\omega^2 - 1)} \right| = \frac{\sqrt{2}}{2},$$

$$\left| \frac{1}{1 + j(L/R\omega - 1/RC\omega)} \right| = \frac{1}{\sqrt{1 + (L/R\omega - 1/RC\omega)^2}} = \frac{1}{\sqrt{2}},$$

which implies:

$$\left(\frac{L}{R}\omega - \frac{1}{RC\omega} \right)^2 = 1 \Longrightarrow \frac{L}{R}\omega - \frac{1}{RC\omega} = \pm 1,$$

$$LC\omega^2 \pm RC\omega - 1 = 0.$$

For:

$$LC\omega^2 + RC\omega - 1 = 0,$$

we obtain:

$$\omega_1^1 = \frac{-RC + \sqrt{(RC)^2 + 4LC}}{2LC},$$

$$\omega_2^1 = \frac{-RC - \sqrt{(RC)^2 + 4LC}}{2LC} < 0,$$

and for:

$$LC\omega^2 - RC\omega - 1 = 0,$$

we obtain:

$$\omega_1^2 = \frac{RC + \sqrt{(RC)^2 + 4LC}}{2LC},$$

$$\omega_2^2 = \frac{RC - \sqrt{(RC)^2 + 4LC}}{2LC} < 0.$$

Finally, the bandwidth is equal to:

$$\omega_1^2 - \omega_1^1 = \frac{R}{L}.$$

2.6. Feedback systems

PROBLEM 2.32. Show that the transfer function:

$$F_1\left(s\right) = \frac{\omega_n^2}{s^2 + 2\zeta\omega_n s + \omega_n^2}$$

is equivalent to a unity-feedback system with a transfer function of the direct path equal to:

$$G\left(s\right) = \frac{\omega_n^2}{s\left(s + 2\zeta\omega_n\right)}.$$

SOLUTION 2.32. The closed-loop transfer function of the unity-feedback system is given by:

$$\begin{aligned}
F\left(s\right) &= \frac{G\left(s\right)}{1 + G\left(s\right)} = \frac{\omega_n^2/\left[s\left(s + 2\zeta\omega_n\right)\right]}{1 + \omega_n^2/\left[s\left(s + 2\zeta\omega_n\right)\right]} \\
&= \frac{\omega_n^2}{s^2 + 2\zeta\omega_n s + \omega_n^2} = F_1\left(s\right),
\end{aligned}$$

which corresponds to the desired result.

PROBLEM 2.33. For the system depicted in Figure 2.17, calculate the Laplace transform of the system output $y\left(t\right)$, as a function of the Laplace transforms of the input $y_r\left(t\right)$ and the disturbance $d\left(t\right)$.

SOLUTION 2.33. There exist two methods for solving this problem. The first method consists of introducing secondary variables and writing equations related to the behavior of the system. The second method is based on the superposition principle, which holds for linear systems. This principle can be stated as follows. Consider a linear system with input u and output y. If an output y_i results from an input u_i, i.e., $u_i \rightarrow y_i$, then:

$$\sum_i u_i \rightarrow \sum_i y_i.$$

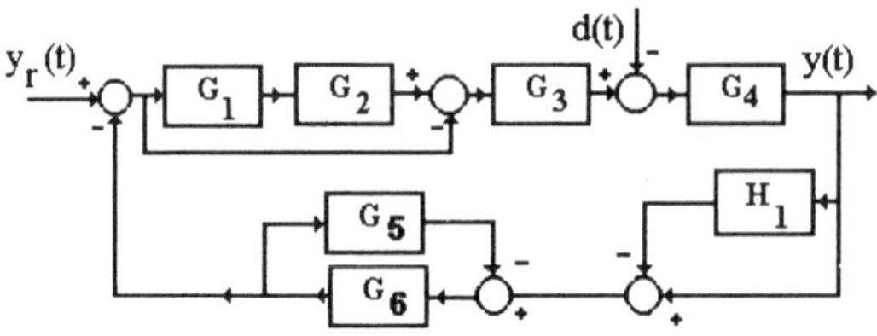

Figure 2.17. *Schematic diagram of a disturbed system*

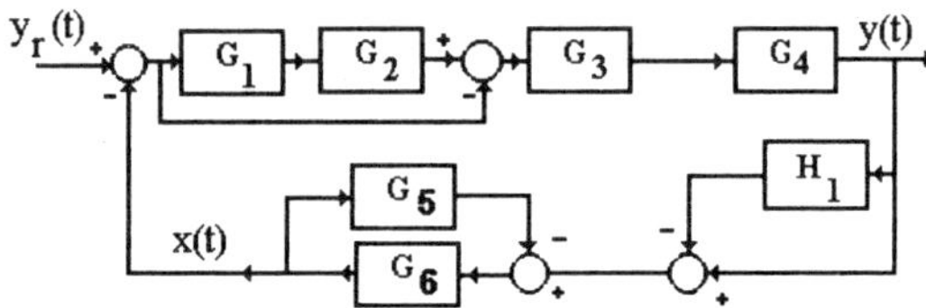

Figure 2.18. *Block diagram in the absence of the perturbation $d(t)$*

We shall solve this problem by using the second method. On the basis of the superposition principle, we derive:

$$y\left(t\right) = y\left(y_r\left(t\right), d\left(t\right)\right) = y_{d(t)\equiv 0}\left(t\right) + y_{y_r(t)\equiv 0}\left(t\right),$$

$$Y\left(s\right) = Y_{d(t)\equiv 0}\left(s\right) + Y_{y_r(t)\equiv 0}\left(s\right).$$

This decomposition of the output into two terms permits us to consider two block diagrams (Figures 2.18 and 2.19).

Observe that the input of the block constituted by $G_1\left(s\right)$ and $G_2\left(s\right)$ corresponds to $\varepsilon\left(t\right) = y_r\left(t\right) - x(t)$. Taking into account the fact that in Figure 2.19, we consider only the effect of $d\left(t\right)$ under the assumption that $y_r\left(t\right) \equiv 0$, the input of this block is $\varepsilon\left(t\right) = -x(t)$. This explains why we have introduced a block of transfer function -1. In order to obtain a classical feedback structure, we use the fact that adding is equivalent to subtracting the opposite quantity and vice versa (see Figure 2.20).

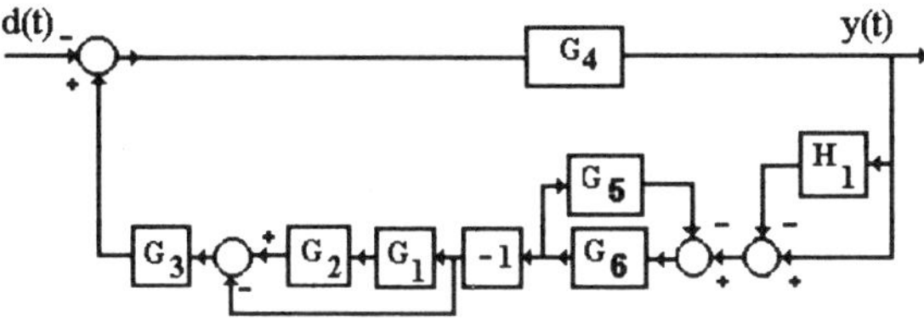

Figure 2.19. *Effect of a disturbance on the system output*

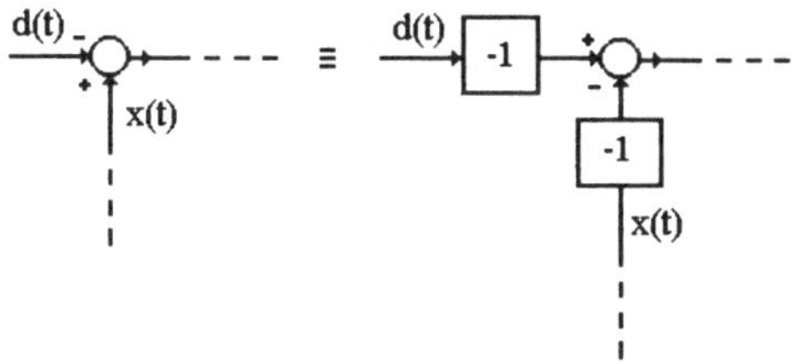

Figure 2.20. *Modification of the feedback inputs*

By simplifying the schemes depicted in Figures 2.18 and 2.19, we obtain:

$$Y(s) = \frac{(G_1 G_2 - 1)(1 + G_5 G_6) G_3 G_4}{1 + G_5 G_6 + G_3 G_4 G_6 (G_1 G_2 - 1)(1 - H_1)} Y_r(s)$$
$$- \frac{G_4 (1 + G_5 G_6)}{1 + G_5 G_6 + G_3 G_4 G_6 (G_1 G_2 - 1)(1 - H_1)} D(s).$$

REMARK 2.1. Block diagrams can also be reduced using signal flow graphs and Mason's rule.

PROBLEM 2.34. Consider a speed regulation system, namely the Ward–Leonard system. The objective is to rotate a mass of moment of inertiainertia J_m at a desired

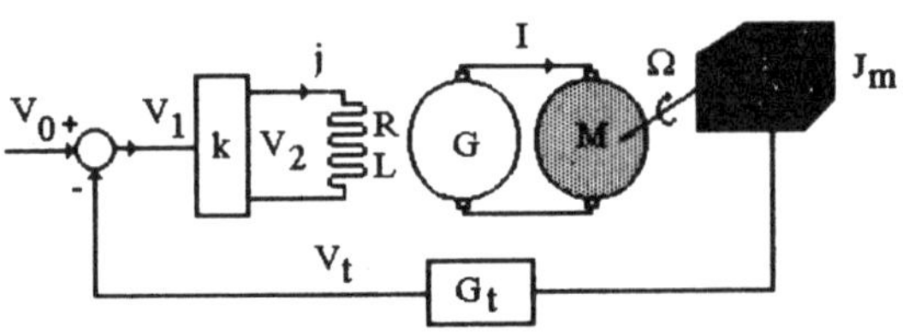

Figure 2.21. *Schematic diagram of the Ward–Leonard system*

speed Ω_d. To achieve this objective, the mass is connected to an electric motor, which is supplied by a generator, the excitation current of which is controlled by manipulating the excitation voltage. This excitation voltage is provided by another generator. A schematic diagram of the Ward–Leonard system is shown in Figure 2.21. The model of this system is as follows.

Electrical phenomena. The electromotive force E_G of the generator is equal to the back electromotive force E_M of the motor plus the voltage drop across the resistance $R_G + R_M$ and the inductor $L_G + L_M$. The electromotive force E_G is proportional to the excitation current j, and the back electromotive force E_M is proportional to the angular velocity Ω:

$$E_G \;=\; E_M + (R_G + R_M)\, I + (L_G + L_M)\,\frac{dI}{dt},$$

$$E_G \;=\; k_2 j, \quad V_2 = Rj + L\frac{dj}{dt}, \quad V_2 = k_1 V_1,$$

$$V_1 \;=\; V_0 - V_t, \quad E_M = k_4 \Omega, \quad V_t = k_5 \Omega,$$

where the indexes G and M correspond to the generator and the motor, respectively. The angular velocity is measured by a tacho-generator which produces an electromotive force proportional to the angular velocity.

Mechanical phenomena. The driving torque C_M is proportional to the current I, and the difference between the driving torque and the opposing torque is equal to the moment of inertia J_m multiplied by the angular acceleration $d\Omega/dt$:

$$C_M \;=\; k_3 I,$$

$$C_M \;=\; J_m \frac{d\Omega}{dt} + C_r.$$

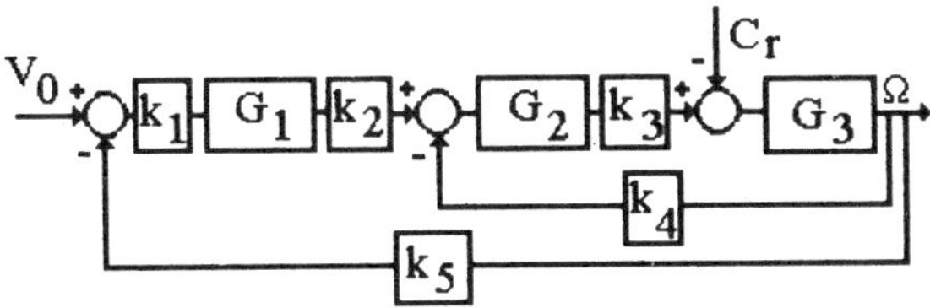

Figure 2.22. *Block diagram of the Ward–Leonard system*

1. Draw a block diagram of the system using the transfer functions of each part of the system.

2. Determine an expression for the Laplace transform of the output $\Omega(t)$ as a function of the input $V_0(t)$ and the perturbation $C_r(t)$.

SOLUTION 2.34. 1. The transfer functions associated with these equations are:

$$I(s) = \frac{E_G(s) - E_M(s)}{(R_G + R_M) + (L_G + L_M)s}, \tag{2.36}$$

$$J(s) = \frac{V_2}{R + Ls},$$

$$C_M(s) = k_3 I(s), \tag{2.37}$$

$$\Omega(s) = \frac{C_M(s) - C_r(s)}{J_m s}. \tag{2.38}$$

We start with Equation (2.36), draw a block the output of which is $I(s)$, and then we use Equation (2.37) and obtain a block the output of which is $C_r(s)$. On the basis of this output and equation (2.38), we obtain the process output and so on, and we construct the feedback using the remaining equations. The resulting block diagram of the Ward–Leonard system is depicted in Figure 2.22, where:

$$G_1(s) = \frac{1}{R + Ls}, \quad G_2(s) = \frac{1}{(R_G + R_M) + (L_G + L_M)s}, \quad G_3(s) = \frac{1}{J_m s}.$$

2. The system (model) is linear. We apply the superposition principle (see Problem 2.33). It follows that:

$$\Omega(s) = \frac{k_1 k_2 k_3 G_1 G_2 G_3}{1 + k_3 k_4 G_2 G_3 + k_1 k_2 k_3 k_5 G_1 G_2 G_3} V_0(s)$$
$$- \frac{G_3}{1 + k_3 G_2 G_3 (k_4 + k_1 k_2 k_5 G_1)} C_r(s).$$

PROBLEM 2.35. Consider a structure consisting of N systems in series. Each system has the following transfer function:

$$F_i(s) = \frac{k_i}{1 + T_i s}, \quad \forall i k_i = k \text{ and } T_i = T, \quad i = 1, \cdots, N., \quad y(0) = 0.$$

1. Calculate the derivatives $d^i y(t)/dt^i$, $i = 1, 2, 3$, of the step response for $t = 0$. For $k = 1$ and $T = 2$, draw the step responses for $N = \overline{1, 6}$.
2. For $k_i = k = 1$ and $T_i = T/N$, $i = 1, \cdots, N$, show that when $N \rightarrow \infty$, the step response of this structure is equivalent to a time delay. Draw the step response of this system for $T = 2\ s$ and $N = 150$.

SOLUTION 2.35. 1. The transfer function of this structure is:

$$F(s) = (F_i(s))^N = \frac{k}{1 + Ts} \frac{k}{1 + Ts} \cdots \frac{k}{1 + Ts} = \left(\frac{k}{1 + Ts}\right)^N,$$

and the Laplace transform of its step response is given by:

$$Y(s) = \frac{1}{s} \left(\frac{k}{1 + Ts}\right)^N \tag{2.39}$$

Let us now consider the partial fraction expansion of equation (2.39),

$$Y(s) = \frac{\alpha_0}{s} + \frac{\alpha_1}{1 + Ts} + \frac{\alpha_2}{(1 + Ts)^2} + \cdots + \frac{\alpha_N}{(1 + Ts)^N}, \tag{2.40}$$

where[16]:

$$\alpha_0 = k^N, \quad \alpha_i = -Tk^N, \quad i = 1, \cdots, N.$$

---

16. The following result can proved by induction:

$$\frac{b}{x(1 + ax)^{N-1}} = \frac{b}{x} - \frac{ab}{(1 + ax)} - \cdots - \frac{ab}{(1 + ax)^{N-1}}.$$

Let us calculate the derivatives of the step response. The Laplace transform of the derivative $d^i y(t)/dt^i$ is given by:

$$\mathcal{L}\left(\frac{d^i y(t)}{dt^i}\right) = s^i Y(s) - s^{i-1}y(0) - s^{i-2}\left.\frac{dy(t)}{dt}\right|_{t=0} - \cdots - \left.\frac{d^{i-1}y(t)}{dt}\right|_{t=0}.$$

Taking into account the initial condition $y(t)_{t=0} = y(0) = 0$ and Tauber's Theorem, we obtain the following.

First derivative:

$$\mathcal{L}\left(\frac{dy(t)}{dt}\right) = \left(\frac{k}{1+Ts}\right)^N - y(0) = \left(\frac{k}{1+Ts}\right)^N.$$

$$\lim_{t\to 0}\frac{dy(t)}{dt} = \lim_{s\to\infty} s\left(\frac{k}{1+Ts}\right)^N = y'(0)\begin{cases} = \dfrac{k}{T} & \text{for } N=1, \\ \to 0 & \text{for } N\ge 2.\end{cases}$$

Second derivative:

$$\mathcal{L}\left(\frac{d^2y(t)}{dt^2}\right) = s^2\frac{1}{s}\left(\frac{k}{1+Ts}\right)^N - sy(0) - y'(0)$$

$$= \begin{cases} s\left(\dfrac{k}{1+Ts}\right) - \dfrac{k}{T} & \text{for } N=1, \\[2ex] s\left(\dfrac{k}{1+Ts}\right)^N & \text{for } N\ge 2.\end{cases}$$

For $N = 1$:

$$\lim_{t\to 0}\frac{d^2y(t)}{dt^2} = \lim_{s\to\infty} s\left[s\left(\frac{k}{1+Ts}\right) - \frac{k}{T}\right]$$

$$= \lim_{s\to\infty} s\left[\frac{k}{T} - \frac{k}{T(1+Ts)} - \frac{k}{T}\right]$$

$$= \lim_{s\to\infty}\left[-\frac{ks}{T(1+Ts)}\right] = -\frac{k}{T^2}.$$

This leads to:

$$\frac{b}{x(1+ax)^N} = \frac{1}{(1+ax)}\frac{b}{x(1+ax)^{N-1}}$$

$$= \frac{1}{(1+ax)}\left[\frac{b}{x} - \frac{ab}{(1+ax)} - \cdots - \frac{ab}{(1+ax)^{N-1}}\right]$$

$$= \frac{b}{x(1+ax)} - \frac{ab}{(1+ax)^2} - \cdots - \frac{ab}{(1+ax)^N}$$

$$= \frac{b}{x} - \frac{ab}{(1+ax)} - \frac{ab}{(1+ax)^2} - \cdots - \frac{ab}{(1+ax)^N},$$

which corresponds to the desired result.

For $N = 2$:

$$\lim_{t \to 0} \frac{d^2 y\,(t)}{dt^2} = \lim_{s \to \infty} s\left[s\left(\frac{k}{1 + Ts} \right)^2 \right] = \frac{k^2}{T^2}.$$

For $N > 2$:

$$\lim_{t \to 0} \frac{d^2 y\,(t)}{dt^2} = \lim_{s \to \infty} s\left[s\left(\frac{k}{1 + Ts} \right)^N \right] = 0.$$

Third derivative:

$$\mathcal{L}\left(\frac{d^3 y\,(t)}{dt^3} \right) = s^3 \frac{1}{s} \left(\frac{k}{1 + Ts} \right)^N - s^2 y\,(0) - s\left. \frac{dy\,(t)}{dt} \right|_{t=0} - \left. \frac{d^2 y\,(t)}{dt^2} \right|_{t=0}.$$

For $N = 1$ and

$$y\,(0) = 0, \quad \left. \frac{dy\,(t)}{dt} \right|_{t=0} = \frac{k}{T} \quad \text{and} \quad \left. \frac{d^2 y\,(t)}{dt^2} \right|_{t=0} = -\frac{k}{T^2},$$

$$\mathcal{L}\left(\frac{d^3 y\,(t)}{dt^3} \right) = s^2 \left(\frac{k}{1 + Ts} \right) - \frac{k}{T}s + \frac{k}{T^2}$$

$$= \frac{k}{T}s - \frac{k}{T^2} + \frac{k}{T^2}\frac{1}{1 + Ts} - \frac{k}{T}s + \frac{k}{T^2} = \frac{k}{T^2}\frac{1}{1 + Ts},$$

$$\lim_{t \to 0} \frac{d^3 y\,(t)}{dt^3} = \lim_{s \to \infty} \left(\frac{k}{T^2}\frac{s}{1 + Ts} \right) = \frac{k}{T^3}.$$

For $N = 2$ and:

$$y\,(0) = 0, \quad \left. \frac{dy\,(t)}{dt} \right|_{t=0} = 0 \quad \text{and} \quad \left. \frac{d^2 y\,(t)}{dt^2} \right|_{t=0} = \frac{k^2}{T^2},$$

$$\mathcal{L}\left(\frac{d^3 y\,(t)}{dt^3} \right) = s^2 \left(\frac{k}{1 + Ts} \right)^2 - \frac{k^2}{T^2} = -\frac{k^2\,(1 + 2Ts)}{T^2\,(1 + Ts)^2},$$

$$\lim_{t \to 0} \frac{d^3 y\,(t)}{dt^3} = \lim_{s \to \infty} \left(-\frac{k^2\,(1 + 2Ts)\,s}{T^2\,(1 + Ts)^2} \right) = -\frac{2k^2}{T^3}$$

For $N > 2$ and:

$$y\,(0) = 0, \quad \left. \frac{dy\,(t)}{dt} \right|_{t=0} = 0 \quad \text{and} \quad \left. \frac{d^2 y\,(t)}{dt^2} \right|_{t=0} = 0,$$

$$\mathcal{L}\left(\frac{d^3 y(t)}{dt^3}\right) = s^2 \left(\frac{k}{1+Ts}\right)^N,$$

$$\lim_{t \to 0} \frac{d^3 y(t)}{dt^3} = \lim_{s \to \infty} s^3 \left(\frac{k}{1+Ts}\right)^N = \begin{cases} \dfrac{k^2}{T^3} \text{ for } N = 3, \\ 0 \text{ for } N > 3. \end{cases}$$

Recall that:

$$L\left[t^n \exp(-\lambda t)\right] = \frac{\Gamma(n+1)}{(s+\lambda)^{n+1}}, \quad \Gamma(n+1) = n!$$

where $\Gamma(.)$ represents the gamma function. The inverse of the Laplace transform of Equation (2.40) is given by:

$$Y(s) = \frac{k^N}{s} - \frac{k^N}{1/T+s} - \frac{k^N/T}{(1/T+s)^2} - \cdots - \frac{k^N/T^{N-1}}{(1/T+s)^N},$$

$$\begin{aligned} y(t) \\ = k^N - k^N \left[+1 + \frac{1}{T}t + \cdots + \frac{1}{(N-1)!T^{N-1}}t^{N-1}\right] \exp\left(-\frac{t}{T}\right). \end{aligned}$$

For $k = 1$ and $T = 2$, the step responses for $N = \overline{1,5}$ are depicted in Figure 2.23.

2. For $k_i = k = 1$ and $T_i = T/N$, the transfer function and the step response of this system are given by:

$$F(s) = \frac{1}{(1+(T/N)s)^N}, \quad Y(s) = \frac{1}{s(1+(T/N)s)^N}.$$

The partial fraction expansion of the step response gives:

$$Y(s) = \frac{1}{s} - \sum_{i=1}^{N} \frac{1}{T^{i-1}} \frac{N^{i-1}}{(T/N+s)^i}.$$

The inverse Laplace transform is:

$$y(t) = \left[1 - \sum_{i=1}^{N} \frac{1}{(i-1)!} \left(\frac{N}{T}\right)^{i-1} t^{i-1} \exp\left(-\frac{N}{T}t\right)\right] 1(t).$$

Because the Laplace transform is defined for functions equal to zero for negative time, we have multiplied the second term of the right-hand side by $1(t)$ which, is equal to zero for $t < 0$.

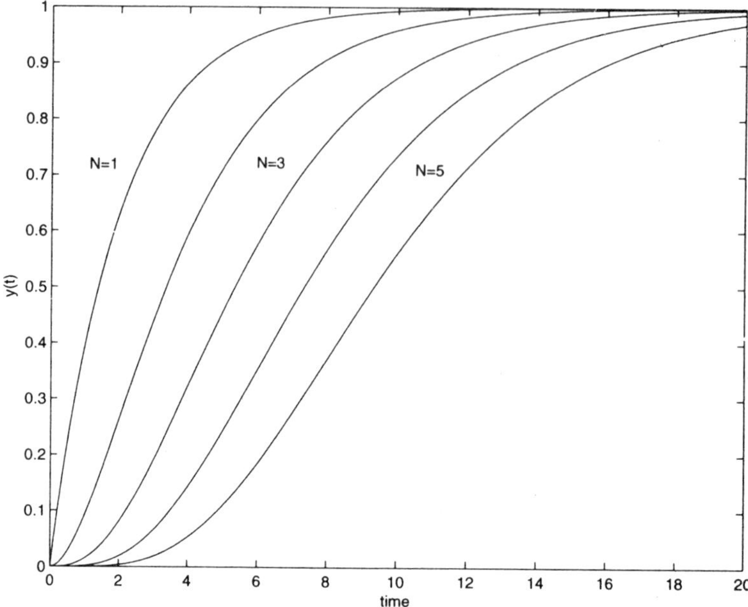

Figure 2.23. *Step responses for $N = \overline{1,5}$*

Recall that the transfer function of the system considered[17] is:

$$\frac{\dfrac{1}{\left(1 + T/Ns\right)^N}}{} = \frac{1}{1 + Ts + \left(N\left(N-1\right)/2N^2\right)T^2s^2 + \left(N\left(N-1\right)\left(N-2\right)/3!N^3\right)T^3s^3 + \cdots}.$$

For $N \rightarrow \infty$, we obtain:

$$\frac{1}{\left(1 + (T/N)\,s\right)^N} \rightarrow \frac{1}{1 + Ts + (1/2)\,T^2s^2 + (1/3!)\,T^3s^3 + \cdots} \qquad (2.41)$$

---

17. $(a+b)^n = \sum\limits_{i=0}^{n} C_n^i a^{n-i}b^i = \sum\limits_{i=0}^{n} C_n^i a^i b^{n-i}, \quad C_n^i = \frac{n!}{i!(n-i)!}.$

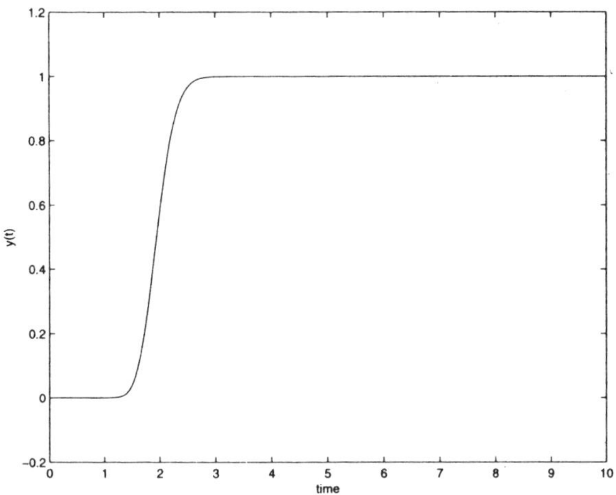

Figure 2.24. *Step response of N first order systems in series with $N = 50$,*
$T = 2 \sec$ and $k = 1$

which corresponds to the Taylor expansion of a time delay:

$$\frac{1}{\exp(\tau s)} = \frac{1}{1 + \tau s + (1/2!)\,\tau^2 s^2 + (1/3!)\,\tau^3 s^3 + \cdots + (1/n!)\,\tau^n s^n + \cdots}. \tag{2.42}$$

Expressions (2.41) and (2.42) are similar and show that a series of N first-order systems with time constants equal to T/N behaves like a system with a time delay equal to T. The step response of a system for $N = 50$ is plotted in Figure 2.24. This response corresponds to a time delay very close to $\tau = 2\,s$.

Note that there exist other approximations to a time delay, such as the Padé approximations. A simple approximation is:

$$G(s) = \frac{P(s)}{Q(s)} = \frac{1 - (\tau/2)\,s}{1 + (\tau/2)\,s}. \tag{2.43}$$

Long division of $G(s)$ leads to:

$$G(s) = 1 - \tau s + \frac{\tau^2}{2}s^2 - \frac{\tau^3}{4}s^3 + \frac{\tau^4}{8}s^4 + \cdots, \tag{2.44}$$

and the Taylor expansion of $\exp(-\tau)$ gives:

$$\exp(-\tau) = 1 - \tau s + \frac{\tau^2}{2}s^2 - \frac{\tau^3}{3!}s^3 + \frac{\tau^4}{4!}s^4 + \cdots. \tag{2.45}$$

Note that only the first three terms are identical. Nevertheless, Equation (2.43) represents a good approximation of the time delay τ. Other approximations which lead to high accuracy can be obtained by considering polynomials $P(s)$ and $Q(s)$ of high degree. Another approximation is given by:

$$\exp(-\tau) = \left(\frac{-\tau s + 2n}{\tau s + 2n}\right)^n, \quad n = 1, 2, 3, \cdots.$$

In order to obtain a desired reaction time τ by using reactors of sizes commonly available in the market, chemical engineers use a cascade of reactors of the same size with residence times τ_i such that the total residence time is greater than or equal to τ.

PROBLEM 2.36. Show that for the non-linear system:

$$\frac{dx_1}{dt} = f_1(x_1, x_2) + f_2(x_1, x_2) u,$$
$$\frac{dx_2}{dt} = f_3(x_1, x_2), \tag{2.46}$$
$$y = x_1,$$

the feedback controller:

$$u = \frac{v - f_1(x_1, x_2)}{f_2(x_1, x_2)} \tag{2.47}$$

leads to a linear system.

SOLUTION 2.36. There exist various approaches for transforming a non-linear system into a linear system [NIJ 90]. One approach consists of transforming the control input into a non-linear feedback. This is the case here. Substitution of Equation (2.47) into (2.46) gives:

$$\frac{dx_1}{dt} = f_1(x_1, x_2) + v - f_1(x_1, x_2) = v, \quad y = x_1,$$

which is linear.

PROBLEM 2.37. Consider the position regulation system depicted in Figure 2.25.

1. Calculate the closed-loop transfer function $Y(s)/Y_r(s)$.

2. For what values of the parameters A_1 and A_2 does this system have a damping factor ζ equal to 0.7?

3. For $A_1 A_2 = 20$, compute the damping factor, the resonant frequency and the height of the resonant peak (the maximum magnitude of the frequency response).

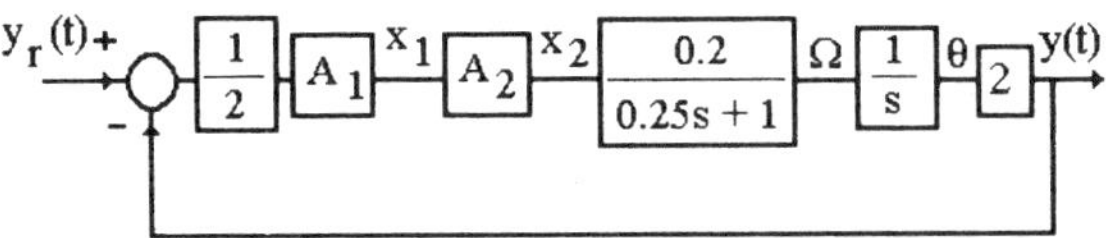

Figure 2.25. *Position regulation system*

SOLUTION 2.37. 1. The open-loop transfer function is equal to:

$$T\left(s\right) = \frac{0.4A_1 A_2}{2s\left(1. + 0.25s\right)}.$$

The closed-loop transfer function is:

$$F\left(s\right) = \frac{T\left(s\right)}{1 + T\left(s\right)} = \frac{0.4A_1 A_2/\left[2s\left(1. + 0.25s\right)\right]}{1 + 0.4A_1 A_2/\left[2s\left(1. + 0.25s\right)\right]},$$

and we obtain:

$$F\left(s\right) = \frac{0.4A_1 A_2}{0.5s^2 + 2s + 0.4A_1 A_2}.$$

2. The transfer function of a second-order system is:

$$\frac{k\omega_n^2}{s^2 + 2\zeta\omega_n s + \omega_n^2}$$

By identification, we obtain:

$$\frac{0.8A_1 A_2}{s^2 + 4s + 0.8A_1 A_2} = \frac{k\omega_n^2}{s^2 + 2\zeta\omega_n s + \omega w_n^2},$$

$$k\omega_n^2 = 0.8A_1 A_2, \quad \zeta\omega_n = 2, \quad \omega_n^2 = 0.8A_1 A_2. \tag{2.48}$$

Substituting $\zeta = 0.7$, we obtain:

$$\omega_n = \frac{2}{0.7} = 2.85, \quad k = 1,$$

and finally:

$$A_1 A_2 = \frac{\omega_n^2}{0.8} = 10.2.$$

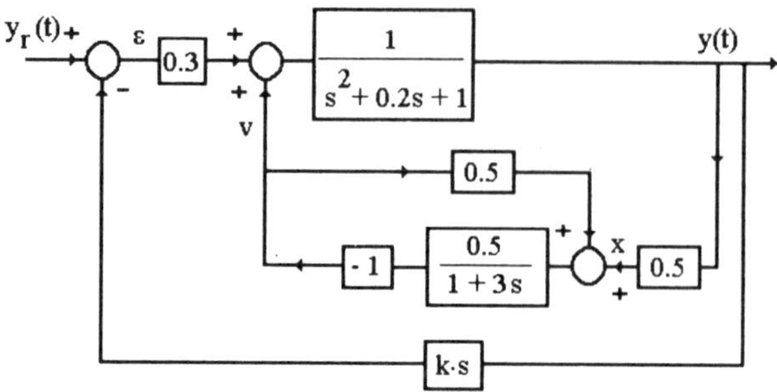

Figure 2.26. *Control system for the blade pitch of the propellers of an aircraft*

3. For $A_1 A_2 = 20$, we obtain from Equation (2.48):

$$\omega_n^2 = 0.8 A_1 A_2 = 16, \quad \omega_n = 4, \quad \zeta = \frac{2}{\omega_n} = 0.5.$$

The resonant frequency and the height of the resonant peak are given by:

$$\omega_r = \omega_n \sqrt{1 - 2\zeta^2} = 2.82, \quad M_p = \frac{1}{2\zeta \sqrt{1 - \zeta^2}} = 1.15.$$

PROBLEM 2.38. A control system for the blade pitch of the propellers of an aircraft is illustrated in Figure 2.26.

1. Calculate the closed-loop transfer function.

2. Calculate the steady-state error (position error, $\varepsilon\,(\infty)$) for $y_r\,(t) = 1\,(t)$ (a unit step).

SOLUTION 2.38. 1. For a system with input $x\,(t)$ and output $v\,(t)$, we derive:

$$-\frac{0.5}{1 + 3s} \left[X\,(s) + 0.5V\,(s) \right] = V\,(s),$$

and obtain the transfer function:

$$\frac{V\,(s)}{X\,(s)} = -\frac{0.5}{1.25 + 3s}.$$

The transfer function of the internal loop ($\mathcal{L}\left[y_r\left(t\right)\right]/\mathcal{L}\left[0.3\varepsilon\left(t\right)\right]$) is:

$$
\begin{aligned}
G_1\left(s\right) &= \frac{1/\left(s^2+0.2s+1\right)}{1+0.25/\left[\left(s^2+0.2s+1\right)\left(1.25+3s\right)\right]} \\[2mm]
&= \frac{1.25+3s}{3s^3+1.85s^2+3.25s+1.5}.
\end{aligned}
$$

Let us denote ks and $0.3G_1\left(s\right)$ by $H\left(s\right)$ and $G\left(s\right)$, respectively. The closed-loop transfer function is:

$$
F\left(s\right) = \frac{Y\left(s\right)}{Y_r\left(s\right)} = \frac{G\left(s\right)}{1+G\left(s\right)H\left(s\right)} = \frac{0.3G_1\left(s\right)}{1+0.3G_1\left(s\right)H\left(s\right)}
$$

$$
= \frac{0.3\left(1.25+3s\right)}{3s^3+\left(1.85+0.9k\right)s^2+\left(3.25+0.375k\right)s+1.5}.
$$

2. The Laplace transform of the error $\varepsilon\left(t\right)$ is given by:

$$
\Xi\left(s\right) = \frac{1}{1+G\left(s\right)H\left(s\right)}Y_r\left(s\right) = \frac{1}{1+0.3G_1\left(s\right)H\left(s\right)}Y_r\left(s\right)
$$

$$
= \frac{1.85s^2+3s^3+3.25s+1.5}{3s^3+\left(1.85+0.9k\right)s^2+\left(3.25+0.375k\right)s+1.5}\frac{1}{s}.
$$

Using the final-value theorem, we obtain:

$$
\varepsilon\left(\infty\right) = \lim_{s\to0}s\frac{1.85s^2+3s^3+3.25s+1.5}{3s^3+\left(1.85+0.9k\right)s^2+\left(3.25+0.375k\right)s+1.5}\frac{1}{s},
$$

$$
\varepsilon\left(\infty\right) = \frac{1.5}{1.5} = 1.
$$

REMARK 2.2. In the simplification of transfer functions, it is important to avoid simplification of unstable zeros and poles. In fact, for initial conditions different from zero, the system output tends to infinity.[18] In other words, the transfer function does not display all the modes of the system.

---

18. Let us consider the system governed by the following differential equation:

$$
y\left(t\right)+a\frac{d}{dt}y\left(t\right) = u\left(t\right)+a\frac{d}{dt}u\left(t\right), \quad y\left(0\right)=y_0, \quad u\left(0\right)=0.
$$

A Laplace transform yields:

$$
Y\left(s\right) = \frac{1+as}{1+as}U\left(s\right)+\frac{y_0}{1/a+s}.
$$

PROBLEM 2.39. Calculate the error coefficients (the error series) for a unity-feedback system with:

$$G(s) = \frac{k(1+s)}{(s+3)(s+5)}, \quad H(s) = 1.$$

SOLUTION 2.39. The Laplace transform of the error is given by:

$$\Xi(s) = \frac{1}{1+G(s)H(s)} Y_r(s) = W(s) Y_r(s)$$

$$= \frac{(s+3)(s+5)}{(s+3)(s+5)+k(1+s)} Y_r(s),$$

and, on the basis of the definition of the convolution,

$$W(s) Y_r(s) = \mathcal{L}\left[w(t) * y_r(t)\right],$$

we obtain:

$$\varepsilon(t) = \int_0^t w(\tau) y_r(t-\tau) d\tau, \quad w(t) = \mathcal{L}^{-1}[W(s)].$$

where $W(s)$ represents the sensitivity transfer function. If the nth derivative of $y_r(t)$ exists, then the Taylor expansion of the error leads to:

$$y_r(t-\tau) = y_r(t) - \tau \frac{d}{dt} y_r(t) + \frac{\tau^2}{2!} \frac{d^2}{dt^2} y_r(t) + \cdots + (-1)^n \frac{\tau^n}{n!} \frac{d^n}{dt^n} y_r(t) + \dots.$$

It follows that:

$$\varepsilon(t) = y_r(t) \int_0^t w(\tau) d\tau - \frac{d}{dt} y_r(t) \int_0^t \tau w(\tau) d\tau + \frac{d^2}{dt^2} y_r(t) \int_0^t \frac{\tau^2}{2!} w(\tau) d\tau + \cdots$$

If we ignore the initial condition y_0, we obtain the following step response:

$$Y(s) = \frac{1+as}{1+as} U(s), \quad y(t) = 1,$$

Taking the initial condition into account, we derive:

$$y(t) = 1 + y_0 \exp\left(-\frac{t}{a}\right) \to \infty \quad \text{for } a < 0.$$

The steady-state error is given by:

$$\varepsilon\left(\infty\right) = \lim_{t \to \infty} \varepsilon\left(t\right)$$

$$= y_r\left(t\right) \int_0^\infty w\left(\tau\right) d\tau - \frac{d}{dt} y_r\left(t\right) \int_0^\infty \tau w\left(\tau\right) d\tau$$

$$+ \frac{d^2}{dt^2} y_r\left(t\right) \int_0^\infty \frac{\tau^2}{2!} w\left(\tau\right) d\tau + \cdots .$$

Let us denote by $\alpha_0, \alpha_1, \cdots, \alpha_n$ the following terms:

$$\alpha_0 = \int_0^\infty w\left(\tau\right) d\tau, \quad \alpha_1 = -\int_0^\infty \tau w\left(\tau\right) d\tau, \quad \cdots, \alpha_n = \left(-1\right)^n \int_0^\infty \tau^n w\left(\tau\right) d\tau.$$

Finally, we obtain:

$$\varepsilon\left(\infty\right) = \alpha_0 y_r\left(t\right) + \frac{\alpha_1}{1!} \frac{d}{dt} y_r\left(t\right) + \cdots + \frac{\alpha_n}{n!} \frac{d^n}{dt^n} y_r\left(t\right) + \cdots .$$

The coefficients α_i, $i = 0, 1, \cdots$ represent the error coefficients. They permit the calculation of the steady-state error for any input.

Let us now relate these coefficients to the transfer function $W\left(s\right) = 1/\left(1 + G\left(s\right) H\left(s\right)\right)$ between the error and the input. From the definition of the Laplace transform, we derive:

$$W\left(s\right) = \int_0^\infty w\left(t\right) \exp\left(-st\right) dt, \quad \frac{dW\left(s\right)}{ds} = -\int_0^\infty tw\left(t\right) \exp\left(-st\right) dt,$$

$$\frac{d^2 W\left(s\right)}{ds^2} = \int_0^\infty t^2 w\left(t\right) \exp\left(-st\right) dt, \quad \cdots,$$

$$\frac{d^n W\left(s\right)}{ds^n} = \left(-1\right)^n \int_0^\infty t^n w\left(t\right) \exp\left(-st\right) dt,$$

which yields:

$$\lim_{s \to 0} W\left(s\right) = \int_0^\infty w\left(t\right) dt = \alpha_0, \quad \lim_{s \to 0} \frac{dW\left(s\right)}{ds} = \alpha_1, \quad \lim_{s \to 0} \frac{d^2 W\left(s\right)}{ds^2} = \alpha_2,$$

$$\cdots \quad \lim_{s \to 0} \frac{d^n W\left(s\right)}{ds^n} = \alpha_n.$$

For the system considered, we obtain:

$$\alpha_0 = \lim_{s \to 0} W\left(s\right) = \lim_{s \to 0} \frac{\left(s + 3\right)\left(s + 5\right)}{\left(s + 3\right)\left(s + 5\right) + k\left(1 + s\right)} = \frac{15}{15 + k},$$

$$\alpha_1 = \lim_{s \to 0} \frac{dW\left(s\right)}{ds} = \lim_{s \to 0} k \frac{s^2 + 2s - 7}{\left(s^2 + 8s + ks + k + 15\right)^2} = \frac{-7k}{\left(15 + k\right)^2},$$

etc.

Observe that the error coefficients can be used to calculate the steady-state output. We obtain:

$$\varepsilon\left(\infty\right) = y_r\left(\infty\right) - y\left(\infty\right) \Rightarrow y\left(\infty\right) = y_r\left(\infty\right) - \varepsilon\left(\infty\right).$$

PROBLEM 2.40. Let us consider again the electrical circuit considered in Problem 26 in Chapter 1.

1. Calculate the inverse Laplace transform of the current $I_k\left(s\right)$.

2. Derive an expression for the current $i_k\left(t\right) = \mathcal{L}^{-1}\left[I_k\left(s\right)\right]$ for $n \to \infty$.

SOLUTION 2.40. 1. Let us recall the expression for the current $I_k\left(s\right)$,

$$I_k(s) = \frac{V \cosh\left(n - k\right)\lambda}{z \, \sinh \lambda \sinh \lambda n}, \quad V\left(s\right) = \frac{v_0}{s}, \quad z = \frac{1}{Cs}, \tag{2.49}$$

and the equation for the parameter λ,

$$s^2 + \frac{R}{L}s + \frac{2}{CL}\left(1 - \cosh \lambda\right) = 0. \tag{2.50}$$

The poles of $I_k\left(s\right)$ are given by[19]:

$$\sinh \lambda n = 0 \Rightarrow \lambda_l = j\frac{l\pi}{n}, l = 1, \cdots, n. \tag{2.51}$$

For $l = 0$, we derive from Equation (2.50):

$$l = 0 \Rightarrow \lambda_0 = 0 \Rightarrow \cosh 0 = 1 \Rightarrow s^2 + \frac{R}{L}s = 0, \tag{2.52}$$

---

19. Recall that:

$$\sin s = \frac{1}{2j}\left[\exp\left(js\right) - \exp\left(-js\right)\right],$$

$$\cos s = \frac{1}{2}\left[\exp\left(js\right) + \exp\left(-js\right)\right],$$

$$\sinh s = -j\sin\left(js\right) \quad \text{and} \quad \cosh s = \cos\left(js\right),$$

and the roots of $\sin s$ are given by:

$$\sin s = 0 \Rightarrow \exp\left(js\right) = \exp\left(-js\right),$$

or

$$\exp\left(2js\right) = \exp\left(js\right)\exp\left(-js\right) = 1.$$

Using the result:

$$\exp\left(2jn\pi\right) = \cos 2n\pi + j\sin 2n\pi = 1,$$

we obtain the roots of $\sin s$,

$$s = n\pi, \quad n \in \mathcal{Z},$$

from which we obtain the roots of $\sinh s$, i.e., $s = jn\pi$.

and the poles are:

$$s = 0 \quad \text{and } s = -\frac{R}{L}. \tag{2.53}$$

For $l \neq 0$, we derive from equation (2.50):

$$s^2 + \frac{R}{L}s + \frac{2}{CL}\left(1 - \cosh j\frac{l\pi}{n}\right) = 0, \tag{2.54}$$

since:

$$\cosh j\frac{l\pi}{n} = \frac{1}{2}\left(\exp\left(j\frac{l\pi}{n}\right) + \exp\left(-j\frac{l\pi}{n}\right)\right)$$
$$= \frac{1}{2}\left(\cos\frac{l\pi}{n} + j\sin\frac{l\pi}{n} + \cos\frac{l\pi}{n} - j\sin\frac{l\pi}{n}\right) = \cos\frac{l\pi}{n}.$$

Let us write Equation (2.54) in a more convenient form:

$$s^2 + \frac{R}{L}s + \frac{2}{CL}\left(1 - \cos\frac{l\pi}{n}\right) = 0,$$
$$\left(s + \frac{R}{2L}\right)^2 - \frac{R^2}{4L^2} + \frac{2}{CL}\left(1 - \cos\frac{l\pi}{n}\right). \tag{2.55}$$

Therefore, the corresponding poles are:

$$s = -\delta \pm jw_l, \tag{2.56}$$

where:

$$\delta = \frac{R}{2L} \quad \text{and } w_l = \sqrt{\frac{2}{LC}\left(1 - \cos\frac{l\pi}{n}\right) - \frac{R^2}{4L^2}}. \tag{2.57}$$

Recall that for a Laplace transform of the form:

$$Y(s) = \frac{N(s)}{D(s)} = \frac{N(s)}{\displaystyle\prod_{i=1}^{n}(s - p_i)} = \sum_{i=1}^{n}\frac{\mu_i}{s - p_i},$$

where $N(s)$ and $D(s)$ are polynomials, the coefficients μ_i of the partial fraction expansion of $Y(s)$ are given by:

$$\mu_i = \frac{N(p_i)}{D'(p_i)}, \quad D'(p_i) = \frac{d}{ds}D(s)\Big|_{s=p_i}$$

Before we calculate the inverse of the Laplace transform $I_k(s)$ of the current I_k, let us calculate the derivative[20] of the denominator of equation (2.49) with respect to s.

Derivatives. From Equation (2.50), we derive:

$$\cosh \lambda = \frac{1}{2}\left(LCs^2 + RCs\right) + 1,$$

$$\cosh \lambda = \frac{\exp \lambda + \exp\left(-\lambda\right)}{2} = \frac{1}{2}\left(LCs^2 + RCs\right) + 1,$$

$$\exp \lambda + \exp\left(-\lambda\right) = LCs^2 + RCs + 2,$$

$$\frac{d}{ds}\left(\exp \lambda + \exp\left(-\lambda\right)\right) = \left(\frac{d\lambda}{ds}\right)\exp \lambda - \left(\frac{d\lambda}{ds}\right)\exp\left(-\lambda\right) = 2LCs + RC$$

$$= \left(\exp \lambda - \exp\left(-\lambda\right)\right)\frac{d\lambda}{ds} = 2\left(\sinh \lambda\right)\frac{d\lambda}{ds} = 2LCs + RC,$$

$$\frac{d\lambda}{ds} = \frac{1}{\sinh \lambda}\left[LCs + \frac{RC}{2}\right] \tag{2.58}$$

and:

$$\frac{d}{ds}\cosh \lambda = LCs + \frac{RC}{2}. \tag{2.59}$$

We need to calculate $\dfrac{d}{ds}\sinh \lambda$:

$$\frac{d}{ds}\sinh \lambda = \left(\frac{d}{d\lambda}\sinh \lambda\right)\frac{d\lambda}{ds}.$$

From the definition of $\sinh \lambda$, we obtain:

$$\frac{d}{d\lambda}\sinh \lambda = \frac{d}{d\lambda}\left(\frac{\exp \lambda - \exp\left(-\lambda\right)}{2}\right) = \frac{\exp \lambda + \exp\left(-\lambda\right)}{2} = \cosh \lambda. \tag{2.60}$$

---

20. Recall the expressions for the hyperbolic functions:

$$\cosh s \;=\; \frac{1}{2}\left(\exp s + \exp\left(-s\right)\right) = \cos js,$$

$$\sinh s \;=\; \frac{1}{2}\left(\exp s - \exp\left(-s\right)\right) = -j \sin js,$$

$$\tanh s \;=\; \frac{\sinh s}{\cosh s} = \frac{\exp s - \exp\left(-s\right)}{\exp s + \exp\left(-s\right)} = -j \tan js,$$

$$\coth s \;=\; \frac{\cosh s}{\sinh s} = \frac{\exp s + \exp\left(-s\right)}{\exp s - \exp\left(-s\right)} = j \cot js.$$

Finally, combining this expression with equation (2.58), we obtain:

$$\frac{d}{ds}\sinh\lambda = \left(\frac{d}{d\lambda}\sinh\lambda\right)\frac{d\lambda}{ds} = \cosh\lambda\frac{1}{\sinh\lambda}\left[LCs + \frac{RC}{2}\right],$$

$$\frac{d}{ds}\sinh\lambda = \coth\lambda\left[LCs + \frac{RC}{2}\right]. \tag{2.61}$$

Now, we have to calculate the following derivative:

$$\frac{d}{ds}\sinh\lambda n = \left(\frac{d}{d\lambda}\sinh\lambda n\right)\frac{d\lambda}{ds} = (n\cosh\lambda n)\frac{d\lambda}{ds}.$$

From Equations (2.60) and (2.58), we obtain:

$$\frac{d}{ds}\sinh\lambda n = n\cosh(\lambda n)\frac{1}{\sinh\lambda}\left[LCs + \frac{RC}{2}\right]. \tag{2.62}$$

Now, we are ready to calculate the derivative of the denominator of $I_k(s)$:

$$\frac{d}{ds}\left(\sinh\lambda\sinh\lambda n\right) = \sinh(\lambda n)\frac{d}{ds}\sinh\lambda + \sinh\lambda\frac{d}{ds}\sinh(\lambda n)$$

$$= \sinh(\lambda n)\coth\lambda\left[CLs + \frac{RC}{2}\right] + n\sinh\lambda\cosh(\lambda n)\frac{1}{\sinh\lambda}\left[LCs + \frac{RC}{2}\right],$$

$$\frac{d}{ds}\left(\sinh\lambda\sinh\lambda n\right) = (\sinh\lambda n\coth\lambda + n\cosh\lambda n)\left[LCs + \frac{RC}{2}\right]. \tag{2.63}$$

This result will enable us to calculate the partial fraction expansion of $I_k(s)$.

Partial fraction expansion. For the poles $s = 0$ and $s = -R/L$ corresponding to $\lambda \to \lambda_0 = 0$, we obtain[21]:

$$\lim_{\lambda\to 0}\cosh\lambda n = \lim_{\lambda\to 0}\frac{\exp\lambda n + \exp(-\lambda n)}{2} = 1,$$

$$\lim_{\lambda\to 0}\sinh\lambda n\coth\lambda = \lim_{\lambda\to 0}\cosh\lambda\frac{\sinh\lambda n}{\sinh\lambda}$$

---

21. Recall that:

$$a^n - b^n = (a - b)\left(\sum_{i=0}^{n-1} a^{n-i-1}b^i\right).$$

For $a = \exp\lambda$ and $b = \exp(-\lambda)$, we obtain:

$$\exp(n\lambda) - \exp(-n\lambda) = (\exp\lambda - \exp(-\lambda))\left(\sum_{i=0}^{n-1}\exp((n-i-1)\lambda)\exp(-\lambda i)\right)$$
$$= (\exp\lambda - \exp(-\lambda))\left(\sum_{i=0}^{n-1}\exp((n-2i-1)\lambda)\right).$$

$$= \lim_{\lambda \to 0} \cosh \lambda \frac{\exp \lambda n - \exp(-\lambda n)}{\exp \lambda - \exp(-\lambda)}$$

$$= \lim_{\lambda \to 0} \cosh \lambda \left(\sum_{i=0}^{n-1} \exp(-\lambda(n - 2i - 1)) \right) = n.$$

Therefore, for $s = 0$ and $s = -R/L$, we obtain:

$$s = 0 \implies \frac{d}{ds} \left(\sinh \lambda \sinh \lambda n \right) \bigg|_{\lambda=0} = nRC,$$

$$s = -\frac{R}{L} \implies \frac{d}{ds} \left(\sinh \lambda \sinh \lambda n \right) \bigg|_{\lambda=0} = -nRC.$$

For $\lambda \to \lambda_n = j\pi$, we obtain:

$$\cosh j\pi = \frac{\exp(j\pi) + \exp(-j\pi)}{2} = -1,$$

$$\cosh j\pi n = \frac{\exp(j\pi n) + \exp(-j\pi n)}{2} \doteq (-1)^n,$$

$$\sinh j\pi n \coth j\pi = \cosh j\pi \frac{\sinh j\pi n}{\sinh j\pi} = -1 \frac{\sinh j\pi n}{\sinh j\pi}$$

$$= -1 \frac{\exp(j\pi n) - \exp(-j\pi n)}{\exp(j\pi) - \exp(-j\pi)} = -\left(\sum_{i=0}^{n-1} \exp(j(n - 2i - 1)\pi) \right)$$

$$- \left(\sum_{i=0}^{n-1} \cos((n - 2i - 1)\pi) \right) = -n(-1)^{n-1} = (-1)^n n,$$

$$\frac{d}{ds} \left(\sinh \lambda \sinh \lambda n \right) \bigg|_{\lambda=j\pi} = 2(-1)^n n \left[LCs + \frac{RC}{2} \right]. \tag{2.64}$$

If we insert Equations (2.56) and (2.57) into Equation (2.64), we obtain:

$$\frac{d}{ds} \left(\sinh \lambda \sinh \lambda_n \right) \bigg|_{\lambda=j\pi} = 2(-1)^n n \left[LC \left(-\frac{R}{2L} \pm jw_n \right) + \frac{RC}{2} \right]$$

$$= \pm 2j(-1)^n nLCw_n. \tag{2.65}$$

For $\lambda \to \lambda_l = jl\pi/n \ (l = 1, 2, \cdots, n - 1)$, we obtain:

$$n \cosh n\lambda_l = n \cosh jl\pi = n \frac{\exp(jl\pi) + \exp(-jl\pi)}{2} = (-1)^l n,$$

$$\cosh j\frac{l\pi}{n} = \frac{\exp jl\pi/n + \exp(-jl\pi/n)}{2} = \cos \frac{l\pi}{n} \quad \text{and} \quad \sinh j\frac{l\pi}{n} = j \sin \frac{l\pi}{n}.$$

Observe that:

$$\sinh jl\pi = 0 \Rightarrow \sinh jl\pi \coth j\frac{l\pi}{n} = \cos \frac{l\pi}{n} \frac{\sinh jl\pi}{j \sin l\pi/n} = 0.$$

Combining the above with Equations (2.56) and (2.57), we obtain:

$$\frac{d}{ds}\left(\sinh\lambda\sinh\lambda n\right)\Big|_{\lambda_l=jl\pi/n} = n\left(-1\right)^l\left[LC\left(-\delta\pm jw_l\right)+\frac{RC}{2}\right]$$

$$= \frac{d}{ds}\left(\sinh\lambda\sinh\lambda n\right)\Big|_{\lambda_l=jl\pi/n} = n\left(-1\right)^l\left[-LC\left(\frac{R}{2L}\right)\pm jLCw_l+\frac{RC}{2}\right],$$

$$\frac{d}{ds}\left(\sinh\lambda\sinh\lambda n\right)\Big|_{\lambda_l=jl\pi/n} = \pm jn\left(-1\right)^l LCw_l.$$

Inverse Laplace transform. Finally, the inverse Laplace transform of $I_k\left(s\right)$ is:

$$i_k\left(t\right) = \frac{V_0}{nR} - \frac{V_0}{nR}\exp\left(-\frac{R}{L}t\right) + \frac{\left(-1\right)^k V_0}{nLw_n}\exp\left(-\delta t\right)\sin\left(w_n t\right) \qquad (2.66)$$

$$+ \frac{2V_0}{nL}\exp\left(-\delta t\right)\sum_{l=1}^{n-1}\frac{\sin\left(w_l t\right)}{w_l}\cos\frac{kl\pi}{n},$$

where the first, second, third and fourth terms of the right-hand side of this equation correspond to the poles $s = 0$, $s = R/L$, $s = -R/2L \pm jw_n$ and $s = -R/2L \pm jw_l$, respectively.

2. Let us consider the case where $n = \infty$. From Equation (2.49), we derive:

$$I_k(s) = \lim_{n\to\infty}\frac{V\cosh\left(n-k\right)\lambda}{z\ \sinh\lambda\sinh\lambda n} = \frac{V\exp\left(-k\lambda\right)}{z\ \sinh\lambda}. \qquad (2.67)$$

Recall that[22]:

$$\sinh \lambda = \sqrt{\cosh^2 \lambda - 1} \quad \text{and} \quad \exp(\pm\lambda) = \cosh \lambda \pm \sqrt{\cosh^2 \lambda - 1}. \qquad (2.68)$$

From the expressions for the current and $\cosh \lambda$ (see Problem 1.26), we thus obtain:

$$I_k(s) = \frac{V(s)}{\sqrt{zz_1}} \frac{\left[(1 + z_1/2z) - \sqrt{(1 + z_1/2z)^2 - 1}\right]^k}{\sqrt{1 + z_1/4z}}.$$

For $z = 1/Cs$, $z_1 = R$, and $V(s) = v_0/s$, we obtain:

$$I_k(s) = \frac{2v_0}{R\mu^k} \frac{\left(s + \mu - \sqrt{s^2 + 2\mu s}\right)^k}{\sqrt{s^2 + +2\mu s}},$$

where $\mu = 2/RC$. From a table of Laplace transforms, we obtain:

$$i_k(t) = \frac{2v_0}{R} \exp(-\mu t) J_k(\lambda t),$$

where:

$$J_k(t) = \frac{J_k(jt)}{j^k}$$

represents the modified Bessel function of order k.[23]

---

22. From the definitions of the hyperbolic functions, we obtain:

$$\cosh^2 \lambda = \frac{(\exp \lambda + \exp(-\lambda))^2}{4} = \frac{\exp(2\lambda) + \exp(-2\lambda) + 2}{4},$$

$$\sinh^2 \lambda = \frac{(\exp \lambda - \exp(-\lambda))^2}{4} = \frac{\exp(2\lambda) + \exp(-2\lambda) - 2}{4},$$

$$\sinh \lambda = \sqrt{\cosh^2 \lambda - 1}$$

and

$$\cosh^2 \lambda - 1 = \frac{\exp(2\lambda) + \exp(-2\lambda) - 2}{4} = \frac{\exp(2\lambda) + \exp(-2\lambda) - 2\exp(\lambda)\exp(-\lambda)}{4}$$

$$= \frac{[\exp(\lambda) - \exp(-\lambda)]^2}{4}$$

$$\Rightarrow \sqrt{\frac{[\exp(\lambda) - \exp(-\lambda)]^2}{4}} = \frac{\exp(\lambda) - \exp(-\lambda)}{2},$$

$$\cosh \lambda + \sqrt{\cosh^2 \lambda - 1} = \frac{\exp(\lambda) - \exp(-\lambda)}{2} + \frac{\exp(\lambda) + \exp(-\lambda)}{2} = \exp(\lambda).$$

23. The solutions of the differential equation:

We shall end this chapter by presenting a general method for calculating the coefficients of the partial fraction expansion of a rational function.

PROBLEM 2.41. Determine the partial fraction expansions of:

$$F_1\left(s\right) = \frac{s^4 + 3s^2 + 1}{\left(s + 2\right)^3 \left(s + 3\right)^2 \left(s + 4\right) \left(s^2 + 2s + 2\right)}$$

and

$$F_2\left(s\right) = \frac{s^4 + 59s^2 + 60}{13s^3 + 107s}.$$

SOLUTION 2.41. The partial fraction expansion of the rational function [GRA 00]:

$$\frac{P\left(s\right)}{Q\left(s\right)} = \frac{P\left(s\right)}{\left(s - p_1\right)^\alpha \left(s - p_2\right)^\beta \cdots \left(s - p_m\right)^\mu}$$

is given by:

$$\frac{P\left(s\right)}{Q\left(s\right)} = \frac{A_\alpha}{\left(s - p_1\right)^\alpha} + \frac{A_{\alpha-1}}{\left(s - p_1\right)^{\alpha-1}} + \cdots + \frac{A_1}{\left(s - p_1\right)}$$

$$+ \frac{B_\beta}{\left(s - p_2\right)^\beta} + \frac{B_{\beta-1}}{\left(s - p_2\right)^{\beta-1}} + \cdots + \frac{B_1}{\left(s - p_2\right)} + \cdots$$

$$+ \frac{M_\mu}{\left(s - p_m\right)^\mu} + \frac{M_{\mu-1}}{\left(s - p_m\right)^{\mu-1}} + \cdots + \frac{M_1}{\left(s - p_m\right)},$$

$$x^2 \frac{d^2 y}{dx^2} + x \frac{dy}{dx} + \left(x^2 - k^2\right) y = 0$$

which is called Bessel's differential equation, are called Bessel functions of order k. They are given by:

$$J_k\left(x\right) = \left(\frac{x}{2}\right)^k \sum_{i=0}^{\infty} \frac{\left(-1\right)^i \left(x/2\right)^{2i}}{i!\,\Gamma\left(k + i + 1\right)}, \quad u \in C, \quad k \neq -1, -2, \cdots,$$

where $\Gamma\left(.\right)$ represents the gamma function. This expression may derived by looking for solutions of the form:

$$y = x^\alpha \sum_{n=0}^{\infty} a_n x^n, \quad a_0 \neq 0.$$

The Bessel functions obey the following recurrence:

$$J_{k+1}\left(x\right) = \frac{2k}{x} J_k\left(x\right) - J_{k-1}\left(x\right).$$

The Bessel functions are used in the solution of potential and wave propagation problems.

where:

$$A_{\alpha-k+1} = \frac{R_1^{(k-1)}(p_1)}{(k-1)!}, \quad B_{\beta-k+1} = \frac{R_2^{(k-1)}(p_2)}{(k-1)!}, \cdots,$$

$$M_{\mu-k+1} = \frac{R_m^{(k-1)}(p_m)}{(k-1)!},$$

$$R_1(s) = \frac{P(s)(s-p_1)^\alpha}{Q(s)}, \quad , \quad R_m(s) = \frac{P(s)(s-p_m)^\mu}{Q(s)},$$

and

$$R_i^{(k)}(s) = \frac{d^k}{ds^k} R_i(s).$$

If p_i, $i = \overline{1, m}$, are simple roots, i.e., $\alpha_i = 1$, $i = \overline{1, m}$, we obtain:

$$A = \frac{P(p_1)}{Q'(p_1)}, \cdots, \quad M = \frac{P(p_m)}{Q'(p_m)}.$$

We shall describe in detail the computations related to the triple pole $p_1 - 2$. We have $\alpha = 3$:

$$\frac{A_3}{(s-p_1)^3} + \frac{A_2}{(s-p_1)^2} + \frac{A_1}{(s-p_1)},$$

$$A_3 = \frac{R_1^{(0)}(-2)}{(0)!} = R_1(-2) = \left. \frac{P(s)(s+2)^3}{Q(s)} \right|_{p=-2} = \frac{29}{4},$$

$$A_2 = \frac{R_1^{(1)}(-2)}{(1)!} = \left. \frac{d}{ds} \frac{P(s)(s+2)^3}{Q(s)} \right|_{p=-2} = -\frac{175}{8},$$

$$A_1 = \frac{R_1^{(2)}(-2)}{(2)!} + \left. \frac{d^2}{ds^2} \frac{P(s)(s+2)^3}{Q(s)} \right|_{p=-2} = \frac{663}{16}.$$

The partial fraction expansions of $F_1(s)$ and $F_2(s)$ are given by:

$$F_1(s) = \frac{29}{4(s+2)^3} - \frac{175}{8(s+2)^2} + \frac{633}{16(s+2)} - \frac{109}{5(s+3)^2} - \frac{896}{25(s+3)}$$
$$- \frac{61}{16(s+4)} + \frac{3}{100} \frac{7+3s}{s^2+2s+2},$$

$$F_2(s) = \frac{s^4 + 59s^2 + 60}{13s^3 + 107s} = \frac{1}{13}s + \frac{(59 - 107/13)s^2 + 60}{13s^3 + 107s}$$
$$= \frac{1}{13}s + \frac{60}{107s} + \frac{60480}{1391} \frac{s}{13s^2 + 107}.$$

Observe that for complex poles, the relations given earlier remain valid. We have only to regroup the fractions associated with conjugate poles. Notice also that:

$$F_2(s) = \frac{f_1(s)}{f_2(s)}, \quad f(s) = s^4 + 13s^3 + 59s^2 + 107s + 60,$$

where $f_1(s)$ and $f_2(s)$ correspond to the decomposition of the polynomial $f(s)$ into odd and even terms (see the proof of the Routh–Hurwitz criterion given in Appendix A).

Chapter 3

Analysis

3.1. Introduction

The main objective of transient and frequency analysis is to convey information about the process considered. For a given system:

1) the transfer function is equal to the Laplace transform of the impulse response, i.e.:

$$Y(s) = F(s)\,\mathcal{L}\left[\delta(t)\right] = F(s), \quad g(t) = \mathcal{L}^{-1}\left[F(s)\right];$$

2) the transfer function is equal to the Laplace transform of the derivative of the step response, i.e.:

$$Y(s) = F(s)\,\frac{1}{s}, \quad F(s) = sY(s);$$

3) the frequency response is a function of the magnitude and argument of the transfer function, i.e.:

$$|F(j\omega)|\,u_0 \sin(\omega t + \varphi), \quad \varphi = \arg F(j\omega),$$

where the input is given by:

$$u(t) = u_0 \sin \omega t.$$

Transient and frequency analysis can be compared to a recruitment interview where applicants have to answer specific questions ("skilful" questioning). On the basis of the applicant's responses, the recruiter is "able" to evaluate the applicants and to select a qualified applicant for the position offered.

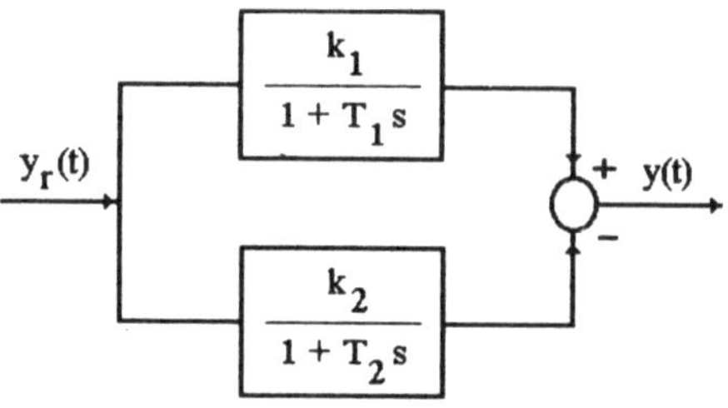

Figure 3.1. *Two parallel systems*

3.2. Step responses

PROBLEM 3.1. Consider the system depicted in Figure 3.1. Determine for what values of the gains k_1 and k_2 the derivative at the origin of the step response of the system is positive.

SOLUTION 3.1. The Laplace transform of the step response is given by:

$$Y(s) = \left(\frac{k_1}{1 + T_1 s} - \frac{k_2}{1 + T_2 s} \right) \frac{1}{s}.$$

The inverse of this Laplace transform gives expressions for the step responses of the systems:

$$y(t) = k_1 \left[1 - \exp\left(-\frac{t}{T_1} \right) \right] - k_2 \left[1 - \exp\left(-\frac{t}{T_2} \right) \right].$$

The derivative is given by:

$$\frac{dy(t)}{dt} = \frac{k_1}{T_1} \exp\left(-\frac{t}{T_1} \right) - \frac{k_2}{T_2} \exp\left(-\frac{t}{T_2} \right).$$

At the origin, we obtain:

$$\left. \frac{dy(t)}{dt} \right|_{t=0} = \frac{k_1}{T_1} - \frac{k_2}{T_2}$$

which is positive for:

$$\frac{k_1}{T_1} > \frac{k_2}{T_2}.$$

Observe that for:

$$\frac{k_1}{T_1} < \frac{k_2}{T_2},$$

we obtain a non-minimum phase system.

PROBLEM 3.2. 1. Determine the step response of the following system:

$$F(s) = \frac{3(2+s)}{(s+1)(s^2+0.8s+0.15)} = \frac{N(s)}{D(s)}.$$

2. Calculate the steady-state error (position error).

SOLUTION 3.2. 1. The Laplace transform of the step response is given by:

$$Y(s) = \frac{3(2+s)}{(s+1)(s^2+0.8s+0.15)}Y_r(s), \quad Y_r(s) = \frac{1}{s}.$$

A partial fraction expansion[1] of $Y(s)$ leads to:

$$Y(s) = \frac{\alpha}{s} + \frac{\beta}{(s+1)} + \frac{\delta}{(s+0.3)} + \frac{\gamma}{(s+0.5)}. \tag{3.1}$$

There exist two main methods for calculating the parameters of this expansion, namely α, β, δ and γ. We now present these methods.

First method. The common denominator in Equation (3.1) is:

$$D(s) = s(s+1)(s+0.3)(s+0.5)$$

which leads to:

$$Y(s) = \frac{\alpha(s+1)(s+0.3)(s+0.5) + \beta s(s+0.3)(s+0.5)}{D(s)}$$
$$+ \frac{\delta s(s+1)(s+0.5) + \gamma s(s+1)(s+0.3)}{D(s)}.$$

Equating coefficients of identical powers of s, we obtain a set of algebraic equations:

$$0.15\alpha = 6,$$

$$0.95\alpha + 0.15\beta + 0.5\delta + 0.3\gamma = 3,$$

$$1.8\alpha + 0.8\beta + 1.5\delta + 1.3\gamma = 0,$$

$$\alpha + \beta + \delta + \gamma = 0.$$

---

1. The partial fraction expansion is suitable for inversion.

Solving these equations for α, β, δ and γ, we obtain:

$$\alpha = 40.0, \quad \beta = -8.5714, \quad \delta = -121.43 \quad \text{and } \gamma = 90.0.$$

Second method (Heaviside expansion[2]). Let us consider Equation (3.1) again. We multiply both sides by s and set $s = 0$, and obtain:

$$\frac{3\,(2+0)}{(0+1)\,(0^2 + 0.8 \times 0 + 0.15)} = \alpha,$$

$$\alpha = \frac{6}{0.15} = 40.$$

By multiplying by $(s + 1)$, $(s + 0.3)$ and $(s + 0.5)$ and setting s equal to -1, -0.3 and -0.5, respectively, we derive:

$$\frac{3\,(2-1)}{(-1)\left((-1)^2 + 0.8\,(-1) + 0.15\right)} = \beta = -8.5714,$$

$$\frac{3\,(2-0.3)}{-0.3\,(-0.3+1)\,(-0.3+0.5)} = \delta = -121.43,$$

$$\frac{3\,(2-0.5)}{-0.5\,(-0.5+1)\,(-0.5+0.3)} = \gamma = 90.$$

In the case of multiple zeros, however, this method can not lead to all the coefficients of the partial fraction expansion.[3]

---

2. This method consists of successively multiplying both sides of the expansion by the factors of the denominator and to setting s equal to the pole corresponding to the factor considered.

3. Consider the following example:

$$\frac{k}{s\,(s+1)^2}.$$

Its expansion is:

$$\frac{k}{s\,(s+1)^2} = \frac{\alpha}{s} + \frac{\beta}{(s+1)} + \frac{\delta}{(s+1)^2}.$$

Multiplying both sides of this expression by s and $(s+1)^2$ and setting $s = 0$ and $s = -1$ respectively, we obtain:

$$\alpha = k, \quad \delta = -k.$$

One simple solution for calculating β consists of selecting any value for s. For example, if s is selected to be equal to 2, we obtain:

$$\frac{k}{2\,(2+1)^2} = \frac{k}{2} - \frac{k}{(2+1)} + \frac{\delta}{(2+1)^2},$$

which leads to:

$$\delta = -k.$$

The inverse Laplace transform of $Y(s)$ is:

$$y(t) \;=\; [40.0 - 8.571 \exp(-t) - 121.43 \exp(-0.3t) + 90.0 \exp(-0.5t)]\, 1(t)$$

where $1(t)$ represents the unit step.

Observe that to calculate β and similarly α, δ and γ, we multiplied $Y(s) = N(s)/D(s)$ by $(s+1)$ (and s, $(s+0.3)$ and $(s+0.5)$, respectively,), i.e.:

$$(s+1)\frac{N(s)}{D(s)} \;=\; (s+1)\frac{N(s)}{D(s) - D(-1)}$$

$$=\; \frac{N(s)}{[D(s) - D(-1)]/[s - (-1)]},$$

which leads to:

$$\beta = \lim_{n \to \infty} (s+1)\frac{N(s)}{D(s)} = \frac{N(-1)}{D'(-1)}, \quad D'(s) = \frac{d}{ds}D(s).$$

In general, the coefficients of the partial fraction expansion of $N(s)/D(s)$ which are associated with simple poles p_i are given[4] by:

$$\frac{N(p_i)}{dD(p_i)/ds}.$$

2. The Laplace transform of the error is given by:

$$\Xi(s) \;=\; Y_r(s) - Y(s) = (1 - F(s))\,Y_r(s) = \frac{1}{s}(1 - F(s))$$

$$=\; \frac{1}{s}\left(\frac{(s+1)(s^2 + 0.8s + 0.15) - 3(2+s)}{(s+1)(s^2 + 0.8s + 0.15)}\right)$$

---

4. Note that, if for $m \neq 0$,

$$\lim_{s \to p_0} (s - p_0)^m f(s)$$

exists and is equal to $a \neq 0$, then p_0 is an m-tuple pole of f, i.e.:

$$f(s) = \frac{A(s)}{(s - p_0)^m B(sz)}.$$

and, using Tauber's (initial-and final-value) theorem, we obtain:

$$\varepsilon\left(\infty\right) = \lim_{s\to 0} s\frac{1}{s}\left(\frac{\left(s+1\right)\left(s^2+0.8s+0.15\right)-3\left(2+s\right)}{\left(s+1\right)\left(s^2+0.8s+0.15\right)}\right)$$

$$= \frac{0.15-6}{0.15} = -39.$$

PROBLEM 3.3. Calculate the step response of the following system:

$$F\left(s\right) = \frac{P\left(s\right)}{\left(s-\sigma\right)^3\left(s+5\right)}, \qquad \sigma = -1, \quad P\left(s\right) = \left(s+2\right).$$

SOLUTION 3.3. The Laplace transform of the step response is given by:

$$Y\left(s\right) = \frac{1}{s}\frac{P\left(s\right)}{\left(s+1\right)^3\left(s+5\right)} = \frac{P\left(s\right)}{\left(s+1\right)^3 Q\left(s\right)}, \qquad Q\left(s\right) = s\left(s+5\right).$$

This step response may be decomposed as follows:

$$Y\left(s\right) = \frac{\alpha}{s} + \frac{\beta}{\left(s+1\right)} + \frac{\delta}{\left(s+1\right)^2} + \frac{\gamma}{\left(s+1\right)^3} + \frac{\varepsilon}{\left(s+5\right)}.$$

Multiplying both side of this equation by $\left(s+5\right)$ and setting $s=-5$, we obtain:

$$\varepsilon = -\frac{3}{320}.$$

Multiplying both side of the equation by $\left(s+1\right)^3$ and setting $s=-1$, we obtain:

$$\gamma = -\frac{1}{4}.$$

Multiplying by s and setting $s=0$, we obtain:

$$\alpha = \frac{2}{5}.$$

We shall present a method for the multiple-pole case.[5] To obtain the values of the coefficients β, δ and γ, we make the following variable change:

$$s = \sigma + z.$$

---

5. Let a function $f\left(s\right)$ be analytic on an annulus C around p, defined by:
$$0 < r \le \left|s-p\right| \le R,$$

We obtain:

$$\frac{P(s)}{(s-\sigma)^3 Q(s)} = \frac{P(\sigma+z)}{z^3 Q(\sigma+z)}.$$

This ratio leads to:

$$\frac{P(\sigma+z)}{Q(\sigma+z)} = \frac{\dot{2}+\sigma+z}{\sigma(\sigma+5)+(2\sigma+5)z+z^2}$$
$$= \gamma+\delta z+\beta z^2+z^3\frac{P^*(\sigma+z)}{Q(\sigma+z)}.$$

The polynomial:

$$\gamma+\delta z+\beta z^2$$

represents the quotient resulting from the division of $P(\sigma+z)$ by $Q(\sigma+z)$ when the term z^3 appears in the remainder. This division leads to:

$$z^3\left(\frac{P(\sigma+z)}{Q(\sigma+z)}\right) = -\frac{1}{4}-\frac{7}{16}z-\frac{25}{64}z^2+z^3(\cdots).$$

It follows that:

$$\gamma=-\frac{1}{4},\quad \delta=-\frac{7}{16},\quad \beta=-\frac{25}{64}.$$

Finally, we obtain:

$$Y(s) = \frac{2}{5s}-\frac{25}{64(s+1)}-\frac{7}{16(s+1)^2}-\frac{1}{4(s+1)^3}-\frac{3}{320(s+5)}.$$

PROBLEM 3.4. Calculate the step responses and their derivatives at $t=0$ of the following systems:

$$F_1(s) = \frac{k}{s(s+5)},\quad F_2(s) = \frac{k(s+7)}{s(s+5)},\quad y_i(0)=0, i=1,2.$$

$f(s)$ then has a Laurent expansion in this region,

$$f(s) = \sum_{n=-\infty}^{+\infty} c_n(s-p)^n$$

If the Laurent expansion about p has a finite number of negative powers, i.e.:

$$f(s) = \frac{c_{-m}}{(s-p)^m}+\cdots+\frac{c_{-1}}{(s-p)}+\sum_{n=0}^{+\infty}c_n(s-p)^n,\quad m>0,\quad c_{-m}\neq 0,$$

then $f(s)$ has a pole of order m at $s=p$.

SOLUTION 3.4. The step responses are given by:

$$Y_1\left(s\right) = \frac{k}{s^2\left(s+5\right)}, \quad Y_2\left(s\right) = \frac{k\left(s+7\right)}{s^2\left(s+5\right)},$$

$$Y_1\left(s\right) = -\frac{1}{25}\frac{k}{s} + \frac{1}{5}\frac{k}{s^2} + \frac{1}{25}\frac{k}{s+5},$$

$$Y_2\left(s\right) = -\frac{2}{25}\frac{k}{s} + \frac{7}{5}\frac{k}{s^2} + \frac{2}{25}\frac{k}{s+5},$$

$$y_1\left(t\right) = k\left[-\frac{1}{25} + \frac{1}{5}t + \frac{1}{25}\exp\left(-5t\right)\right]1\left(t\right),$$

$$y_2\left(t\right) = k\left[-\frac{2}{25} + \frac{7}{5}t + \frac{2}{25}\exp\left(-5t\right)\right]1\left(t\right).$$

Using the initial- and final-value theorem (Tauber's theorem), we derive the initial values of the derivatives:

$$\mathcal{L}\left(\frac{dy_1\left(t\right)}{dt}\right) = sY_1\left(s\right) = \frac{k}{s\left(s+5\right)},$$

$$\mathcal{L}\left(\frac{dy_2\left(t\right)}{dt}\right) = sY_2\left(s\right) = \frac{k\left(s+7\right)}{s\left(s+5\right)},$$

$$\left.\frac{dy_1\left(t\right)}{dt}\right|_{t=0} = \lim_{s\to\infty} s\frac{k}{s\left(s+5\right)} = 0,$$

$$\left.\frac{dy_2\left(t\right)}{dt}\right|_{t=0} = \lim_{s\to\infty} s\frac{k\left(s+7\right)}{s\left(s+5\right)} = k.$$

Observe that the introduction of a zero in the first system ($F_1\left(s\right)$) makes $dy_1\left(t\right)/dt|_{t=0} \neq 0$. In other words, if $y_1\left(t\right)$ represents the step response and $dy_1\left(t\right)/dt|_{t=0} \neq 0$ there is a zero present in the system.

PROBLEM 3.5. Determine the logarithmic sensitivity coefficient with respect to the static gain k of an open-loop system with transfer function $kG\left(s\right)$.

SOLUTION 3.5. The logarithmic sensitivity coefficient is given by:

$$\sigma_k = \frac{d\ln kG}{d\ln k} = \frac{d\,kG/kG}{dk/k} = \frac{G\,dk/Gk}{dk/k} = 1.$$

PROBLEM 3.6. Consider a feedback control system where the transfer functions of the forward and feedback paths are $G\left(s\right)$ and $H\left(s\right)$, respectively.

1. Assume that $G(s)$ depends on a parameter k as follows: $G(s) = kG^*(s)$. Calculate the sensitivity coefficient of the closed-loop transfer function with respect to k.

2. Suppose now that $H(s)$ depends on a parameter k as follows: $H(s) = kH^*(s)$. Calculate the sensitivity coefficient of the closed-loop transfer function with respect to k.

SOLUTION 3.6. 1. The sensitivity coefficient is given by:

$$\sigma_k = \frac{dF/F}{dk/k},$$

$$\frac{dF}{dk} = \frac{G^*(s)\left[1 + kG^*(s)H(s)\right] - kG^*(s)G^*(s)H(s)}{\left[1 + kG^*(s)H(s)\right]^2}$$

$$= \frac{G^*(s)}{\left[1 + kG^*(s)H(s)\right]^2},$$

$$\frac{dF}{F} = \left[\frac{1}{1 + kG^*(s)H(s)}\right]\frac{dk}{k},$$

$$\sigma_k = \frac{1}{1 + kG^*(s)H(s)}.$$

If $kG^*(s)H(s) \gg 1$, we derive:

$$\sigma_k \simeq \frac{1}{kG^*(s)H(s)}.$$

2. In the case where the parameter appears in the feedback path, we obtain:

$$\frac{dF}{k} = -\frac{G(s)G(s)H^*(s)}{\left[1 + kG(s)H^*(s)\right]^2},$$

$$\frac{dF}{F} = -\frac{kG(s)H^*(s)}{\left[1 + kG(s)H^*(s)\right]}\frac{dk}{k},$$

$$\sigma_k = -\frac{kG(s)H^*(s)}{1 + kG(s)H^*(s)}.$$

If $kG(s)H^*(s) \gg 1$, we obtain:

$$\sigma_k \simeq -1.$$

PROBLEM 3.7. Consider a unity-feedback system where the forward path consists of an integrator in series with the system:

$$G(s) = \frac{k\prod_{i=1}^{m}(s - z_i)}{\prod_{i=1}^{n}(s - p_i)}.$$

Show that the steady-state error for a step input is equal to zero.

SOLUTION 3.7. The Laplace transform of the error is given by:

$$\Xi(s) = \frac{1}{1 + (1/s)\,G(s)}\frac{1}{s} = \frac{\prod\limits_{i=1}^{n}(s - p_i)}{s\prod\limits_{i=1}^{n}(s - p_i) + k\prod\limits_{i=1}^{m}(s - z_i)}.$$

Using the final-value theorem, we obtain:

$$\varepsilon(\infty) = \lim_{s \to 0} s\,\Xi(s) = 0.$$

It is easy to understand this result. Let us consider the steady-state behavior, $u(t) = u_s$ and $\varepsilon(t) = \varepsilon_s$, where $u(t)$ represents the integrator output. We obtain:

$$u_s = \int \varepsilon_s\,dt,$$

which implies $\varepsilon_s = \varepsilon(\infty) = 0$.

Many processes can be correctly modeled by a first- or second-order system with a time delay [NAJ 88, NAJ 89]. The next two problems are related to the application of some simple identification techniques for the estimation of the parameters of first- and second-order systems. General identification techniques are treated in, for instance, in [LJU 83, IKO 02].

3.3. System identification

PROBLEM 3.8. Derive the parameters of a first-order system from its impulse and step responses.

SOLUTION 3.8. The transfer function of a first-order system and its impulse response are given by:

$$F(s) = \frac{k}{1 + Ts}, \quad Y(s) = \frac{k}{1 + Ts}, \quad y(t) = \frac{k}{T}\exp\left(-\frac{t}{T}\right).$$

Observe that:

$$\int_0^{+\infty} \frac{k}{T}\exp\left(-\frac{t}{T}\right)dt = k.$$

Let us determine the equation of the tangent to the impulse response at a point A of abscissa t_a (t_a, y_a):

$$\frac{dy(t)}{dt} = -\frac{k}{T^2} \exp\left(-\frac{t}{T}\right),$$

$$f = \left.\frac{dy(t)}{dt}\right|_{t=t_a} t + b.$$

Recall that the point A belongs to the impulse response. We obtain:

$$y(t_a) = \frac{k}{T} \exp\left(-\frac{t_a}{T}\right), \quad y_a = \left.\frac{dy(t)}{dt}\right|_{t=t_a} t_a + b$$

$$\implies b = \frac{k}{T} \exp\left(-\frac{t_a}{T}\right)\left[1 + \frac{t_a}{T}\right],$$

$$f(t) = -\frac{k}{T^2} \exp\left(-\frac{t_a}{T}\right) t + \frac{k}{T} \exp\left(-\frac{t_a}{T}\right)\left[1 + \frac{t_a}{T}\right]. \tag{3.2}$$

From this expression, we derive the intersection of the tangent with the time axis, where $f(t_b) = 0$:

$$-\frac{k}{T^2} \exp\left(-\frac{t_a}{T}\right) t_b + \frac{k}{T} \exp\left(-\frac{t_a}{T}\right)\left[1 + \frac{t_a}{T}\right] = 0,$$

$$\frac{k}{T} \exp\left(-\frac{t_a}{T}\right) \frac{T - (t_b - t_a)}{T} = 0$$

$$\implies t_b - t_a = T.$$

Therefore, the difference between the abscissa of the point A (any point belonging to the impulse response) and the abscissa of the intersection of the tangent to the impulse response at this point with the time axis gives the time constant. The impulse response and its tangent at the origin $(0, k/T)$ for a first-order system ($k = 30, T = 5\,s$) are plotted in Figure 3.2.

Let us now consider the step response:

$$Y(s) = \frac{k}{s(1 + Ts)} = \frac{k}{s} - \frac{k}{(1/T + s)},$$

$$y^*(t) = \mathcal{L}^{-1}(Y(s)) = k\left[1 - \exp\left(-\frac{t}{T}\right)\right]. \tag{3.3}$$

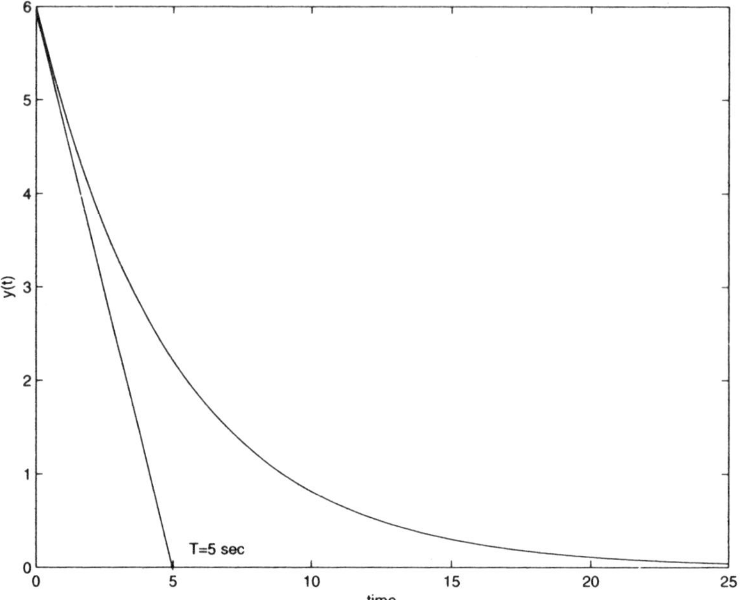

Figure 3.2. *Identification of the static gain and the time constant from the impulse response*

The equation of the tangent to the step response $y(t)$ at a given point A, of coordinates t_a and y_a is given by:

$$\frac{dy^*(t)}{dt} = \frac{k}{T}\exp\left(-\frac{t}{T}\right), \qquad \left.\frac{dy^*(t)}{dt}\right|_{t=t_a} = \frac{k}{T}\exp\left(-\frac{t_a}{T}\right),$$

$$f = \frac{k}{T}\exp\left(-\frac{t_a}{T}\right)t + const.$$

The point A belongs to both the step response and the tangent considered. We obtain:

$$y_a = k\left[1-\exp\left(-\frac{t_a}{T}\right)\right] = \frac{1}{T}\exp\left(-\frac{t_a}{T}\right)t_a + const$$

$$\implies const = y_a - \frac{1}{T}\exp\left(-\frac{t_a}{T}\right)t_a.$$

Finally, the equation of the tangent to the step response at the point A is:

$$f = \frac{k}{T} \exp\left(-\frac{t_a}{T}\right) t + y_a - \frac{k}{T} \exp\left(-\frac{t_a}{T}\right) t_a.$$

Let us now determine the intersection $B(t_b, y_b)$ of this tangent with the horizontal straight line ($f = k$) corresponding to the steady-state behavior of the system:

$$
\begin{aligned}
f &= \frac{k}{T} \exp\left(-\frac{t_a}{T}\right) t_b + y_a - \frac{k}{T} \exp\left(-\frac{t_a}{T}\right) t_a = k, \\
t_b &= \frac{T}{k}(k - y_a)\exp\left(\frac{t_a}{T}\right) - t_a = T - t_a.
\end{aligned}
$$

Therefore, we have:

$$T = t_b - t_a. \tag{3.4}$$

Observe that a similar result was derived for the impulse response.

The quantities that we are interested in estimating are the time constant T and the static gain k. The steady-state regime ($\lim\limits_{t\to\infty} y(t) = \alpha k$, where α represents the magnitude of the step input) leads directly to the static gain, and the time constant is equal to the difference between the abscissa of the intersection of the tangent to the step response at any point with the straight line representing the steady-state behavior and the abscissa of the point considered. From the step response ($\alpha = 1$) depicted in Figure 3.3, where the segment (straight line) AB represents the tangent at the point A of abscissa $t_a = 5\,s$, we deduce that the time constant T and the static gain k are equal to:

$$T = t_b - t_a = 10 - 5 = 5\,s \quad \text{and}\quad k = 30.$$

In order to obtain good accuracy, it is necessary to draw many tangents and to take the mean value of the results for the time constant obtained.

The time constant can also be computed by considering the point $C(T, y_c)$ of abscissa T and where $y_c = k\left[1 - e^{-1}\right] \simeq 0.632k$, or the points[6] $D(3T, 0.95k)$ and $E(4T, 0.98k)$. Finally, note that this identification method is also applicable to first-order systems with a time delay.

---

6. For a first-order system, the response times to within 5% and 2% are equal to $3T$ and $4T$, respectively.

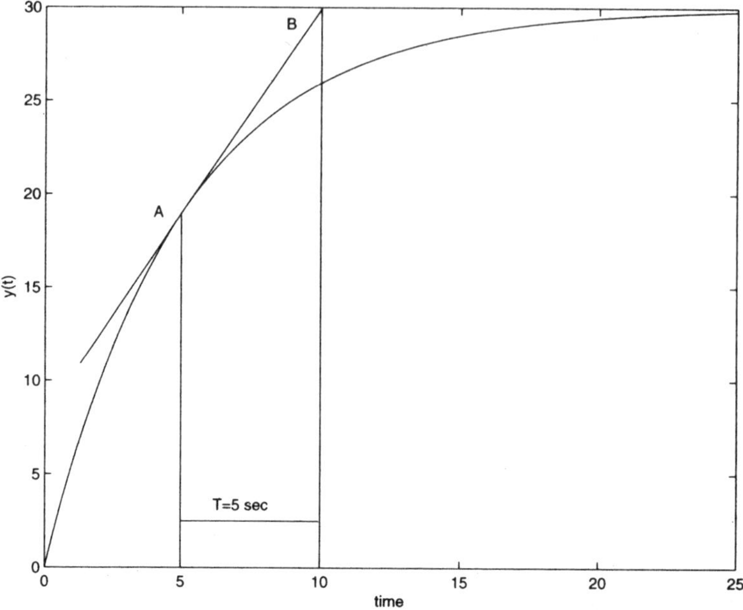

Figure 3.3. *Identification of the parameters of a first-order system*

REMARK 3.1. Observe also that the impulse response and the translated step response are given by a term of the form $\lambda \exp(-t/T)$. A semi-logarithmic plot of this term consists of a straight line such that:

$$y(t) = \lambda \exp\left(-\frac{t}{T}\right),$$

$$y(t + 2.3T) = \lambda \exp\left(-\frac{t + 2.3T}{T}\right) = \lambda \exp\left(-\frac{t}{T}\right)\exp(2.3) \simeq \frac{1}{10}y(t) \quad (3.5)$$

$$\text{and } y(t + T) = \lambda \exp\left(-\frac{t + T}{T}\right) = \frac{1}{e}y(t) = \frac{1}{2.72}y(t).$$

These expressions lead to another identification technique for estimating the time constant T (see Figure 3.4).

REMARK 3.2. The time delays introduced by RC circuit networks in microelectronic technology have induced many studies and the development of RC delay metrics. The impulse responses of such system satisfy the following conditions: $g(t) \geq 0 \ \forall t$ and $\int_0^\infty g(t)\,dt = 1 \quad k = 1$. This impulse response may be considered as a probability

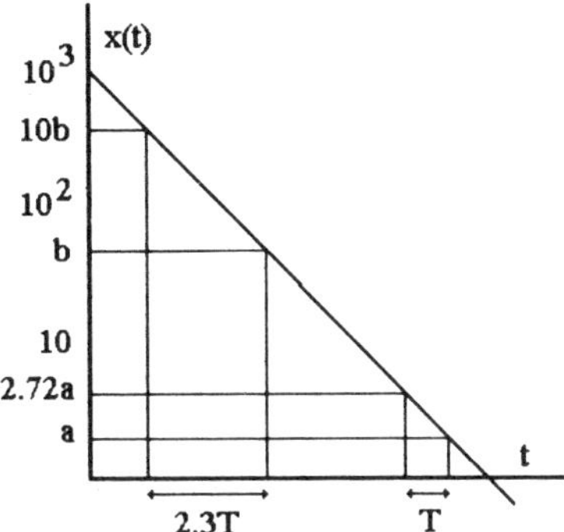

Figure 3.4. *Representation of the impulse response (or the translated step response) of a first-order system in a semi-logarithmic plot*

density function[7] [NAJ 04]. Therefore, the step response, which corresponds to the integral of the impulse response, represents the corresponding probability distribution.

PROBLEM 3.9. Derive the parameters of a second-order system from its step response.

____

7. A probability distribution gives the relationship between all possible values of a given random variable x and their associated probabilities. It is defined by:

$$F(x) = \Pr(X < x).$$

The probability that a continuous random variable X belongs to the interval $[x, x + \Delta x]$ is equal to the variation of the probability distribution in this interval, i.e.:

$$\Pr(x < X \leq x + \Delta x) = F(x + \Delta x) - F(x).$$

If $F(x)$ is continuous and differentiable, then when $\Delta x \to 0$, we obtain:

$$\lim_{\Delta x \to 0} \frac{F(x + \Delta x) - F(x)}{\Delta x} = \frac{dF(x)}{dx} = f(x),$$

where $f(x)$ represents the probability density function.

SOLUTION 3.9. The Laplace transform of the step response is given by:

$$Y(s) = \frac{1}{s} \frac{k}{(1/\omega_n^2)\, s^2 + (2\zeta/\omega_n)\, s + 1} = \frac{1}{s} \frac{k\omega_n^2}{s^2 + 2\zeta\omega_n s + \omega_n^2}$$
$$= \frac{1}{s} \frac{k\omega_n^2}{(s + \zeta\omega_n)^2 + \omega_n^2 (1 - \zeta^2)}.$$

The static gain is given by the final value of $y(t)$,

$$\lim_{t \to \infty} y(t) = y(\infty) = \lim_{s \to 0} sY(s) = k.$$

In order to derive the step response, we have to consider three cases (given any quadratic equation, there are three categories that the roots could fall into).

Case 1 (over-damped), $\zeta > 1$. In this case, the poles of the system are real.

$$p_1 = -\zeta\omega_n + \omega_n\sqrt{\zeta^2 - 1}, \quad p_2 = -\zeta\omega_n - \omega_n\sqrt{\zeta^2 - 1}.$$

The partial fraction expansion of $Y(s)$ leads to:

$$Y(s) = k\left[\frac{1}{s} - \frac{s + 2\zeta\omega_n}{(s + \zeta\omega_n)^2 - \omega_n^2(\zeta^2 - 1)}\right]$$

$$= k\left[\frac{1}{s} - \frac{\zeta + \sqrt{\zeta^2 - 1}}{2\sqrt{\zeta^2 - 1}\,(s - p_1)} - \frac{\sqrt{\zeta^2 - 1} - \zeta}{2\sqrt{\zeta^2 - 1}\,(s - p_2)}\right],$$

$$y(t) = k\left[1 - \lambda\exp\left(\left(-\zeta\omega_n + \omega_n\sqrt{\zeta^2 - 1}\right)t\right) - \mu\exp\left(\left(-\zeta\omega_n - \omega_n\sqrt{\zeta^2 - 1}\right)t\right)\right].$$

The term $\omega_n\sqrt{\zeta^2 - 1}$ is usually denoted by ω_d and represents the damped natural frequency. Now it is easy to check that the step response is given by:

$$y(t) = k\left[1 - \lambda\exp(p_1 t) - \mu\exp(p_2 t)\right], \tag{3.6}$$

where:

$$\lambda = \frac{\zeta + \sqrt{\zeta^2 - 1}}{2\sqrt{\zeta^2 - 1}}, \quad \mu = \frac{\sqrt{\zeta^2 - 1} - \zeta}{2\sqrt{\zeta^2 - 1}}.$$

For $\zeta > 1$, it is simpler to write the transfer function of a second-order system in the following form:

$$F(s) = \frac{k}{(1 + T_1 s)(1 + T_2 s)}.$$

The Laplace transform of the step response and its inverse are given by:

$$Y(s) = \frac{k}{s\,(1+T_1 s)\,(1+T_2 s)} = \frac{k}{s} - \frac{k}{(T_1 - T_2)}\left(\frac{T_1^2}{(1+T_1 s)} - \frac{T_2^2}{(1+T_2 s)}\right),$$

$$y(t) = k\left[1 - \frac{1}{(T_1 - T_2)}\left(T_1 \exp\left(-\frac{t}{T_1}\right) - T_2 \exp\left(-\frac{t}{T_2}\right)\right)\right] 1(t). \qquad (3.7)$$

Comparing equations (3.6) and (3.7), we derive:

$$\lambda = \frac{T_1}{(T_1 - T_2)}, \quad \mu = -\frac{T_2}{(T_1 - T_2)}.$$

Harriott [HAR 64] has shown that:

$$T_1 + T_2 = t_{73}/1.3,$$

where t_{73} represents the time at which the process output is equal to 73% of the final steady-state value. For the step response depicted in Figure 3.5 corresponding to $k = 30$, $\zeta = 5$ and $\omega_n = 25$, we derive $T_1 = 0.396\,s$ and $T_2 = 0.004\,s$. On the basis of the Harriott method, we obtain $t_{73} = 0.526\,s$ and $T_1 + T_2 = 0.402\,s$. The difference between the exact value of $T_1 + T_2$ and the estimate is negligible, it is equal to 0.002.

Let us now consider the following function:

$$f(t) = \log\left(1 - \frac{y(t)}{ku_0}\right)$$

$$= \log\left(\frac{T_1}{T_1 - T_2}\exp\left(-\frac{t}{T_1}\right) + \frac{T_2}{T_2 - T_1}\exp\left(-\frac{t}{T_2}\right)\right),$$

where u_0 represents the magnitude of the step input. If $T_2 \gg T_1$, we obtain:

$$f(t) \simeq \log\left(\frac{T_2}{T_2 - T_1}\exp\left(-\frac{t}{T_2}\right)\right) = -\frac{t}{T_2} + \log\left(\frac{T_2}{T_2 - T_1}\right).$$

The slope of this straight line is approximately equal to $-0.43/T_2$. Let α be the intersection of this straight line with the vertical axis. It follows that:

$$f(0) = \alpha = \log\left(\frac{T_2}{T_2 - T_1}\right).$$

In summary, we draw the function $f(t)$ and determine its slope and its intersection with the vertical axis in order to obtain $-0.43/T_2$ and α. From the slope, we derive the time constant T_2, and from α, we obtain the value of the time constant T_1.

There exists a wide range of constructive numerical methods for finding minima of non-smooth functions of several variables. We advise the reader to implement one of them in order to identify the time constants T_1 and T_2 by making use of filtered data. High-pass filtering makes it possible to eliminate DC offsets, load disturbances, etc., and low-pass filtering is used to eliminate high-frequency noise, etc. The rule of thumb governing the design of the filter is that the upper frequency should be about twice the desired system bandwidth and the lower frequency should be about one-tenth of the desired bandwidth.

Case 2 (critically damped), $\zeta = 1$. In this case, the system has a double pole:

$$Y(s) = \frac{1}{s} \frac{k\omega_n^2}{(s+\omega_n)(s+\omega_n)}.$$

The partial fraction expansion of $Y(s)$ leads[8] to:

$$Y(s) = k \left[\frac{1}{s} - \frac{1}{(s+\omega_n)} - \frac{\omega_n}{(s+\omega_n)^2} \right]$$

$$y(t) = k \left[1 - (1 + \omega_n t) \exp(-\omega_n t) \right].$$

Let us now calculate the second derivative of the step response:

$$
\begin{aligned}
y(t) &= k \left[1 - (1 + \omega_n t) \exp(-\omega_n t) \right], \\
\frac{dy(t)}{dt} &= -\omega_n \exp(-\omega_n t) + \omega_n (1 + \omega_n t) \exp(-\omega_n t), \\
\frac{d^2 y(t)}{dt^2} &= \omega_n^2 \left[1 - \omega_n t \right] \exp(-\omega_n t).
\end{aligned}
$$

We thus obtain:

$$\frac{d^2 y(t)}{dt^2} = 0 \quad \text{for } t = \frac{1}{\omega_n}. \tag{3.8}$$

This will enable us to determine the natural frequency by considering the abscissa of the inflection point of the step response. For $t = 1/\omega_n$, the step response is equal to:

$$y\left(\frac{1}{\omega_n}\right) = k \left[1 - \frac{2}{e} \right] = 0.264k.$$

---

8. $\mathcal{L}\left[\exp(-\lambda t) t^n\right] = \dfrac{n!}{(s+\lambda)^{n+1}}.$

Case 3 (under-damped), $\zeta < 1$. In this case, the system has complex poles:

$$Y(s) = \frac{1}{s} \frac{k\omega_n^2}{(s + \zeta\omega_n)^2 + \omega_n^2(1 - \zeta^2)}$$

The partial fraction expansion of $Y(s)$ leads[9] to:

$$Y(s) = k\left[\frac{1}{s} - \frac{s + \zeta\omega_n}{(s + \zeta\omega_n)^2 + \omega_n^2(1 - \zeta^2)} - \frac{\zeta\omega_n}{(s + \zeta\omega_n)^2 + \omega_n^2(1 - \zeta^2)}\right]$$

$$y(t) = k - k\exp(-\zeta\omega_n t)\cos\left(\omega_n\sqrt{1 - \zeta^2}\,t\right)$$

$$- \frac{k\zeta}{\sqrt{1 - \zeta^2}}\exp(-\zeta\omega_n t)\sin\left(\omega_n\sqrt{1 - \zeta^2}\,t\right).$$

$$(3.9)$$

Let us denote ζ and $\sqrt{1 - \zeta^2}$ by $\cos\varphi$ and $\sin\varphi$, respectively. Performing simple computations in equation (3.9), we derive[10]:

$$y(t) = k\left[1 - \frac{\exp(-\zeta\omega_n t)}{\sqrt{1 - \zeta^2}}\sin\left(\omega_n\sqrt{1 - \zeta^2}\,t + \varphi\right)\right].$$

Let us now look at the first maximum (overshoot) of the step response:

$$\frac{dy(t)}{dt} = k\omega_n\exp(-\zeta\omega_n t)\left[\frac{\zeta}{\sqrt{1 - \zeta^2}}\sin\left(\omega_n\sqrt{1 - \zeta^2}\,t + \varphi\right)\right.$$

$$\left. - \cos\left(\omega_n\sqrt{1 - \zeta^2}\,t + \varphi\right)\right]$$

$$= k\omega_n\exp(-\zeta\omega_n t)\left[\frac{\cos\varphi}{\sin\varphi}\sin\left(\omega_n\sqrt{1 - \zeta^2}\,t + \varphi\right)\right.$$

$$\left. - \cos\left(\omega_n\sqrt{1 - \zeta^2}\,t + \varphi\right)\right]$$

$$= k\omega_n\exp(-\zeta\omega_n t)\frac{1}{\sin\varphi}\left[\cos\varphi\sin\left(\omega_n\sqrt{1 - \zeta^2}\,t + \varphi\right)\right.$$

$$\left. - \sin\varphi\cos\left(\omega_n\sqrt{1 - \zeta^2}\,t + \varphi\right)\right] = 0.$$

---

9. $\mathcal{L}[\exp(-\lambda t)\sin(\omega t + \varphi)] = \dfrac{\omega\cos\varphi + (s + \lambda)\sin\varphi}{(s + \lambda)^2 + \omega^2}$,

$\mathcal{L}[\exp(-\lambda t)\cos(\omega t + \varphi)] = \dfrac{(s + \lambda)\cos\varphi - \omega\sin\varphi}{(s + \lambda)^2 + \omega^2}$.

10. $\sin(a + b) = \sin a\cos b + \cos a\sin b$,

$\sin(a - b) = \sin a\cos b - \cos a\sin b$.

We derive:

$$\sin\left(\omega_n\sqrt{1-\zeta^2}t\right) = 0 \Longrightarrow \omega_n\sqrt{1-\zeta^2}t = \pi,$$

$$t_m = \frac{\pi}{\omega_n\sqrt{1-\zeta^2}} \Rightarrow y(t_m) = k\left[1+\exp\left(-\frac{\zeta\pi}{\sqrt{1-\zeta^2}}\right)\right], \qquad (3.10)$$

where t_m represents the time to first peak. The height of the resonant peak is equal to:

$$M_p = k\left(1+\exp\left(-\frac{\zeta\pi}{\sqrt{1-\zeta^2}}\right)\right).$$

From Equation (3.10), we derive the damping factor and the natural frequency:

$$\log\frac{k}{k-y(t_m)} = \frac{\zeta\pi}{\sqrt{1-\zeta^2}} = \sigma,$$

$$\left(\sigma^2+\pi^2\right)\zeta^2 = \sigma^2, \quad \zeta = \frac{\sigma}{\sqrt{\sigma^2+\pi^2}}, \quad \omega_n = \frac{\pi}{t_m\sqrt{1-\zeta^2}}.$$

For $\omega_n = 25\ rad/s$ and three values ($\zeta = 0.2, 1$ and 5) of the damping factor, the step responses are depicted in Figure 3.5. For $\zeta < 1$, the translated step response $(y(t) - k)$ corresponds to an exponentially decreasing harmonic function. This figure shows also that the parameter ζ represents a measure of the severity of the damping. When the damping factor becomes small, the behavior of a second-order system can be compared to the behavior of a car when the shock absorbers are old and worn.

The method described above for the identification of the parameters of a first-order system can be extended to second-order systems with two different time constants $T_1 \gg T_2$. For example, if we consider an electric motor, we can associate a time constant T_1 with the mechanical phenomena and a time constant T_2 with the electrical phenomena. Of course, the dynamics associated with the mechanical phenomena are slower than those associated with the electrical phenomena.

The impulse response of this kind of system is given by:

$$y(t) = \frac{k}{T_1-T_2}\left[\exp\left(-\frac{t}{T_1}\right)-\exp\left(-\frac{t}{T_2}\right)\right]$$

and:

$$\int_0^{+\infty} y(t)\,dt = \frac{k}{T_1-T_2}\int_0^{+\infty}\left[\exp\left(-\frac{t}{T_1}\right)-\exp\left(-\frac{t}{T_2}\right)\right]dt = k.$$

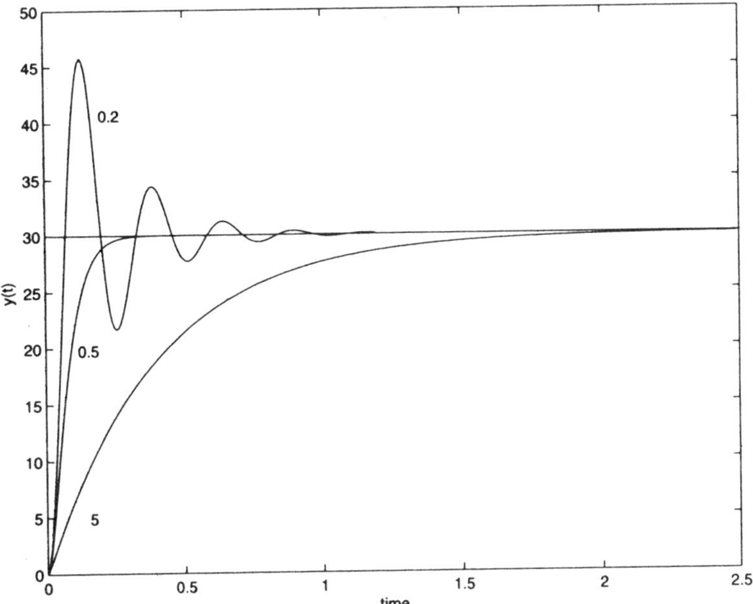

Figure 3.5. *Step responses of second-order systems*

For $T_1 \gg T_2$ (in this case T_2 represents the dominant time constant) and small values of t, $\exp\left(-t/T_1\right) \gg \exp\left(-t/T_2\right)$ and:

$$y\left(t\right) \simeq \frac{k}{T_1 - T_2} \exp\left(-\frac{t}{T_1}\right),$$

which looks like the impulse response of a first-order system.

In order to identify the remaining parameters, we just have to consider the data (the values of $y\left(t\right)$) for large values of t, and then calculate the term $\left[k/\left(T_2 - T_1\right)\right]$ $\exp\left(-\frac{t}{T_2}\right)$ by subtracting the term $\left[k/\left(T_1 - T_2\right)\right]\exp\left(-\frac{t}{T_1}\right)$ from the values of the impulse response and using the same identification technique. In other words, for small values of t, the fast dynamics are dominant, while for large values of t, the slow dynamics are dominant.

PROBLEM 3.10. The step response of the system:

$$G\left(s\right) = \frac{k}{\left(1 + Ts\right)^2}, \quad k = 2, \quad T = 1,$$

is depicted in Figure 3.6. Derive the parameters k and T as a function of the area A and the final value $y\left(\infty\right)$.

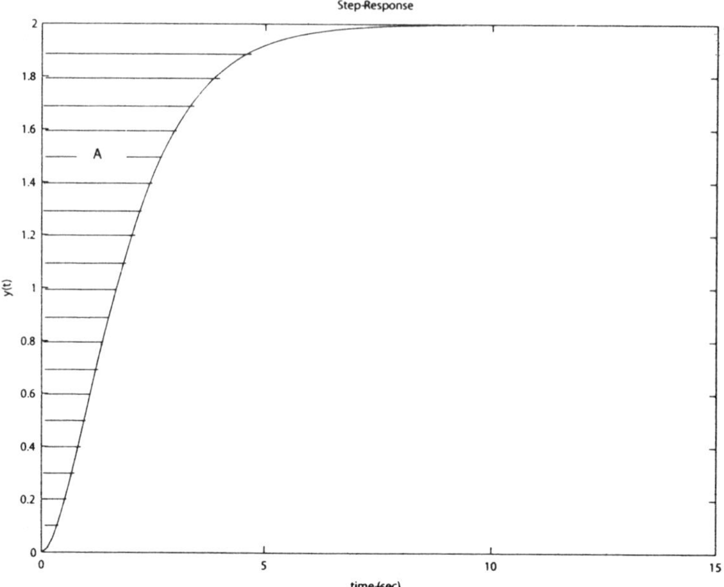

Figure 3.6. *Step response of the system*

SOLUTION 3.10. The Laplace transform of the step response is given by:

$$Y(s) = \frac{k}{s(1+Ts)^2} = \frac{k}{s} - \frac{kT}{(1+Ts)} - \frac{kT}{(1+Ts)^2},$$

and its inverse Laplace transform is:

$$y(t) = k\left(1 - \exp(-\frac{t}{T}) - \frac{t}{T}\exp(-\frac{t}{T})\right).$$

For $b = 2$ and $T = 1\,s$, we obtain:

$$y(t) = 2\left(1 - \exp(-t) - t\exp(-t)\right).$$

Therefore, the final value of the step response is equal to:

$$y(\infty) = k.$$

Let us now calculate the area A:

$$A = \int_0^\infty (y(\infty) - y(t))\,dt = k\int_0^\infty \exp(-\frac{t}{T})\,dt + k\int_0^\infty \frac{t}{T}\exp(-\frac{t}{T})\,dt.$$

Integration of the second integral by part[11] leads to:

$$A = 2kT,$$

which implies:

$$T = \frac{A}{2k}.$$

PROBLEM 3.11. Consider the following system:

$$G(s) = \frac{k}{(1 + T_1 s)(1 + T_2 s)}.$$

Express the parameters k, T_1 and T_2 as functions of the following integrals:

$$\int_0^\infty g(t)\, dt, \quad \int_0^\infty tg(t)\, dt \quad \text{and} \quad \int_0^\infty t^2 g(t)\, dt,$$

where $g(t)$ represents the impulse response of the system.

SOLUTION 3.11. The Laplace transform of the impulse response (see Appendix A) is given by:

$$\mathcal{L}(g(t)) = G(s) = \int_0^\infty g(t) \exp(-st)\, dt.$$

For $s = 0$, we obtain the static gain,

$$G(0) = \int_0^\infty g(t)\, dt = k.$$

Let us now consider the first derivative of the transfer function $G(s)$:

$$\frac{dG(s)}{ds} = -\int_0^\infty tg(t) \exp(-st)\, dt = -k\frac{T_1(1 + T_2 s) + T_2(1 + T_1 s)}{(1 + T_1 s)^2 (1 + T_2 s)^2}.$$

---

11. Recall that:

$$\int P(x) \exp(\alpha x)\, dx = \frac{\exp(\alpha x)}{\alpha} \sum_{i=0}^n (-1)^i \frac{P^{(i)}(x)}{\alpha^i},$$

where $\deg P(x) = n$ and $P^{(i)}(x) = d^i P(x)/dx^i$.

For $s = 0$, we obtain:

$$\frac{dG(s)}{ds}\bigg|_{s=0} = -\int_0^\infty tg(t)\,dt = -k(T_1 + T_2).$$

It follows that:

$$T_1 + T_2 = a = -\frac{dG(s)/d|_{s=0}}{G(0)} = \frac{\int_0^\infty tg(t)\,dt}{\int_0^\infty g(t)\,dt}. \tag{3.11}$$

Finally, consider the second derivative:

$$\frac{d^2}{ds^2}G(s) = 2k\frac{T_1^2 + 3T_1^2T_2s + 3T_1T_2^2s + 3T_1^2T_2^2s^2 + T_1T_2 + T_2^2}{(1+T_1s)^3(1+T_2s)^3},$$

which leads to:

$$\frac{d^2}{ds^2}G(s)\bigg|_{s=0} = 2k\left(T_1^2 + T_1T_2 + T_2^2\right)$$

and

$$T_1^2 + T_1T_2 + T_2^2 = b = \frac{d^2G(s)/ds^2\big|_{s=0}}{2G(0)}.$$

Recall that:

$$\frac{d^2G(s)}{ds^2} = \int_0^\infty t^2g(t)\exp(-st)\,dt \quad \text{and} \quad \frac{d^2}{ds^2}G(s)\bigg|_{s=0} = \int_0^\infty t^2g(t)\,dt.$$

Therefore, we obtain:

$$T_1^2 + T_1T_2 + T_2^2 = b = \frac{\int_0^\infty t^2g(t)\,dt}{2\int_0^\infty g(t)\,dt}. \tag{3.12}$$

From Equations (3.11) and (3.12), we derive:

$$T_1 + T_2 = a \quad \text{and } T_1T_2 = a^2 - b.$$

T_1 and T_2 are the solutions of the following algebraic equation:

$$x^2 - ax + a^2 - b = 0.$$

Observe that:

$$\int_0^\infty \frac{g(t)}{\int_0^\infty g(t)\,dt}\,dt = \int_0^\infty \frac{g(t)}{k}\,dt = 1.$$

The function:

$$\frac{g(t)}{\int_0^\infty g(t)\,dt}\,dt$$

can be considered as a probability density function, and the term:

$$\int_0^\infty t^n \frac{g(t)}{\int_0^\infty g(t)\,dt}\,dt$$

as the moment of order n [NAJ 04].

PROBLEM 3.12. Consider Problem 2.12 again. Calculate the final values of the Fresnel integrals.

SOLUTION 3.12. The Laplace transforms of the Fresnel integrals are given by:

$$\mathcal{L}[C(t)] = \frac{\sqrt{s + \sqrt{s^2 + 1}}}{2s\sqrt{s^2 + 1}}, \quad \mathcal{L}[S(t)] = \frac{\sqrt{\sqrt{s^2 + 1} - s}}{2s\sqrt{s^2 + 1}}.$$

Using the final-value theorem, we obtain:

$$\lim_{t \to \infty} C(t) = C(\infty) = \lim_{s \to 0} s\frac{\sqrt{s + \sqrt{s^2 + 1}}}{2s\sqrt{s^2 + 1}} = \frac{1}{2},$$

$$\lim_{t \to \infty} S(t) = S(\infty) = \lim_{s \to 0} s\frac{\sqrt{\sqrt{s^2 + 1} - s}}{2s\sqrt{s^2 + 1}} = \frac{1}{2}.$$

3.4. Frequency response

Teaching frequency analysis to chemical-engineering students is not an easy task. They have trouble considering the dynamics of a process because they are familiar with working with static models and using static descriptions for the design of chemical processes. In order to introduce frequency analysis, we can start by presenting some examples.

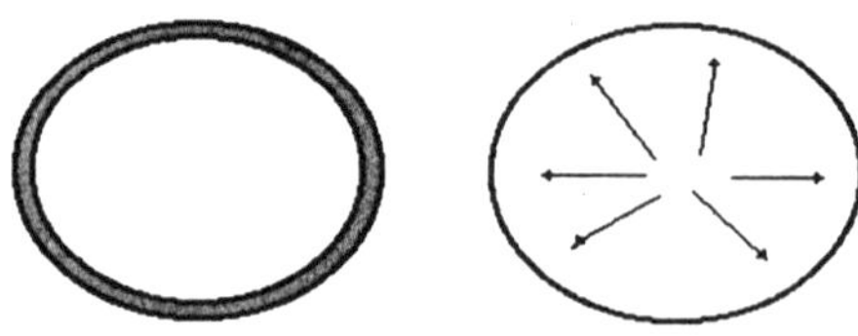

Figure 3.7. *Pseudo-cyclic drying. The left side of the figure shows a prune and its wet superficial layer. The right side shows the diffusion of water from the inside of the prune to its external surface*

The first example is this: consider a very big apple tree from which you want to pick some apples. You have no ladder or pole. What do you do? The answer is to shake the tree. The apple tree can be considered as a system, and shaking it is equivalent to applying a sinusoidal input to this system, the output of which is apples. The application of a sinusoidal function permits one to get information about the dynamics of a system.

The second example concerns the cyclic drying of an agricultural product such as prunes. The objective here is to dry the prunes without affecting their quality (i.e., without burning them). One solution would be the following: heat the prunes until the water contained in the superficial layer has been removed, decrease the temperature and wait until the water contained inside the prunes moves to the surface, increase the temperature again, and so on. This procedure is adequate because the diffusion of water inside the prunes is the limiting factor. In other words, the dynamics associated with the diffusion of water are very slow compared with the dynamics associated with the drying process (see Figure 3.7). This drying procedure leads to a sinusoidal variation of the temperature.

Finally, let us recall Fourier series. Sinusoidal functions can be compared to Lego^{TM} bricks: we can construct practically any function with the basic elements. In order to obtain the response of a system to a given input, it is enough to decompose the input into a sum of sinusoidal functions a Fourier series, and to use the superposition principle. In other words, frequency analysis can be considered as analogous to a photographic scan of the process.

PROBLEM 3.13. Consider a system described by a transfer function $F(s)$. Derive its response to a sinusoidal input $u(t) = u_0 \sin \omega t$.

SOLUTION 3.13. There exist various approaches for deriving the frequency response of a given system. The Laplace transform of the system output for a sinusoidal input is given by:

$$Y(s) = \frac{u_0 \omega}{s^2 + \omega^2} F(s), \quad F(s) = \frac{b_0 \prod_{i=1}^{n} (s - z_i)}{a_0 \prod_{i=1}^{n} (s - p_i)}.$$

Approach 1. Let us assume that $Re p_i < 0$, and consider the partial fraction expansion of the Laplace transform $Y(s)$,

$$Y(s) = \sum_{i=1}^{n} \frac{\alpha_i}{s - p_i} + \frac{\delta s + \gamma}{s^2 + \omega^2}. \tag{3.13}$$

The inverse Laplace transform[12] of $\alpha_i / (s - p_i)$ is $\alpha_i \exp(p_i t)$, and $\lim_{t \to \infty} \alpha_i \exp(p_i t) = 0$. The stationary response is given by the inverse of $(\delta s + \gamma) / (s^2 + w^2)$. Expression (3.13) yields:

$$\lim_{s \to jw} (s + w^2) \frac{u_0 \omega}{s^2 + \omega^2} F(s) = u_0 \omega F(j\omega) = \delta j\omega + \gamma,$$

$$F(j\omega) = R(j\omega) + jI(j\omega) = \frac{1}{u_0 \omega} (\delta j\omega + \gamma),$$

$$\gamma = u_0 \omega R(j\omega), \delta = u_0 I(j\omega).$$

Recall that for large value of t, we have:

$$Y(s) = \frac{\delta s + \gamma}{s^2 + \omega^2} \implies$$

$$y(t) = \delta \cos(\omega t) + \frac{\gamma}{\omega} \sin(\omega t) =$$

$$u_0 I(j\omega) \cos(\omega t) + u_0 R(j\omega) \sin(\omega t). \tag{3.14}$$

Let us write $F(j\omega)$ in the following form[13]:

$$F(j\omega) = |F(j\omega)| (\cos\varphi + j\sin\varphi), \quad \varphi = \arctan \frac{Y(j\omega)}{X(j\omega)},$$

$$\cos\varphi = \frac{R(j\omega)}{|F(j\omega)|}, \quad \sin\varphi = \frac{I(j\omega)}{|F(j\omega)|}. \tag{3.15}$$

---

12. For complex poles, we obtain terms of the form $\exp(Re p_i t) \sin(\cdot)$.

13. The argument of a complex number $s = x + jy$ (Cartesian form) is equal to:

$$\arg s = \arctan \frac{y}{x}.$$

Combining Equations (3.14) and (3.15), we obtain:

$$y(t) = u_0 |F(j\omega)| [\sin \varphi \cos(\omega t) + \cos \varphi \sin(\omega t)] = u_0 |F(j\omega)| \sin(\omega t + \varphi).$$

Approach 2. The Laplace transform of the frequency response is:

$$Y(s) = \frac{u_0 \omega}{s^2 + \omega^2} F(s).$$

Let us calculate the residues associated with the singularities $p = j\omega$ and $p^* = -j\omega$:

$$\begin{aligned}
r_{j\omega} &= \lim_{s \to j\omega} \frac{u_0 \omega (s - j\omega)}{(s - j\omega)(s + j\omega)} F(s) \exp(st) = \frac{1}{2j} u_0 F(j\omega) \exp(j\omega t), \\[2mm]
r_{-j\omega} &= \lim_{s \to -j\omega} \frac{u_0 \omega (s + j\omega)}{(s - j\omega)(s + j\omega)} F(s) \exp(st) \\[2mm]
&= -\frac{1}{2j} u_0 F(-j\omega) \exp(-j\omega t),
\end{aligned}$$

which yields:

$$y(t) = \frac{1}{2j} u_0 F(j\omega) \exp(j\omega t) - \frac{1}{2j} u_0 F(-j\omega) \exp(-j\omega t).$$

$F(j\omega)$ may be written as follows:

$$\begin{aligned}
F(j\omega) = R(j\omega) + jI(j\omega) = |F(j\omega)|(\cos \varphi + j \sin \varphi), \quad \varphi = \arg F(j\omega) \\
\implies F(-j\omega) = R(j\omega) - jI(j\omega) = |F(j\omega)|(\cos \varphi - j \sin \varphi).
\end{aligned}$$

The following formulae are valid for $x = 0$ and $y \neq 0$:

$$\arctan \frac{y}{x} = \begin{cases} \dfrac{\pi}{2} & \text{if } y > 0, \\[2mm] -\dfrac{\pi}{2} & \text{if } y < 0. \end{cases}$$

The function $\arg s$ is discontinuous for s real and negative: we observe a jump of magnitude 2π, from $+\pi$ to $-\pi$ or from $-\pi$ to $+\pi$, when we cross the negative real axis. The following formulae is used also for the calculation of the argument of a complex number $s = x + jy$, $x \leq 0$ ($s = r(\cos \theta + j \sin \theta)$, polar form):

$$\arg s = 2 \arctan \left(\frac{y}{x + r} \right), r = \sqrt{x^2 + y^2}.$$

To prove this result, recall that:

$$\tan \left(\frac{\theta}{2} \right) = \frac{\sin \theta}{\cos \theta + 1} = \frac{r \sin \theta}{r \cos \theta + r} = \frac{y}{x + r}.$$

It follows that:

$$y\left(t\right) = \frac{1}{2j} u_0 \left|F\left(j\omega\right)\right| \left(\cos\varphi + j\sin\varphi\right)\exp\left(j\omega t\right)$$

$$-\frac{1}{2j} u_0 \left|F\left(j\omega\right)\right| \left(\cos\varphi - j\sin\varphi\right)\exp\left(-j\omega t\right)$$

$$= \frac{1}{2j} u_0 \left|F\left(j\omega\right)\right| \left(\cos\varphi + j\sin\varphi\right)\left[\cos\omega t + j\sin\omega t\right]$$

$$-\frac{1}{2j} u_0 \left|F\left(j\omega\right)\right| \left(\cos\varphi - j\sin\varphi\right)\left[\cos\omega t - j\sin\omega t\right])$$

$$= u_0 \left|F\left(j\omega\right)\right| \left[\sin\varphi\cos\omega t + \cos\varphi\sin\omega t\right]$$

$$= y\left(t\right) = u_0 \left|F\left(j\omega\right)\right| \sin\left(\omega t + \varphi\right).$$

Approach 3. Let us write $F\left(s\right)$ in the following form:

$$F\left(s\right) = \frac{b_0 s^m + b_1 s^{m-1} + \cdots + b_m}{a_0 s^n + a_1 s^{n-1} + \cdots + a_n}$$

This system is governed by the following differential equation:

$$a_0 \frac{d^n}{dt^n} y\left(t\right) + \cdots + a_n y\left(t\right) = b_0 \frac{d^m}{dt^m} u\left(t\right) + \cdots + b_m u\left(t\right) \qquad (3.16)$$

As the system considered is linear, the output may be expressed as follows:

$$y\left(t\right) = A u_0 \sin\left(\omega t + \varphi\right).$$

Let us introduce the complex representation of $\sin\left(.\right)$:

$$\sin\omega t = Im\left(\exp\left(j\omega t\right)\right),$$

which leads to:

$$u\left(t\right) = u_0 Im\left(\exp\left(j\omega t\right)\right), \quad y\left(t\right) = A u_0 Im\left(\exp\left(j\left(\omega t + \varphi\right)\right)\right). \qquad (3.17)$$

Notice that:

$$\frac{d}{dt}\exp\left(j\omega t\right) = j\omega\exp\left(j\omega t\right), \quad \frac{d^k}{dt^k}\exp\left(j\omega t\right) = \left(j\omega\right)^k\exp\left(j\omega t\right).$$

In other words, for sinusoidal functions, $j\omega$ plays the role of the operator d/dt. If we insert Equation (3.17) in (3.16), we obtain:

$$a_0\left(j\omega\right)^n A u_0 \exp\left(j\left(\omega t + \varphi\right)\right) + \cdots + a_n A u_0 \exp\left(j\left(\omega t + \varphi\right)\right)$$

$$= b_0\left(j\omega\right)^m u_0 \exp\left(j\omega t\right) + \cdots + b_m u_0 \exp\left(j\omega t\right),$$

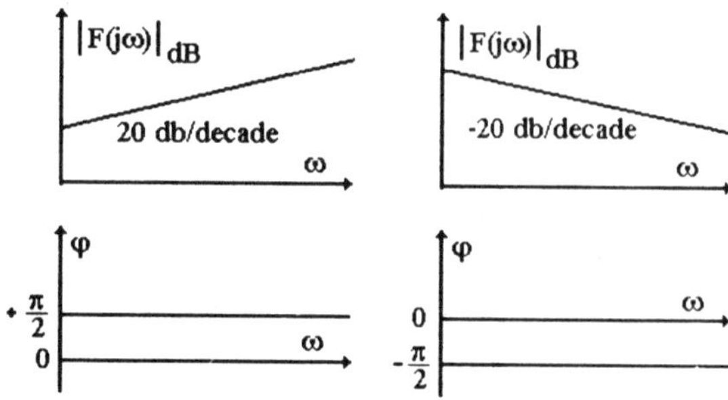

Figure 3.8. *Bode diagrams of s and* $1/s$

which leads to:

$$[a_0 (j\omega)^n + \cdots + a_n] \, A \exp (j\varphi) = [b_0 (j\omega)^m + \cdots + b_m] \, ,$$

$$A \exp (j\varphi) \;=\; \frac{b_0 (j\omega)^m + \cdots + b_m}{a_0 (j\omega)^n + \cdots + a_n} = F (j\omega) \, ,$$

$$A \;=\; |F (j\omega)| \, , \varphi = \arg F (j\omega) \, ,$$

which corresponds to the desired result.

The frequency response can be graphically represented by the Bode, Nyquist and Nichols diagrams. These diagrams correspond to different geometrical representations of the complex number $A \exp (j\varphi)$. Let us recall the Bode diagrams of transfer functions of the forms s (derivative), $1/s$ (integrator), $1 + Ts$ (process lead) and $1/ (1 + Ts)$ (process lag). Asymptotic (approximate) diagrams of these elementary systems are shown in Figures 3.8 and 3.9. The ω axes are logarithmic (i.e., the plots are semi-logarithmic).

The Bode diagram of the term $(1 + jTw)$ has two asymptotes, namely the low- and high-frequency asymptotes:

$$\omega \simeq 0 \Longrightarrow 1 + jT\omega \simeq 1 \quad \text{and} \quad |1 + jT\omega|_{dB} = 0 \; dB, \quad \arg (1 + j\omega) = 0,$$

$$\omega \simeq \infty \Longrightarrow 1 + jT\omega \simeq jT\omega \quad \text{and} \quad |1 + j\omega|_{dB} = 20 \log T\omega, \quad \arg (1 + j\omega) = \frac{\pi}{2}.$$

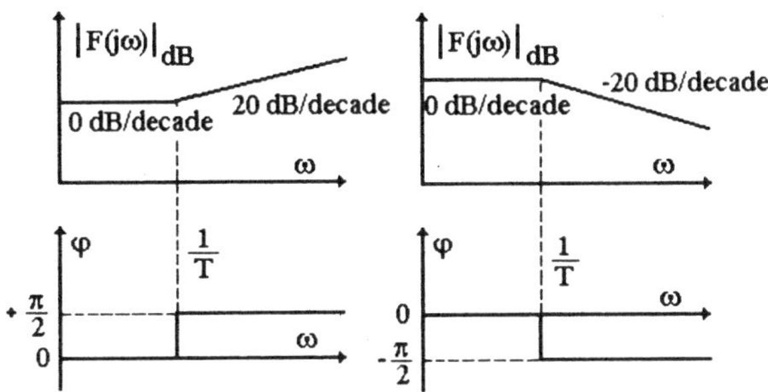

Figure 3.9. *Bode diagrams of* $1 + Ts$ *and* $1/(1+Ts)$

The intersection of these asymptotes occurs at:

$$20 \log T\omega = 0 \Longrightarrow \omega = \frac{1}{T}.$$

Their slopes can be expressed as the variation of the magnitude over a fixed interval. The main intervals used are the decade $[\omega, 10\omega]$ and the octave $[\omega, 2\omega]$. Note that these intervals are constant on a logarithmic scale:

$$\log 2\omega - \log \omega = \log 2, \quad \log 10\omega - \log \omega = \log 10 = 1.$$

For high frequencies, the slope of the asymptote is given by:

$$20 \log 10T\omega - 20 \log T\omega \;=\; 20\, dB/decade,$$

$$20 \log 2T\omega - 20 \log T\omega \;=\; 20 \log 2 \simeq 6\, dB/octave.$$

The real diagrams (see Figure 3.10) are close to these asymptotic diagrams for ω much less than, $1/T$ and for ω much greater than $\dfrac{1}{T}$. In the neighborhood of $\omega = 1/T$, the difference between these diagrams is equal to $3\, dB$. For $\omega = 1/T$, the phase contributions of the process lead and lag are equal to $\pi/4$ and $-\pi/4$, respectively. The phase diagram can be drawn precisely with a template corresponding to the argument of $(1 + Ts)$. Note that $\arg(1 + Ts) = -\arg 1/(1+Ts)$. The contributions of an all-pass element:

$$\frac{1 - as}{1 + as}$$

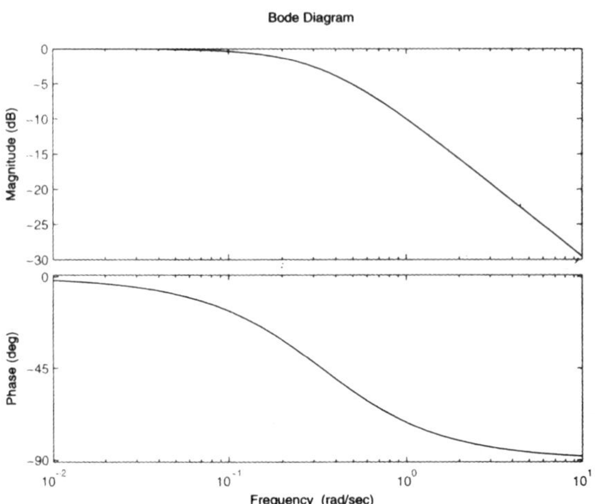

Figure 3.10. *Bode diagram of a first-order system*

to the magnitude expressed in decibels and to the maximal phase shift are equal to zero and $-\pi$, respectively:

$$\left|\frac{1-aj\omega}{1+aj\omega}\right| = 1, \quad \left|\frac{1-aj\omega}{1+aj\omega}\right|_{dB} = 0, \quad \arg_{\omega\to\infty}\frac{1-aj\omega}{1+aj\omega} = -\pi, \quad a > 0.$$

In the Nyquist diagram, the transfer function $G(j\omega) = X(\omega) + jY(\omega)$ is plotted in a two-dimensional coordinate system whose abscissa is the real part $X(\omega)$ and the ordinate is the imaginary part $Y(\omega)$. The Nichols diagram corresponds to the graphical representation of $|G(j\omega)|_{dB}$ versus the argument $\varphi = \arg G(j\omega)$, which corresponds to another representation of the complex number $G(j\omega)$.

PROBLEM 3.14. Draw the Bode diagram of the system:

$$F(s) = \frac{(s+0.7)(s+1)}{s^2(s+0.1)(s+0.6)(s+0.4)}.$$

SOLUTION 3.14. To draw the Bode diagram of this system, we have first to note the break frequencies (corner frequencies), namely 0.1, 0.4, 0.6, 0.7 and 1. We have:

$$F(s) = \frac{(s+0.7)(s+1)}{s^2(s+0.1)(s+0.6)(s+0.4)} = \frac{k(1+T_1 s)(1+T_2 s)}{s^2(1+T_1^* s)(1+T_2^* s)(1+T_3^* s)}.$$

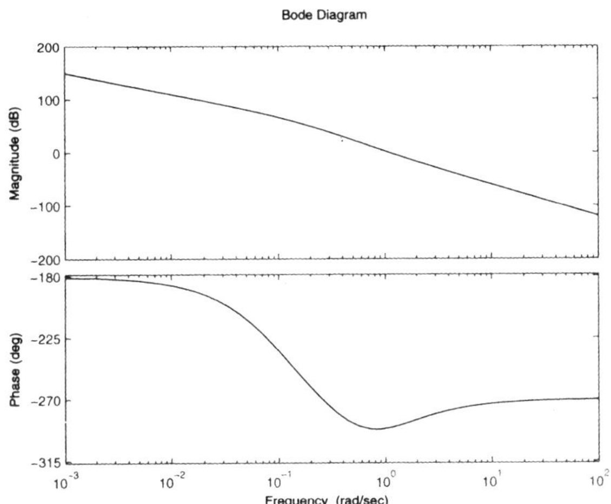

Figure 3.11. *Bode diagram of the system*
$$(s + 0.7)(s + 1) / \left[s^2 (s + 0.1)(s + 0.6)(s + 0.4) \right]$$

The magnitude, expressed in decibels, and the argument are given by:

$$20 \log |F(j\omega)| = |F(j\omega)|_{dB} = \sum_{i=1}^{2} |1 + jT_i\omega|_{dB} + \sum_{i=1}^{3} \left| \frac{1}{1 + jT_i^*\omega} \right|_{dB} + 2 \left| \frac{1}{j\omega} \right|,$$

$$\arg F(j\omega) = \sum_{i=1}^{2} \arg (1 + jT_i\omega) + \sum_{i=1}^{3} \arg \frac{1}{1 + jT_i^*\omega} + 2 \arg \frac{1}{j\omega}.$$

It follows that the magnitude and the argument of $F(j\omega)$ are the sums of the magnitudes and the arguments of terms of the form $(1/s, 1+Ts, 1/(1 + Ts))$, for which the Bode diagrams have been given earlier. The Bode diagram of the system is depicted in Figure 3.11.

Observe that for a time delay $\exp(-j\tau\omega) = \cos \tau\omega + j \sin \tau\omega$, the magnitude is equal to $1 (0\,dB)$, and the argument (phase) is equal to τw. The argument increases with frequency and tends to infinity when $\omega \to \infty$.

PROBLEM 3.15. Draw the Nyquist diagram of a first-order system. What happens as the static gain increases?

SOLUTION 3.15. The transfer function of a first-order system is given by:

$$F(s) = \frac{k}{1 + Ts}, \quad k > 0.$$

The Nyquist diagram corresponds to the parameterized function $x = f_1(\omega) = Re\left(F(j\omega)\right)$, $y = f_2(\omega) = Im\left(F(j\omega)\right)$. It follows that:

$$F(j\omega) = \frac{k}{1 + jT\omega} = \frac{k(1 - jT\omega)}{1 + T^2\omega^2} = x + jy,$$

$$x = \frac{k}{1 + T^2\omega^2}, \quad y = \frac{-kT\omega}{1 + T^2\omega^2}.$$

Observe that:

$$x > 0, \quad y < 0$$

except at the limits:

$$\text{for } \omega = 0, \quad x = k, \quad y = 0 \quad \text{and for } \lim_{\omega \to \infty} x = 0, \quad \lim_{\omega \to \infty} y = 0,$$

we also have that:

$$\arg F(j0) = 0, \quad \arg_{\omega \to \infty} F(j\omega) \to -\frac{\pi}{2} \quad \text{and} \quad \arg F(j\omega) \geq -\frac{\pi}{2}.$$

From these observations, we deduce that the Nyquist diagram lies in the quarter of the plane $x > 0$, $y < 0$. By eliminating ω from the expressions for x and y, we obtain the equation of the Nyquist diagram:

$$x = \frac{k}{1 + T^2\omega^2}, \quad y = \frac{-kT\omega}{1 + T^2\omega^2}$$

$$\frac{y}{x} = -T\omega \Longrightarrow T\omega = -\frac{y}{x},$$

$$x = \frac{k}{1 + y^2/x^2} = \frac{kx^2}{x^2 + y^2} \Rightarrow x^2 + y^2 = kx. \tag{3.18}$$

The function $f(x, y) = x^2 + y^2 - kx = 0$ corresponds to a circle centred at the point $(x = k/2, y = 0)$ with a radius equal to $k/2$. Taking into account the observations made above, the Nyquist diagram consists only of the half of this circle located in the quarter of the plane mentioned above. This diagram is depicted in Figure 3.12.

Note that for systems of order n greater than 1, it is practically impossible to derive analytically the corresponding function $f(x, y) = 0$, i.e., the elimination of ω from the expressions for x and y is practically impossible. For these systems, a computer is needed in order to achieve this objective.

Now, let us consider the distance of a given point $M(x, y)$ in the Nyquist diagram from the origin. From equation (3.18), we derive:

$$|F(j\omega)|^2 = x^2 + y^2 = kx > 0 \quad \text{for } k > 0,$$

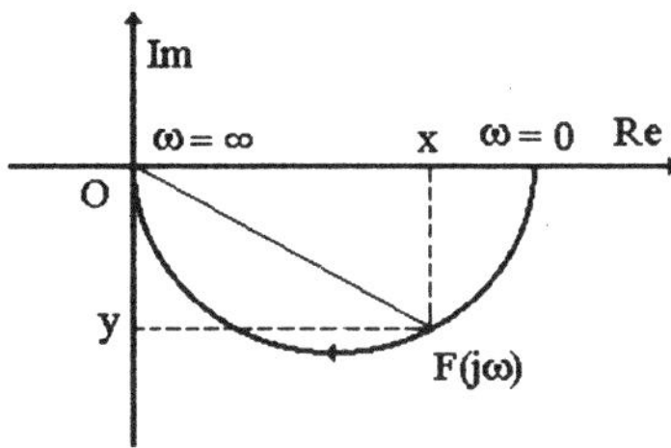

Figure 3.12. *Nyquist diagram of a first-order system*

which shows that this distance is proportional to the static gain k. Note that $\arg F\left(j\omega\right)$ is independent of k, i.e.:

$$\arg F\left(j\omega\right) = \arctan\left(-T\omega\right).$$

PROBLEM 3.16. Derive the magnitude and argument fot a unity-feedback system from the Nichols diagram of the open-loop transfer function.

SOLUTION 3.16. The closed-loop transfer function is given by:

$$F\left(j\omega\right) = \frac{G\left(j\omega\right)}{1 + G\left(j\omega\right)}.$$

The diagram of $F\left(j\omega\right)$ is derived from the diagram of $G\left(j\omega\right)$ by the mapping $G\left(jw\right) / \left(1 + G\left(jw\right)\right)$. From Figure 3.13, we derive:

$$F\left(j\omega\right) \quad \rightarrow \quad \frac{\overrightarrow{OM}}{\overrightarrow{AM}},$$

$$\arg F\left(j\omega\right) \quad = \quad \arg \overrightarrow{OM} - \arg \overrightarrow{AM} = \alpha, \quad \alpha = \varphi - \theta.$$

The Nichols diagram [JAM 47] corresponds to the locus:

$$\frac{OM}{AM} \quad = \quad \lambda = const, \tag{3.19}$$

$$\alpha \quad = \quad const. \tag{3.20}$$

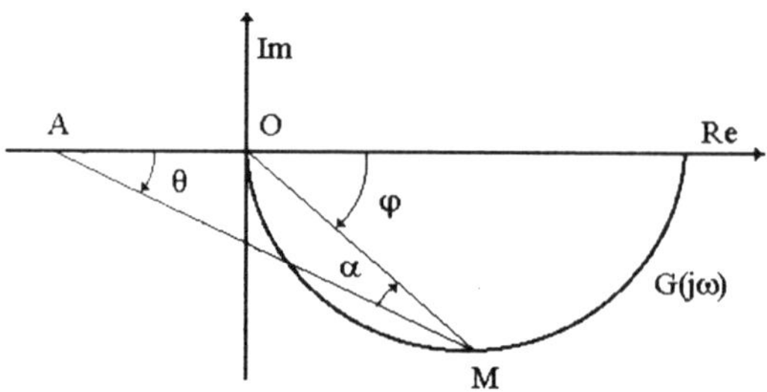

Figure 3.13. *Graphical representation of the magnitude and argument of the closed-loop transfer function*

Loci $\lambda = const$. The loci of points for which the ratio (3.19) are a constant is referred to as M-contours, and the loci of points for which (3.20) is a constant are called N-contours. Recall that, given two fixed points (O and A), the locus of the points M such that the ratio of OM to AM is constant is an Apollonian circle, and there are infinitely many Apollonian circles, one for each value of λ. Each of the circles divides the segment OM harmonically.[14] The set of Apollonian circles is called a hyperbolic pencil. This set includes the bisector of the segment OA. The circles are coaxial, and their centers are located on the line OA. The points O and A represent the limit points. Two circles symmetric around the bisector of the segment OA correspond to ratios λ and $1/\lambda$, respectively.

Loci $\alpha = const$. Taking into account the fact that the inscribed angles are constant, the loci of points M such that the angle α has a constant value are circular arcs that pass through the points O and A. These arcs are symmetric around the segment OA. A set of circles such that $\alpha = const$ is called an elliptic pencil (the locus of points at which a segment subtends a constant angle). They are coaxal circles, i.e., their centres are collinear. They are located on the bisector of the segment OA. The circles of the hyperbolic and elliptic pencils are orthogonal, i.e., they cut one other at right angles. If we allow λ and α to vary, we obtain the Nichols diagram. There exists a similarity

---

14. The points C and D divide a segment AB harmonically if:

$$\frac{|AC|}{|BC|} = \frac{|AD|}{|BD|},$$

where A, B, C and D are collinear. If C and D divide the segment AB harmonically, it follows that the points A and B divide the segment CD harmonically.

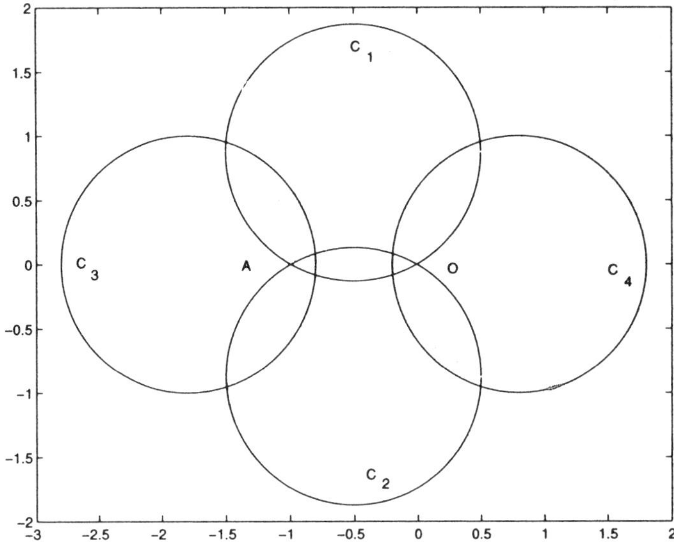

Figure 3.14. *Elliptic and hyperbolic pencils*

between Nichols diagrams and equipotential and gravitational surfaces (fields are a topic of central importance, and they play a prominent role in modern physics).

In Figure 3.14, the circles C_1 and C_2 correpond to α and $-\alpha$, respectively. They belong to an elliptic pencil. The circles C_3 and C_4 correspond to λ and $1/\lambda$, respectively. They belong to a hyperbolic pencil. For each frequency ω_i, the magnitude and phase of the transfer function of the unity-feedback system can be identified as the values of λ and α associated with the magnitude and phase circles, respectively, crossing the open-loop diagram at the point M_i, of coordinates $ReG\,(j\omega_i)$ and $ImG\,(j\omega_i)$. The value of $\max\limits_{\omega} F\,(j\omega)$ is given by the value λ of the magnitude contour which is tangent to the Nyquist diagram of the open-loop transfer function.

Let us consider again the closed-loop transfer function:

$$F\,(j\omega) = \frac{G\,(j\omega)}{1 + G\,(j\omega)}, G\,(j\omega) = \mu \exp\,(j\varphi),$$

where:

$$\mu = |G\,(j\omega)| \quad \text{and} \quad \varphi = \arg G\,(j\omega).$$

It follows that:

$$
\begin{aligned}
|1 + G(j\omega)| &= |1 + \mu\cos\varphi + j\mu\sin\varphi| = \sqrt{1 + \mu^2 + 2\mu\cos\varphi}, \\[2mm]
|F(j\omega)| &= \frac{|G(j\omega)|}{|1 + G(j\omega)|}, \\[2mm]
|F(j\omega)| &= \frac{\mu}{\sqrt{1 + \mu^2 + 2\mu\cos\varphi}}, \\[2mm]
F(j\omega) &= \frac{\mu\cos\varphi + \mu^2 + j\mu\sin\varphi}{(1 + \mu\cos\varphi)^2 + (\mu\sin\varphi)^2}, \\[2mm]
\arg F(j\omega) &= \arctan\frac{\sin\varphi}{\mu + \cos\varphi}.
\end{aligned}
$$

The Hall chart [HAL 43] corresponds to the loci:

$$
20\log\frac{\mu}{\sqrt{1 + \mu^2 + 2\mu\cos\varphi}} = \lambda\, dB = const,
$$

$$
\arctan\frac{\sin\varphi}{\mu + \cos\varphi} = \alpha = const.
$$

These loci are orthogonal.

PROBLEM 3.17. Consider a system with the following transfer function (a lead network or lead compensator):

$$
F(s) = \frac{1 + aTs}{1 + Ts}, \quad a > 1, \quad T > 0 \tag{3.21}
$$

where $a = 2$ and $T = 5\,s$.

1. Draw the Nyquist, Bode and Nichols diagrams of this system.

2. Determine the maximal value of $\arg F(j\omega)$, and show that it does not depend on the time constant T.

3. Show that the transfer function (3.21) corresponds to the system depicted in Figure 3.15.

SOLUTION 3.17. 1. For large values of a, the lead compensator acts as a derivative system and its zero $z = -1/aT$ is close to the imaginary axis. The Nyquist, Bode and Nichols (Black) diagrams are depicted in Figures 3.16, 3.17 and 3.18, respectively.

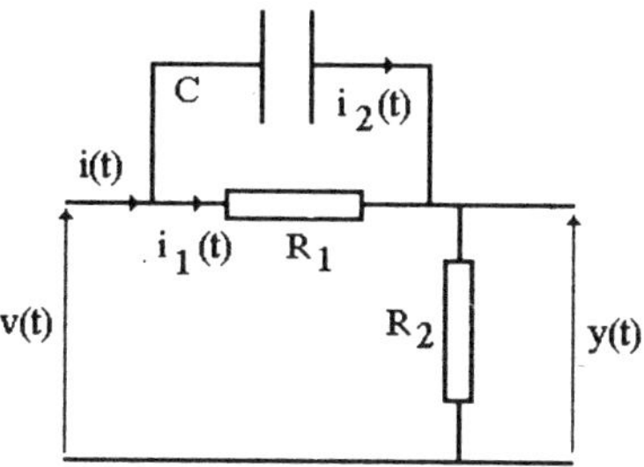

Figure 3.15. *Lead compensator*

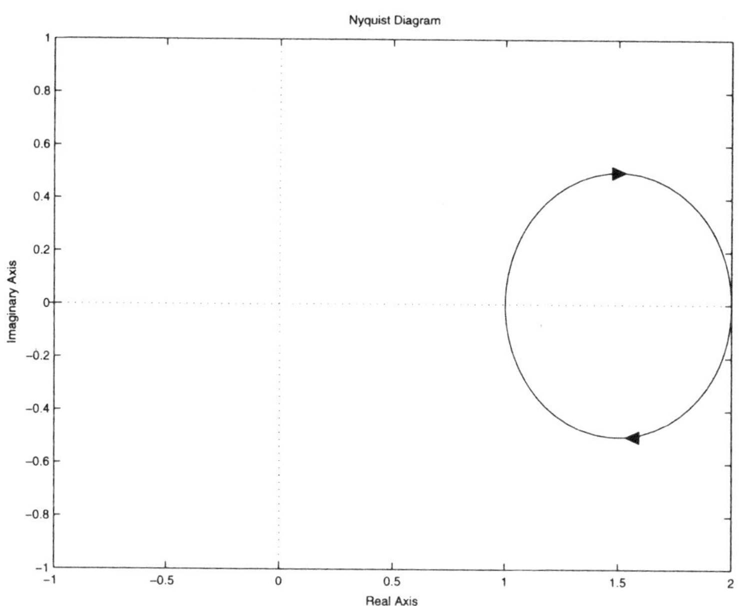

Figure 3.16. *Nyquist diagram of a lead compensator*

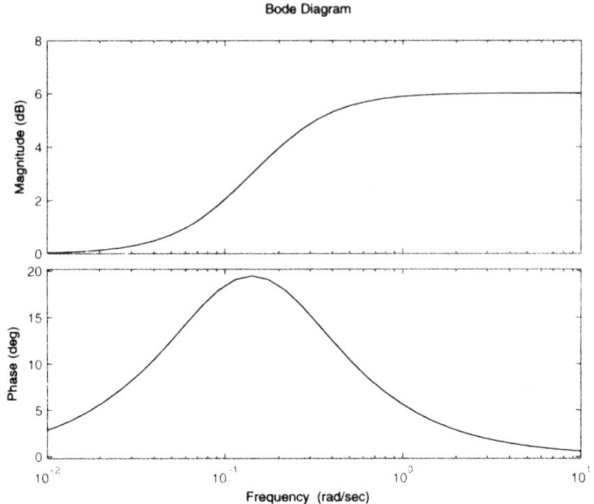

Figure 3.17. *Bode diagram of a lead compensator*

From these plots, we observe that the lead compensator introduces a gain at high frequencies. Therefore, high-frequency noise will be amplified. The lead compensator also introduces a phase margin, for which the maximal value will be calculated in what follows.

The Nyquist diagram consists of a semicircle. We have:

$$F\left(j\omega\right) = \frac{1 + jaT\omega}{1 + jT\omega} = x + jy,$$

where:

$$x = \frac{1 + aT^2\omega^2}{1 + T^2\omega^2}, \quad y = \frac{T\omega\left(a - 1\right)}{1 + T^2\omega^2}. \tag{3.22}$$

By eliminating $T\omega$ from these expressions, we obtain:

$$T^2w^2 = \frac{x - 1}{a - x},$$

$$y = \frac{\left(a - 1\right)\sqrt{\left(x - 1\right)/\left(a - x\right)}}{1 + \left(x - 1\right)/\left(a - x\right)},$$

$$y^2 = \left(x - 1\right)\left(a - x\right),$$

$$y^2 + x^2 - \left(a + 1\right)x + a = y^2 + x^2 - \frac{2\left(a + 1\right)}{2}x + a = 0.$$

This circle is centered at the point of co-ordinates $\left(a + 1\right)/2$. It is clear from Equation (3.22) that when $\omega \to \infty$, then $x \to a$ and $y \to 0$, and when $\omega \to 0$, then $x \to 1$ and

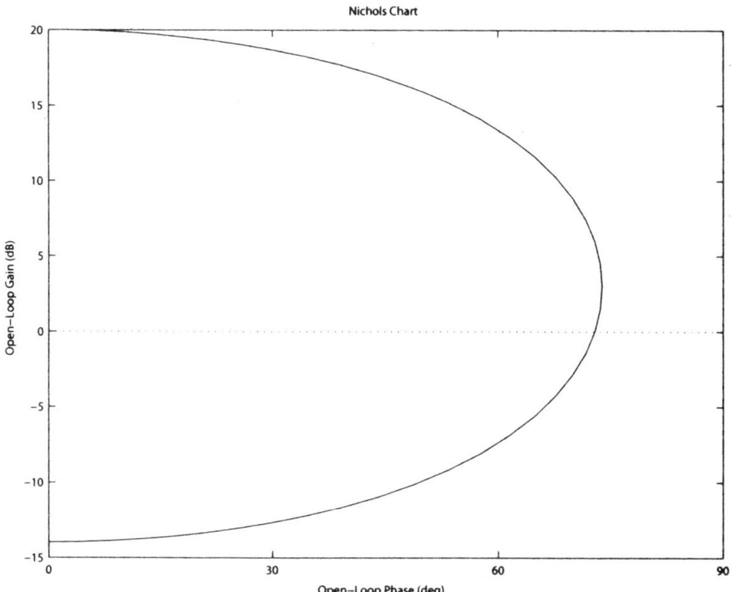

Figure 3.18. *Nichols diagram of a lead compensator*

$y \rightarrow 0$. Notice that y is positive for $\omega \in [0, +\infty[$. Therefore, the Nyquist diagram consists of the semicircle corresponding to $y \geq 0$. This diagram is depicted in Figure 3.16. The other semicircle corresponds to negative values of the frequency.

2. The argument is given by:

$$\arg \frac{1 + jaT\omega}{1 + jT\omega} \quad - \quad \arg \frac{(1 + jaT\omega)(1 - jT\omega)}{1 + T^2\omega^2} = \varphi, \tag{3.23}$$

$$F(jw) \;=\; Re(.) + jIm(.).$$

Let us now calculate its tangent, equal to $Im(.)/Re(.)$:

$$\tan \varphi = \frac{y}{x} = \frac{T\omega(a-1)}{1 + aT^2\omega^2}.$$

The argument φ is equal to the acute angle corresponding to the union of the real axis and OM with the origin, where O and M represent the origin and the point associated with the complex number s, respectively. The derivative of the tangent is:

$$\frac{d\tan \varphi}{d\omega} = \frac{T(1-a)\left(aT^2\omega^2 - 1\right)}{\left(1 + aT^2\omega^2\right)^2}.$$

The maximum value of the argument is obtained by setting this derivative to zero:

$$\frac{T\left(1-a\right)\left(aT^2w^2-1\right)}{\left(1+aT^2w^2\right)^2}=0.$$

Therefore, the maximum occurs at:

$$\omega=\omega_m=\frac{1}{T\sqrt{a}}.$$

From Equation (3.23), using Equation (3.22), we obtain the sine:

$$\sin\varphi=\frac{Tw\left(a-1\right)}{\sqrt{\left(1+aT^2w^2\right)\left(1+T^2w^2\right)}}.$$

For $\omega=\dfrac{1}{T\sqrt{a}}$, we obtain the sine of the maximal value φ_m of the argument φ:

$$\sin\varphi_m=\frac{a-1}{a+1},$$

which depends only on the parameter a. Observe that φ_m is the angle between the real axis and the tangent from the origin to the Nyquist diagram (semi circle).

3. Using Ohm's law, we derive:

$$
\begin{aligned}
V\left(s\right) &= R_1 I_1\left(s\right)+R_2 I\left(s\right), \quad I\left(s\right)=I_1\left(s\right)+I_2\left(s\right), \\
Y\left(s\right) &= R_2 I\left(s\right), \quad R_1 I_1\left(s\right)=\frac{1}{Cs}I_2\left(s\right).
\end{aligned}
$$

Eliminating $I_1\left(s\right)$ and $I_2\left(s\right)$ from these equations, we obtain:

$$F\left(s\right)=\frac{Y\left(s\right)}{V\left(s\right)}=\frac{R_2+R_1R_2Cs}{R_1+R_2+R_1R_2Cs}$$

$$\frac{R_2}{R_1+R_2}=\frac{1+\left[R_1R_2\left(R_1+R_2\right)C\right]s/\left[R_2\left(R_1+R_2\right)\right]}{\left[1+R_1R_2C\right]s/\left[R_1+R_2\right]}$$

$$=\frac{1}{R_1/R_2+1}\frac{1+\left(R_1+R_2\right)R_1R_2Cs/R_2\left(R_1+R_2\right)}{1+R_1R_2Cs/\left[R_1+R_2\right]}$$

$$=\frac{1}{a}\frac{1+aTs}{1+Ts},$$

where:

$$a=1+\frac{R_1}{R_2},\quad T=\frac{R_1R_2}{R_1+R_2}C.$$

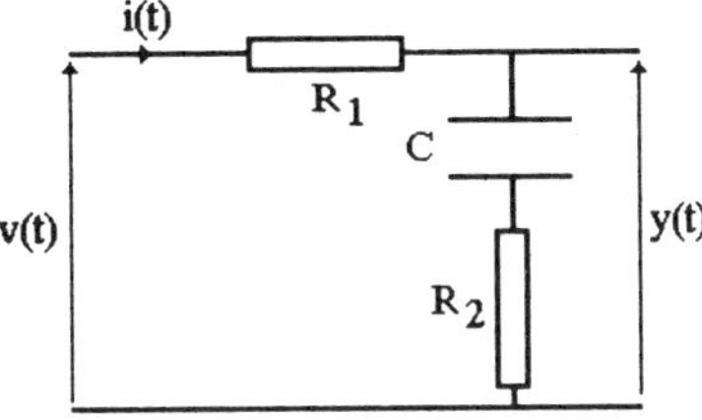

Figure 3.19. *Lag compensator*

Let us recall the expression for the current $i\,(t)$:

$$i\,(t) \;=\; \frac{v\,(t) - y\,(t)}{z}, \quad \frac{1}{z} = Cs + \frac{1}{R_1},$$

$$I\,(s) \;=\; \frac{V\,(s) - Y\,(s)}{R_1}\,(1 + R_1 Cs).$$

For small values of R_1, we obtain:

$$v\,(t) \gg y\,(t), \quad I\,(s) \simeq \frac{V\,(s)}{R_1}\,(1 + R_1 Cs),$$

and

$$Y\,(s) = R_2 I\,(s) \simeq \frac{V\,(s)}{R_1}\left(\frac{R_2}{R_1} + R_2 Cs\right),$$

which corresponds to á proportional and derivative controller.

The synthesis of regulator of this kind (tuning of the parameters a and T) will be presented in Chapter 5.

PROBLEM 3.18. 1. Show that the transfer function:

$$F\,(s) = \frac{1 + Ts}{1 + bTs}, \quad b > 1,\ T > 0,$$

can be realized by the system depicted in Figure 3.19.

2. For $T = 5\,s$ and $b = 2$, draw the Nyquist, Bode and Nichols diagrams of this system.

3. Show that the minimal phase φ_m occurs for $\omega = 1/T\sqrt{b}$. Calculate this phase.

SOLUTION 3.18. From Ohm's law, we derive:

$$v(t) = R_1 i(t) + y(t), \quad y(t) = R_2 i(t) + \frac{1}{C}\int i(t)\,dt,$$

$$V(s) = R_1 I(s) + Y(s), \quad Y(s) = R_2 I(s) + \frac{1}{Cs} I(s),$$

which leads to:

$$F(s) = \frac{Y(s)}{V(s)} = \frac{1 + R_2 C s}{1 + C(R_1 + R_2)s}$$

$$= \frac{1 + R_2 C s}{1 + R_2 C(R_1 + R_2)/R_2 s}.$$

It follows that:

$$b = \frac{(R_1 + R_2)}{R_2} \quad \text{and} \quad T = R_2 C.$$

This transfer function corresponds to a lag compensator. For large values of b, this system acts as an integrator and its pole $p = -1/bT$ is close to the imaginary axis.

Let us recall the expression for the current $i(t)$,

$$i(t) = \frac{v(t) - y(t)}{R_1}.$$

If $y(t) \ll v(t)$, we obtain:

$$i(t) \simeq \frac{v(t)}{R_1}$$

and

$$y(t) \simeq \frac{R_2}{R_1} v(t) + \frac{1}{CR_1}\int v(t)\,dt,$$

which corresponds to a proportional and integral (PI) controller.

2. The Nyquist, Bode and Nichols diagrams are shown in Figures 3.20, 3.21 and 3.22.

3. The frequency response is related to:

$$|F(j\omega)|_{dB} = \left| \frac{1 + jT\omega}{1 + jbT\omega} \right|_{dB},$$

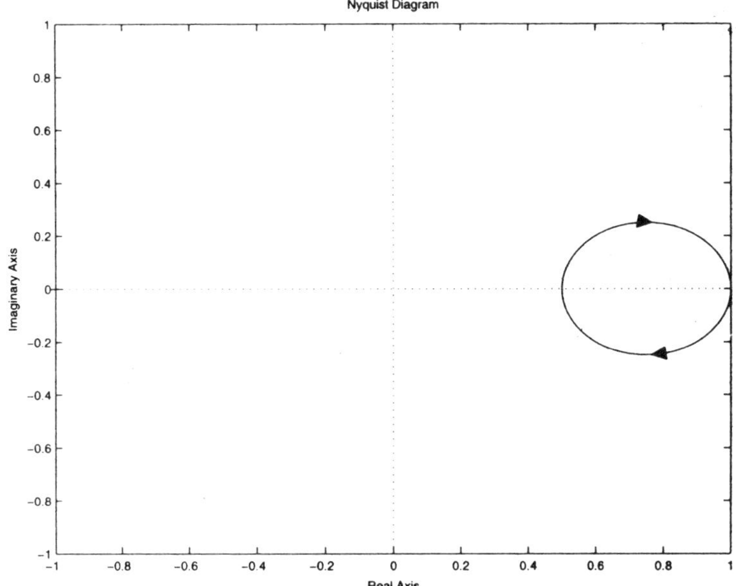

Figure 3.20. *Nyquist diagram of a lag compensator*

$$\varphi = \arg \frac{1 + jT\omega}{1 + jbT\omega} = \arg \frac{(1 + jT\omega)(1 - jbT\omega)}{(1 + jbT\omega)(1 - jbT\omega)},$$

$$\varphi = \arg(x + jy), \quad x = \frac{1 + bT^2\omega^2}{1 + (bT\omega)^2}, y = \frac{T\omega(1 - b)}{1 + (bT\omega)^2}, \tag{3.24}$$

which gives:

$$\tan \varphi = \frac{T\omega(1 - b)}{1 + bT^2\omega^2},$$

and the derivative is:

$$\frac{d\tan \varphi}{d\omega} = T(1 - b)\frac{1 - bT^2\omega^2}{(1 + bT^2\omega^2)^2}.$$

By cancelling this derivative, we obtain:

$$\omega = \frac{1}{T\sqrt{b}}.$$

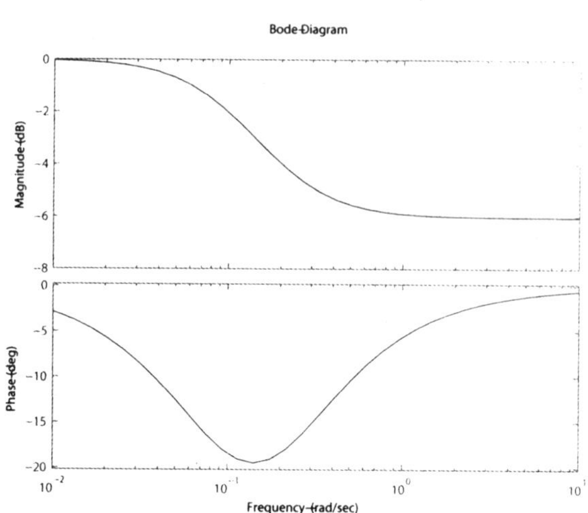

Figure 3.21. *Bode diagram of a lag compensator*

Let us now calculate the value of $\sin \varphi_m$ which corresponds to the frequency $\omega = 1/T\sqrt{b}$:

$$\sin \varphi = \frac{T\omega \left(1 - b\right)}{\sqrt{\left(1 + bT^2\omega^2\right)^2 + T^2\omega^2 \left(1 - b\right)^2}},$$

$$\sin \varphi_m = \frac{\left(1 - b\right)/\sqrt{b}}{\sqrt{\left(1 + 1\right)^2 + \left(1 - b\right)^2/b}} = \frac{1 - b}{b + 1} < 0.$$

Observe that from Equation (3.24), we can derive:

$$x \;\rightarrow\; \frac{1}{b} \quad \text{and } y \rightarrow 0 \quad \text{when } \omega \rightarrow \infty,$$

$$x \;\rightarrow\; 1 \quad \text{and } y \rightarrow 0 \quad \text{when } \omega \rightarrow 0.$$

PROBLEM 3.19. Show that the system depicted in Figure 3.23 corresponds to a lead–lag compensator.

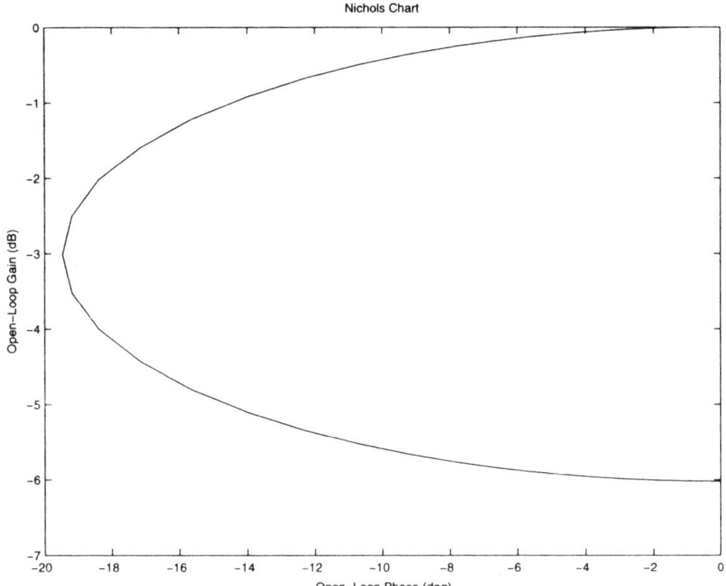

Figure 3.22. *Nichols diagram of a lag compensator*

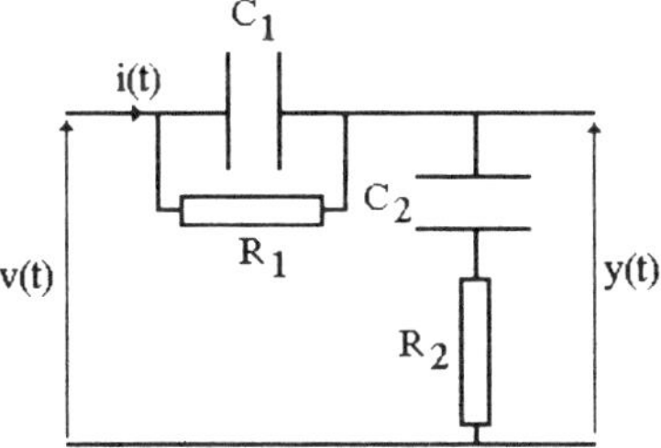

Figure 3.23. *Lead–lag compensator*

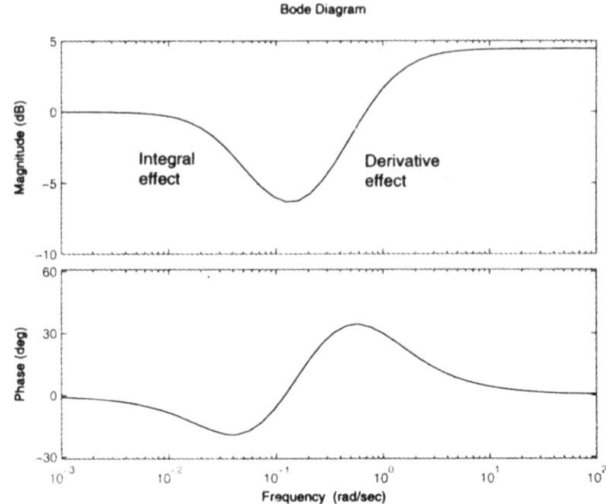

Figure 3.24. *Bode diagram of a lead–lag compensator with*
$$F(s) = k(1 + 5s)(1 + 10s) / (1 + s)(1 + 30s)$$

SOLUTION 3.19. Using Ohm's law, we derive:

$$V(s) = (z_1 + z_2) I(s), \quad Y(s) = z_2 I(s),$$

where z_1 and z_2:

$$z_1 = \frac{R_1}{1 + R_1 C_1 s}, \quad \text{and } z_2 = \frac{1 + R_2 C_2 s}{C_2 s}$$

represent the impedances of the circuits $R_1 C_1$ and $R_2 C_2$, respectively. The circuits of impedance z_2 and z_1 act as a differentiator and an integrator, respectively. It follows that:

$$F(s) = \frac{Y(s)}{V(s)} = \frac{(1 + R_1 C_1 s)(1 + R_2 C_2 s)}{R_1 C_2 s + (1 + R_1 C_1 s)(1 + R_2 C_2 s)}.$$

The Bode diagram of a lead–lag compensator is depicted in Figure 3.24.

PROBLEM 3.20. To control the process:

$$G(s) = \frac{k}{(1 + Ts)^2},$$

the regulator:

$$R(s) = k_c \frac{(s - z_1)(s - z_2)}{s}$$

is used.

1. Determine an expression relating k_c, z_1 and z_2 in order that the steady-state error induced by a ramp input (t) will be less than 0.1.

2. For $z_1 = -1/T$, determine the controller gain k_c and the zero z_2 in order that the compensated system behaves as a second-order system with a natural frequency ω_n and a damping factor equal to ζ.

3. For $T = 1\,s$, $k = 2$, $\zeta = 0.5$ and $\omega_n = 10$, draw the step response of the compensated system and calculate the resonant frequency and the height of the resonant peak.

SOLUTION 3.20. 1. Observe first that the regulator considered here is proportional–integral–derivative (PID) controller. For a ramp input, the Laplace transform of the error is given by:

$$\Xi\left(s\right) = \frac{\left(1+Ts\right)^2}{s\left[s\left(1+Ts\right)^2 + kk_c\left(s-z_1\right)\left(s-z_2\right)\right]}.$$

Using the final-value theorem, we derive:

$$\varepsilon\left(\infty\right) = \lim_{s\to 0} s\frac{\left(1+Ts\right)^2}{s\left[s\left(1+Ts\right)^2 + kk_c\left(s-z_1\right)\left(s-z_2\right)\right]} = \frac{1}{kk_c z_1 z_2} < \frac{1}{10},$$

which implies that:

$$kk_c z_1 z_2 > 10.$$

2. One zero of the PID controller has been used to cancel one of the poles of the system. The closed-loop transfer function is given by:

$$F\left(s\right) = \frac{kk_c\left(s-z_2\right)/T^2}{s^2 + \left(T + kk_c/T^2\right)s - kk_c z_2/T^2}.$$

Therefore,

$$2\zeta\omega_n = \frac{T+kk_c}{T^2} \quad \text{and} \quad \omega_n^2 = -\frac{kk_c}{T^2}z_2,$$

which, on solving, yields:

$$k_c = \frac{T\left(2T\zeta\omega_n - 1\right)}{k}, \quad z_2 = -\frac{T\omega_n^2}{2T\zeta\omega_n - 1}.$$

3. For $T = 1$, $k = 2$, $\zeta = 0.5$ and $\omega_n = 10$, we obtain:

$$k_c = 4.5, \quad z_2 = -11.11.$$

The step response of the compensated system is plotted in Figure 3.25.

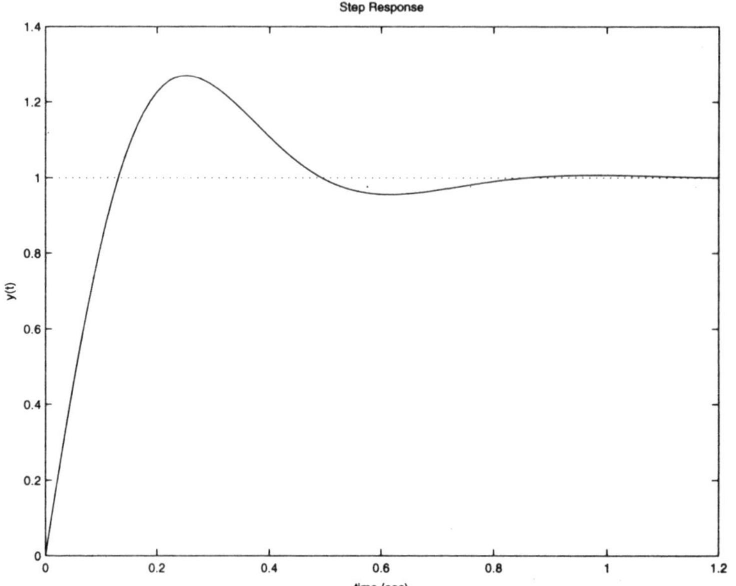

Figure 3.25. *Step response of a compensated system*

From the Bode diagram depicted in Figure 3.26, we obtain the resonant frequency and the height of the resonant peak:

$$\omega_r = 8.3916, \quad M_p = 1.4084 \quad 2.96\,dB$$

PROBLEM 3.21. Consider a second-order system with a damping factor $\zeta < 1$. Let P, P^* and M be the points associated with the poles of this system and with $j\omega$, respectively. Express the magnitude of its transfer function with respect to the angle $\widehat{PMP^*}$. Give a graphical interpretation of the resonant frequency.

SOLUTION 3.21. For a damping factor $\zeta < 1$, the transfer function of a second-order system and the poles are given by:

$$F(s) = \frac{\omega_n^2}{s^2 + 2\zeta\omega_n s + \omega_n^2} = \frac{\omega_n^2}{(s - p)(s - p^*)},$$

$$p = -\zeta\omega_n + j\omega_n\sqrt{1 - \zeta^2}, \quad p^* = -\zeta\omega_n - j\omega_n\sqrt{1 - \zeta^2}.$$

The resonant frequency is given by:

$$\frac{d}{d\omega}|F(j\omega)| = \frac{d}{d\omega}\frac{\omega_n^2}{\sqrt{(\omega_n^2 - w^2)^2 + 4\zeta^2\omega_n^2 w^2}} = 0,$$

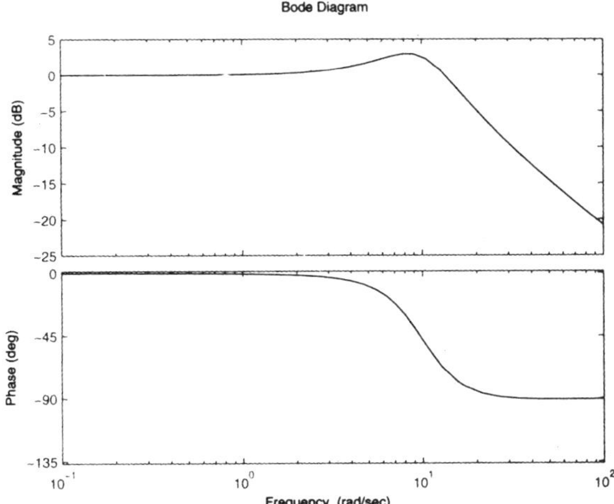

Figure 3.26. *Bode diagram of the compensated system*

which leads to:

$$\omega_r = \omega_n \sqrt{1 - 2\zeta^2}, \quad \zeta < \frac{\sqrt{2}}{2} \simeq 0.7, \quad |F(j\omega_r)| = \frac{1}{2\zeta\sqrt{1 - \zeta^2}}.$$

Let us consider a point $M(0, j\omega)$ located on the imaginary axis $(\overrightarrow{OIm}, \omega > 0)$. It follows from the above that:

$$|F(j\omega)| = \left| \frac{\omega_n^2}{(s - p)(s - p^*)} \right| = \frac{\omega_n^2}{\left| \overrightarrow{PM}\overrightarrow{P^*M} \right|},$$

where P and P^* are the points associated with the poles p and p^*, respectively, i.e., P is $\left(-\zeta\omega_n, \omega_n\sqrt{1 - \zeta^2}\right)$ and P^* is $\left(-\zeta\omega_n, -\omega_n\sqrt{1 - \zeta^2}\right)$. In Figure 3.27, we have represented a semi circle of diameter PP^* and the projection C of the point M onto the segment PP^*. From this figure, we derive:

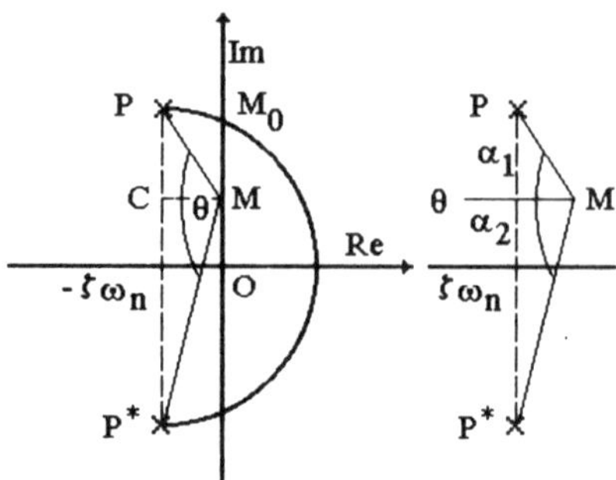

Figure 3.27. *Graphical determination of the resonant frequency of a second-order system*

$$\sin \alpha_1 = \frac{\omega_n \sqrt{1 - \zeta^2} - \omega}{\left|\overrightarrow{PM}\right|} = \frac{\omega_n \sqrt{1 - \zeta^2} - \omega}{\sqrt{\zeta^2 \omega_n^2 + \left(\omega_n \sqrt{1 - \zeta^2} - \omega\right)^2}},$$

$$\cos \alpha_1 = \frac{\zeta \omega_n}{\left|\overrightarrow{PM}\right|} = \frac{\zeta \omega_n}{\sqrt{\zeta^2 \omega_n^2 + \left(\omega_n \sqrt{1 - \zeta^2} - \omega\right)^2}},$$

$$\sin \alpha_2 = \frac{\omega_n \sqrt{1 - \zeta^2} + \omega}{\left|\overrightarrow{P^*M}\right|} = \frac{w_n \sqrt{1 - \zeta^2} + w}{\sqrt{\zeta^2 w_n^2 + \left(w_n \sqrt{1 - \zeta^2} + w\right)^2}},$$

$$\cos \alpha_2 = \frac{\zeta w_n}{\left|\overrightarrow{P^*M}\right|} = \frac{\zeta w_n}{\sqrt{\zeta^2 w_n^2 + \left(\omega_n \sqrt{1 - \zeta^2} + w\right)^2}},$$

and

$$\sin \theta = \sin(\alpha_1 + \alpha_2) = \sin \alpha_1 \cos \alpha_2 + \cos \alpha_1 \sin \alpha_2,$$

$$\sin \theta = \frac{2 \zeta \omega_n^2 \sqrt{1 - \zeta^2}}{\left|\overrightarrow{PM}\right| \left|\overrightarrow{P^*M}\right|}.$$

Let us now express the magnitude $|F(j\omega)|$ as a function of $\sin \theta$. We obtain:

$$|F(j\omega)| = \frac{\omega_n^2}{|\omega_n^2 - \omega^2 + 2j\zeta\omega\omega_n|} = \frac{\omega_n^2}{\left|\overrightarrow{PM}\right|\left|\overrightarrow{P^*M}\right|} = \frac{|\sin \theta|}{2\zeta\sqrt{1 - \zeta^2}}.$$

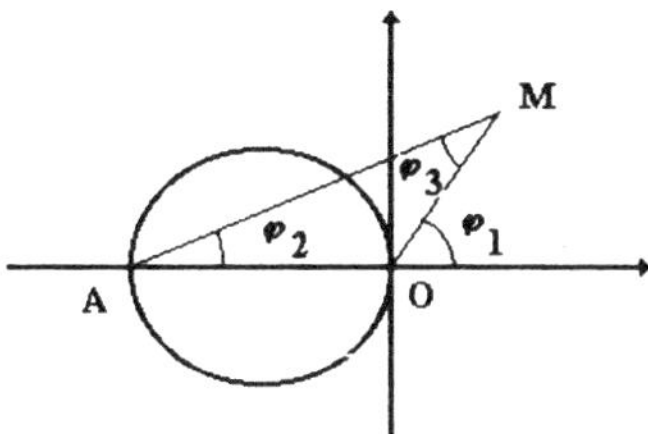

Figure 3.28. *Graphical representation of the argument relation in Problem 3.22*

At the resonant frequency, $|F(j\omega)|$ attains its maximal value, which corresponds to $\sin\theta = 1 \implies \theta = \pi/2$. Observe, from the properties of a circle, that $\theta = \pi/2$ when the point M coincides with the point M_0 $(\widehat{PM_0P^*} = \frac{\pi}{2})$. Consequently, the intersection M_0 of the circle of diameter PP^* with the imaginary axis gives the resonant frequency ω_r.

PROBLEM 3.22. Consider a unity-feedback system such that:

$$\arg \frac{G(j\omega)}{1 + G(j\omega)} = \frac{\pi}{2}, \tag{3.25}$$

where $G(s)$ represent the transfer function of the forward path. Determine the loci of the points M $(x = ReG(j\omega), y = ImG(j\omega))$.

SOLUTION 3.22. From Equation (3.25), we derive:

$$\arg G(j\omega) - \arg(1 + G(j\omega)) = \frac{\pi}{2},$$

which may be written in the following form:

$$\arg \overrightarrow{OM} - \arg \overrightarrow{AM} = \frac{\pi}{2},$$

where O and A represent the origin and the critical point, respectively.

From Figure 3.28, which shows a graphical representation of this expression for the argument, we derive:

$$\varphi_1 - \varphi_2 = \frac{\pi}{2}. \tag{3.26}$$

From the triangle OMA, we obtain the relation:

$$\varphi_1 = \varphi_2 + \varphi_3.$$

Therefore, the points defined by the condition (3.25) belong to a set of points such that:

$$\varphi_3 = \widehat{OMA} = \frac{\pi}{2}. \tag{3.27}$$

It is well known from classical geometry that to fulfil this condition, the point M must lie on a circle C of diameter AO, i.e.:

$$\left| G\left(j\omega\right) + \frac{1}{2} \right| = \frac{1}{2};$$

$G\left(j\omega\right) + 1/2$ is associated with $\overrightarrow{O'M}$, where $O'\left(-1/2, 0\right)$. We have now shown that the points defined by the condition (3.26) lie on the circle C. The points on this circle such that $ImG\left(j\omega\right) > 0$ correspond to the points defined by $\arg \overrightarrow{OM} - \arg \overrightarrow{AM} = \pi/2$, while the points $\in C$ such that $ImG\left(j\omega\right) < 0$ corresponds to the points defined by $\arg \overrightarrow{OM} - \arg \overrightarrow{AM} = -\pi/2$. To verify this result, it is sufficient to consider the point M_1 and M_2 of co-ordinates $\left(-1/2, 1/2\right)$ and $\left(-1/2, -1/2\right)$, respectively. We obtain:

$$\arg \overrightarrow{OM_1} - \arg \overrightarrow{AM_1} \;=\; \arg \frac{\overrightarrow{OM_1}}{\overrightarrow{AM_1}} = \arg \frac{-1/2 + j/2}{1/2 + j/2} = \arg\left(+j\right) = \frac{\pi}{2},$$

$$\arg \overrightarrow{OM_2} - \arg \overrightarrow{AM_2} \;=\; \arg \frac{\overrightarrow{OM_2}}{\overrightarrow{AM_2}} = \arg \frac{-1/2 - j/2}{1/2 - j/2} = \arg\left(-j\right) = -\frac{\pi}{2}.$$

Consequently, the locus defined by the condition (3.25) is the semi circle C located in the second quadrant ($Re\left(\cdot\right) < 0$ and $Im\left(\cdot\right) > 0$).

PROBLEM 3.23. Determine graphically the point M on the Nyquist diagram such that:

$$\left| G\left(j\omega\right) + 1 + jb \right| = b. \tag{3.28}$$

SOLUTION 3.23. The transfer function $G\left(j\omega\right)$ may be written as follows:

$$G\left(j\omega\right) = ReG\left(j\omega\right) + ImG\left(j\omega\right) = x + jy.$$

From Equation (3.28), we derive:

$$\left| x + 1 + j\left(b + y\right) \right| = b.$$

$$(x + 1)^2 + (b + y)^2 = b^2,$$

which corresponds to the equation of a circle centered at $(-1, -b)$ with radius b. Therefore, the point defined by Equation (3.28) is located at the intersection of the Nyquist diagram of $G(j\omega)$ with this circle.

PROBLEM 3.24. The *characteristic equation* is given by:

$$D(s) = \sum_{i=0}^{n} a_i s^i = 0.$$

For $s = \alpha + j\omega$, show that the real and imaginary parts of $s^n = X_n + jY_n$ are solutions of the following recurrence equations:

$$X_{i+2} - 2\alpha X_{i+1} + (\alpha^2 + \omega^2) X_i = 0,$$
$$Y_{i+2} - 2\alpha Y_{i+1} + (\alpha^2 + \omega^2) Y_i = 0.$$

SOLUTION 3.24. For s^{i+1}, we obtain:

$$\begin{aligned}
s^{i+1} &= X_{i+1} + jY_{i+1} = (\alpha + j\omega) s^i = (\alpha + j\omega)(X_i + jY_i) \\
&= \alpha X_i - \omega Y_i + j(\alpha Y_i + \omega X_i),
\end{aligned}$$

which leads to:

$$X_{i+1} = \alpha X_i - \omega Y_i, \quad Y_{i+1} = \alpha Y_i + \omega X_i. \tag{3.29}$$

For s^{i+2}, we derive.

$$\begin{aligned}
s^{i+2} &= X_{i+2} + jY_{i+2} = (\alpha + j\omega)^2 s^i = (\alpha^2 - \omega^2 + 2j\alpha\omega)(X_i + jY_i) \\
&= (\alpha^2 - \omega^2) X_i - 2\alpha\omega Y_i + j\left[(\alpha^2 - \omega^2) Y_i + 2\alpha\omega X_i\right],
\end{aligned}$$

which yields:

$$X_{i+2} = (\alpha^2 - \omega^2) X_i - 2\alpha\omega Y_i, \quad Y_{i+2} = (\alpha^2 - \omega^2) Y_i + 2\alpha\omega X_i. \tag{3.30}$$

Let us consider the following recurrence:

$$X_{i+2} + bX_{i+1} + cX_i = 0.$$

From Equations (3.29) and (3.30), we derive:

$$\left[\left(\alpha^2 - \omega^2\right) + c + \alpha b\right] X_i - \omega\left(b + 2\alpha\right) Y_i = 0,$$

which gives:

$$b = -2\alpha, \quad c = \omega^2 + \alpha^2.$$

Finally, we obtain the following recurrence relationships:

$$X_{i+2} - 2\alpha X_{i+1} + \left(\omega^2 + \alpha^2\right) X_i \;=\; 0,$$
$$Y_{i+2} - 2\alpha Y_{i+1} + \left(\omega^2 + \alpha^2\right) Y_i \;=\; 0,$$

where:

$$X_0 = 1, \quad Y_0 = 0, \quad X_1 = \alpha, \quad Y_1 = \omega.$$

The roots of the system are given by:

$$\sum_{i=0}^{n} a_i s^i = \sum_{i=0}^{n} a_i \left(X_i + jY_i\right) = \sum_{i=0}^{n} a_i X_i + j \sum_{i=0}^{n} a_i Y_i = 0,$$

$$\sum_{i=0}^{n} a_i X_i = 0 \quad \text{and} \quad \sum_{i=0}^{n} a_i Y_i = 0.$$

PROBLEM 3.25. Consider an open-loop system with a transfer function $G\left(s\right)$ such that for the frequency ω_c,

$$\frac{d}{ds} \arg G\left(s\right)\big|_{s=j\omega_c} = 0. \tag{3.31}$$

Show that:

$$\arg \frac{d}{ds} G\left(s\right)\bigg|_{s=j\omega_c} = \arg G\left(s\right)\big|_{s=j\omega_c}.$$

SOLUTION 3.25. Let us write $G\left(j\omega\right)$ in the Cartesian form, i.e.:

$$G\left(j\omega\right) \;=\; X\left(\omega\right) + jY\left(\omega\right),$$
$$\arg G\left(j\omega\right) \;=\; \arctan \frac{Y\left(\omega\right)}{X\left(\omega\right)}, \quad \frac{Y\left(\omega\right)}{X\left(\omega\right)} = \frac{Y}{X}.$$

The derivative of the argument with respect to ω is given by:

$$\frac{d}{d\omega} \arg G(j\omega) = \frac{d}{d\omega} \arctan \frac{Y(\omega)}{X(\omega)}$$
$$= \frac{1}{1 + (Y/X)^2} \left[\frac{dY/d\omega}{X} - \frac{Y\, dX/d\omega}{X^2} \right]$$
$$= \frac{1}{X^2 + Y^2} \left[X\frac{dY}{d\omega} - \frac{Y\, dX}{d\omega} \right] = 0.$$

For $\omega = \omega_c$, we obtain:

$$X\frac{dY}{d\omega} - \frac{Y\, dX}{d\omega} = 0 \implies \frac{Y}{X} = \frac{dY/d\omega}{dX/d\omega}. \tag{3.32}$$

Notice that:

$$\frac{dG(i\omega)}{d\omega} = \frac{dX}{d\omega} + j\frac{dY}{d\omega} \text{ and } \tan\left(\arg \frac{dG(i\omega)}{d\omega}\right) = \frac{dY/d\omega}{dX/d\omega}.$$

From Equation (3.32), it follows that:

$$\tan\left(\arg \frac{dG(i\omega)}{d\omega}\bigg|_{\omega=\omega_c}\right) = \frac{dY/d\omega}{dX/d\omega}\bigg|_{\omega=\omega_c} = \frac{Y}{X}$$
$$= \tan\left(\arg G(i\omega)|_{\omega=\omega_c}\right),$$

which yields:

$$\arg \frac{dG(i\omega)}{d\omega}\bigg|_{\omega=\omega_c} = \arg G(i\omega)|_{\omega=\omega_c}.$$

The condition (3.31) means that at the frequency ω_c, the curve representing the phase versus the frequency is flat. Therefore, around the frequency ω_c, the argument of the system is practically independent of the gain [CHE 05].

PROBLEM 3.26. Determine the frequency for which the argument (phase angle) of the system:

$$G_c(s) = \frac{k(\tau_i s + 1)}{\tau_i s^2} \exp(-\tau s)$$

attains its maximal value.

SOLUTION 3.26. Observe that this system corresponds to an integrator–dead-ime process in series with a PI controller. Let us calculate the argument of $G_c(j\omega)$:

$$\arg G_c(j\omega) = \arg\left[-\frac{k(j\tau_i\omega + 1)}{\tau_i^2\omega^2} \exp(-j\tau\omega)\right] = -\pi - \omega\tau + \arctan(\tau_i\omega).$$

The maximum[15] occurs when:

$$\frac{d}{d\omega} \arg G_c \left(j\omega\right) = -\tau + \frac{\tau_i}{1 + \omega^2 \tau_i^2} = 0,$$

which leads to:

$$\omega_{\max} = \frac{1}{\tau_i} \sqrt{\frac{\tau_i - \tau}{\tau}}.$$

The maximal value of the argument is then given by:

$$\varphi_{\max} = \arg G_c \left(j\omega_{\max}\right) = -\pi - \frac{\tau}{\tau_i} \sqrt{\frac{\tau_i - \tau}{\tau}} + \arctan \left(\sqrt{\frac{\tau_i - \tau}{\tau}}\right),$$

which can be expressed as a function of the ratio $\alpha = \tau/\tau_i$:

$$\varphi_{\max} = -\pi - \alpha\sqrt{\alpha - 1} + \arctan\left(\sqrt{\alpha - 1}\right). \tag{3.33}$$

The Tyreus-Luyben PI tuning technique [TYR 92] is based on the specification of the peak phase angle, the computation of the reset time τ_i from Equation (3.33), and the determination of the controller gain in order to maximize the magnitude of the closed-loop transfer function:

$$|F\left(j\omega\right)|_{dB} = 20 \log \left|\frac{G_c\left(j\omega\right)}{1 + G_c\left(j\omega\right)}\right|.$$

PROBLEM 3.27. Consider a filter of the second-order band-pass type. Its transfer function is given by:

$$G\left(s\right) = \frac{ks}{s^2 + 2\zeta\omega_0 s + \omega_0^2}.$$

1. Show that the circuit depicted in Figure 3.29 corresponds to a band-pass filter.

2. Determine for what value of the frequency the gain $|G\left(j\omega\right)|$ attains its maximal value.

3. Draw the Bode diagram of this filter.

SOLUTION 3.27. 1. The equations governing the behavior of the system represented in Figure 3.29 are:

$$V\left(s\right) = \left(R + z\right)I\left(s\right), \quad Y\left(s\right) = zI\left(s\right),$$

$$V\left(s\right) = \left(R + z\right)\frac{Y\left(s\right)}{z},$$

---

15. $\dfrac{d}{dx} \arctan\left(x\right) = \dfrac{1}{1 + x^2}.$

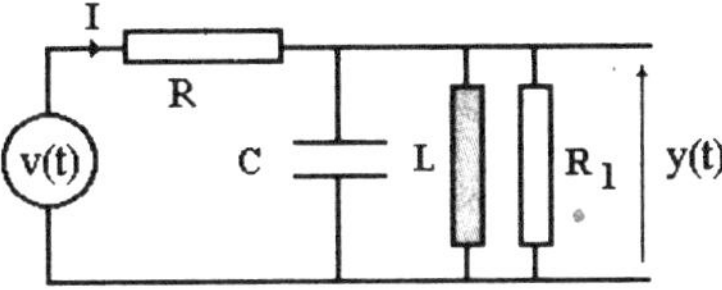

Figure 3.29. *Band-pass filter*

where:
$$\frac{1}{z} = Cs + \frac{1}{Ls} + \frac{1}{R_1}, \quad z = \frac{R_1 Ls}{R_1 + Ls + LR_1 Cs^2}.$$

The transfer function is equal to:

$$G(s) = \frac{Y(s)}{V(s)} = \frac{z}{R+z} = \frac{R_1 Ls}{LRR_1 Cs^2 + (R + R_1)Ls + RR_1}$$

$$= \frac{\dfrac{1}{RC}s}{s^2 + \dfrac{(R + R_1)}{RR_1 C}s + \dfrac{1}{LC}}.$$

By identification, we derive:

$$\omega_0 = \frac{1}{\sqrt{LC}}, \quad \zeta = \frac{(R + R_1)\sqrt{LC}}{2RR_1 C}, \quad k = \frac{1}{RC}.$$

Observe that the equation governing the behavior of an RLC circuit is:

$$V(s) = \left[R + Ls + \frac{1}{Cs}\right] I(s).$$

The transfer function relating the current $I(s)$ to the voltage $V(s)$ is:

$$\frac{I(s)}{V(s)} = \frac{Cs}{LCs^2 + RCs + 1},$$

which corresponds to the transfer function of a band-pass filter.

2. The gain is given by:

$$|G(j\omega)| = \frac{k\omega}{\sqrt{(\omega_0^2 - \omega^2)^2 + 4\zeta^2 \omega_0^2 \omega^2}}.$$

Its derivative with respect to ω is equal to:

$$\frac{d\left|G\left(j\omega\right)/k\right|}{d\omega} = \frac{d}{d\omega}\left(\omega\left(\left(\omega_0^2 - \omega^2\right)^2 + 4\zeta^2\omega_0^2\omega^2\right)^{-1/2}\right)$$

$$= \left(\left(\omega_0^2 - \omega^2\right)^2 + 4\zeta^2\omega_0^2\omega^2\right)^{-1/2}$$

$$-\frac{\omega}{2}\left(\left(\omega_0^2 - \omega^2\right)^2 + 4\zeta^2\omega_0^2\omega^2\right)^{-3/2}\left(-4\omega\left(\omega_0^2 - \omega^2\right) + 8\zeta^2\omega_0^2\omega\right)$$

$$\left(\left(\omega_0^2 - \omega^2\right)^2 + 4\zeta^2\omega_0^2\omega^2 + 2\omega^2\left(\omega_0^2 - \omega^2\right) - 4\zeta^2\omega_0^2\omega^2\right)$$

$$\times\left(\left(\omega_0^2 - \omega^2\right)^2 + 4\zeta^2\omega_0^2\omega^2\right)^{-3/2}$$

$$= \left(\omega_0^2 - \omega^2\right)\left(\omega_0^2 + \omega^2\right)\left(\left(\omega_0^2 - \omega^2\right)^2 + 4\zeta^2\omega_0^2\omega^2\right)^{-3/2}$$

$$= \left(\omega_0^4 - \omega^4\right)\left(\left(\omega_0^2 - \omega^2\right)^2 + 4\zeta^2\omega_0^2\omega^2\right)^{-3/2}.$$

The gain $\left|G\left(j\omega\right)\right|$ attains its maximal value:

$$\left|G\left(j\omega_0\right)\right| = \frac{k}{2\zeta\omega_0}$$

for $\omega = \omega_0$. The argument is equal to:

$$\arg G\left(j\omega_0\right) = 0.$$

Observe that in order to obtain the frequency at which the gain $\left|G\left(j\omega_0\right)\right|$ attains its maximal value, it was not necessary to calculate its derivative; maximization of $\left|G\left(j\omega_0\right)\right|$ implies minimization of the denominator of $\left|G\left(j\omega_0\right)\right|$, which leads to $\omega = \omega_0$. Signals of frequencies outside a band including the frequency ω_0 are attenuated. Observe also that this kind of filter can be constructed using a low-pass and a high-pass filter in series.

3. For $\omega_0 = 1000$ and $\zeta = 0.6, 0.1, 0.01$, the Bode diagrams of the band-pass filter are plotted in Figure 3.30. As ζ decreases, the filter becomes more selective, i.e., practically all signals of frequency different from ω_0 are well attenuated. For system identification purposes, this filter may be used to determine the desired points on the Nyquist diagram.

PROBLEM 3.28. Consider the notch filter described in Problem 1.24.

1. Derive the transfer function:

$$G\left(s\right) = \frac{Y\left(s\right)}{V\left(s\right)}.$$

2. Draw the Bode diagram of this filter.

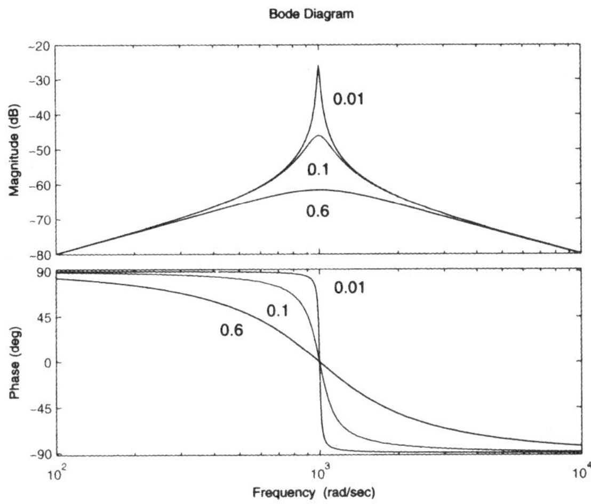

Figure 3.30. *Bode diagrams of a band-pass filter for various values of ζ*

3. Determine the frequencies (rejected band) for which $|G\left(j\omega\right)|_{dB} = 3\,dB$.

SOLUTION 3.28. 1. A notch filter (band reject filter) attenuates (rejects) one frequency band and passes both a lower- and a higher-frequency band. The equations governing the behavior of this filter are:

$$\left[V_{N_1}\left(s\right) - V\left(s\right)\right]Cs + \left[V_{N_1}\left(s\right) - Y\left(s\right)\right]Cs + \frac{V_{N_1}\left(s\right)}{R/2} = 0, \tag{3.34}$$

$$\left[V_{N_2}\left(s\right) - V\left(s\right)\right]\frac{1}{R} + \left[V_{N_2}\left(s\right) - Y\left(s\right)\right]\frac{1}{R} + V_{N_2}\left(s\right)2Cs = 0 \tag{3.35}$$

and

$$\left[Y\left(s\right) - V_{N_1}\left(s\right)\right]Cs + \left[Y\left(s\right) - V_{N_2}\left(s\right)\right]\frac{1}{R} = 0. \tag{3.36}$$

From Equations (3.34), (3.35) and (3.36), we derive:

$$V_{N_1}\left(s\right) = \left[V\left(s\right) + Y\left(s\right)\right]\frac{RCs}{2\left(RCs + 1\right)},$$

$$V_{N_2}\left(s\right) = \left[V\left(s\right) + Y\left(s\right)\right]\frac{1}{2\left(RCs + 1\right)},$$

$$G\left(s\right) = \frac{Y\left(s\right)}{V\left(s\right)} = \frac{1 + \left(RCs\right)^2}{1 + 4RCs + \left(RCs\right)^2}.$$

2. Let us denote RC by $1/\omega_0$. The magnitude of $G\left(j\omega\right)$ is given by:

$$\left|G\left(j\omega\right)\right|^2 = \left|\frac{\omega_0^2 - \omega^2}{\omega_0^2 - \omega^2 + 4j\omega_0\omega}\right|^2 = \frac{\left(\omega_0^2 - \omega^2\right)^2}{\left(\omega_0^2 - \omega^2\right)^2 + 16\omega_0^2\omega^2}.$$

In order to calculate its derivative with respect to ω, let us make the variable change $u = \omega/\omega_0$. We obtain:

$$\frac{d\left|G\left(j\omega\right)\right|^2}{d\omega} = \frac{d}{du}\frac{\left(1 - u^2\right)^2}{\left(1 - u^2\right)^2 + 16u^2} = 32\left(u^2 - 1\right)u\frac{1 + u^2}{\left(1 + 14u^2 + u^4\right)^2}.$$

This derivative is negative for $\omega < \omega_0$, and positive for $\omega > \omega_0$. Therefore, the magnitude $\left|G\left(j\omega\right)\right|$ decreases for $\omega < \omega_0$, increases for $\omega > \omega_0$ and is equal to zero for $\omega = \omega_0$. For $\omega < \omega_0$, $\left(\omega_0^2 - \omega\right)^2 > 0$, we obtain:

$$\arg G\left(j\omega\right) = \arg\left(\frac{\left(\omega_0^2 - \omega^2\right)\left(\omega_0^2 - \omega^2 - 4j\omega_0\omega\right)}{\left(\omega_0^2 - \omega^2\right)^2 + \left(4\omega_0\omega\right)^2}\right) = \arctan\left(-\frac{4\omega_0\omega}{\omega_0^2 - \omega^2}\right).$$

For $\omega = 0$, $\arg G\left(j\omega\right) = 0$, and it tends to $-\pi/2$ when $\omega \to \omega_0$. For $\omega > \omega_0$, $\left(\omega_0^2 - \omega\right)^2 < 0$, we derive:

$$\arg G\left(j\omega\right) = \arg\left(\frac{\left(\omega^2 - \omega_0^2\right)\left(\omega^2 - \omega_0^2 + 4j\omega_0\omega\right)}{\left(\omega_0^2 - \omega^2\right)^2 + \left(4\omega_0\omega\right)^2}\right) = \arctan\left(\frac{4\omega_0\omega}{\omega_0^2 - \omega^2}\right).$$

When $\omega \to \omega_0$ $(\omega > \omega_0)$, $\arg G\left(j\omega\right)$ tends to $\pi/2$, and $\arg G\left(j\omega\right)$ tends to zero when $\omega \to \infty$. Therefore, a singularity (a jump) occurs at the frequency ω_0. A Bode diagram obtained with **MATLAB** for $\omega_0 = 10^3$ is depicted in Figure 3.31. Note the translation of the phase diagram.

3. Recall that $3\,dB$ corresponds to a magnitude of $\sqrt{2}/2$. It follows that:

$$\left|G\left(jw\right)\right|^2 = \frac{1}{2}.$$

From the expression for the transfer function $G\left(s\right)$, we derive:

$$\frac{\left(\omega_0^2 - \omega^2\right)^2}{\left(\omega_0^2 - \omega^2\right)^2 + 16\omega_0^2\omega^2} = \frac{1}{2},$$

$$\left(\omega_0^2 - \omega^2\right)^2 = 16\omega_0^2\omega^2,$$

$$\left(\omega_0^2 - \omega^2\right) = \pm 4\omega_0\omega,$$

which leads to:

$$\omega_1 = \left(-2 + \sqrt{5}\right)\omega_0, \quad \omega_2 = \left(2 + \sqrt{5}\right)\omega_0.$$

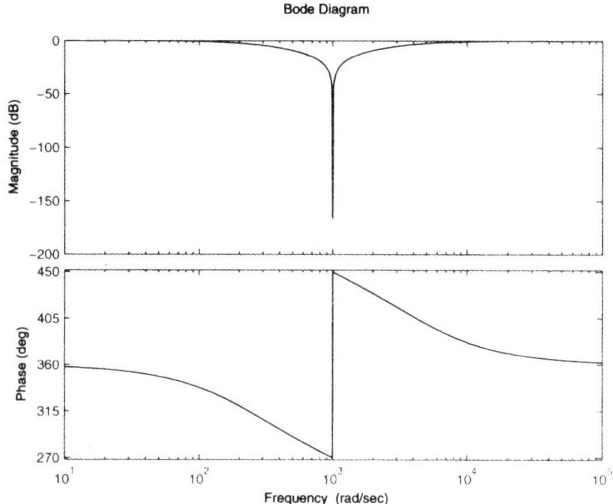

Figure 3.31. *Bode diagram of a notch filter*

The rejected band corresponds to (ω_1, ω_2).

PROBLEM 3.29. Consider a unity-feedback system where the transfer function of the forward path is:

$$G\left(s\right) = \frac{w_n^2}{s^2 + 2\zeta w_n + w_n^2}.$$

1. Determine the maximum of the sensitivity function.

2. Draw the variation of the maximum of the sensitivity function as a function of the damping factor for $\zeta \in \left]0, 1\right]$.

SOLUTION 3.29. 1. The sensitivity function is given by:

$$\frac{1}{1 + G\left(j\omega\right)} = \frac{\omega_n^2 - \omega^2 + 2j\zeta\omega_n\omega}{2\omega_n^2 - \omega^2 + 2j\zeta\omega_n\omega}.$$

In order to simplify the computations, let us make the following variable change:

$$u = \frac{\omega}{\omega_n}.$$

It follows that:

$$M_s = \max_{\omega > 0}\left|\frac{1}{1 + G\left(j\omega\right)}\right| = \max_{u > 0}\sqrt{\frac{\left(1 - u^2\right)^2 + 4\zeta^2 u^2}{\left(2 - u^2\right)^2 + 4\zeta^2 u^2}}.$$

Let M be the point associated with $G\left(j\omega\right)$, i.e., M is $\left(ReG\left(j\omega\right), ImG\left(j\omega\right)\right)$; then $\left[1 + G\left(j\omega\right)\right]$ represents the vector $\overrightarrow{AM}$, where A is the critical point. Therefore,

$$\max_{\omega>0}\left|\frac{1}{1 + G\left(j\omega\right)}\right| = \max_{\omega>0}\frac{1}{\left|\overrightarrow{AM}\right|} \implies \min_{\omega>0}\left|\overrightarrow{AM}\right|.$$

The derivative is given by:

$$\frac{d}{dw}\left|\frac{1}{1 + G\left(ju\right)}\right| = \frac{1}{2}\frac{U'V - V'U}{U^{1/2}V^{3/2}},$$

where:

$$U = \left(1 - u^2\right)^2 + 4\zeta^2 u^2, \quad U' = -4u\left(1 - u^2\right) + 8\zeta^2 u,$$

$$V = \left(2 - u^2\right)^2 + 4\zeta^2 u^2, \quad V' = -4u\left(2 - u^2\right) + 8\zeta^2 u.$$

Simple but lengthy calculations lead to:

$$\frac{d}{dw}\left|\frac{1}{1 + G\left(ju\right)}\right| = -\frac{2u\left(u^4 - 3u^2 - 6\zeta^2 + 2\right)}{\left(u^4 - 4u^2 + 4\zeta^2 u^2 + 4\right)^{\frac{3}{2}}\sqrt{\left(u^4 - 2u^2 + 4\zeta^2 u^2 + 1\right)}}.$$

The maximum value corresponds to $d\left|1/\left(1 + G\left(ju\right)\right)\right|/dw = 0$, which yields:

$$x^2 - 3x - 6\zeta^2 + 2 = 0, x = u^2,$$

The term $\left(x^2 - 3x - 6\zeta^2 + 2\right)$ is negative for $x \in \left]x_1, x_2\right[$ and positive for values of x located outside the segment $\left[x_1, x_2\right]$.

$$\Delta = 9 - 4\left(2 - 6\zeta^2\right) = 1 + 24\zeta^2 > 0,$$

$$\left[\begin{array}{c} x_1 = u_1^2 \\ x_2 = u_2^2 \end{array}\right] = \left[\begin{array}{c} \dfrac{1}{2}\left(3 - \sqrt{1 + 24\zeta^2}\right) \\ \dfrac{1}{2}\left(3 + \sqrt{1 + 24\zeta^2}\right) \end{array}\right].$$

Consequently, the derivative $d\left|1/\left(1 + G\left(ju\right)\right)\right|/dw$ is positive for $u^2 \in \left]u_1^2, u_2^2\right[$ and negative for the values of u^2 located outside the segment $\left[u_1^2, u_2^2\right]$. Therefore, $\left|1/\left(1 + G\left(ju\right)\right)\right|$ decreases when u increases from zero to u_1^2, increases when u increases from u_1^2 to u_2^2, and decreases when u^2 increases from u_2^2 to infinity. Consequently, u_2^2 gives the maximum of $\left|1/\left(1 + G\left(ju\right)\right)\right|$. This maximal value is given by:

$$M_s = \sqrt{\frac{\left(1 - \left(3 + \sqrt{1 + 24\zeta^2}\right)/4\right)^2 + \zeta^2\left(3 + \sqrt{1 + 24\zeta^2}\right)^2}{\left(2 - \left(3 + \sqrt{1 + 24\zeta^2}\right)/4\right)^2 + \zeta^2\left(3 + \sqrt{1 + 24\zeta^2}\right)^2}}$$

$$= \sqrt{\frac{1 + 24\zeta^2 + \left(4\zeta^2 + 1\right)\sqrt{1 + 24\zeta^2}}{1 + 24\zeta^2 + \left(4\zeta^2 - 1\right)\sqrt{1 + 24\zeta^2}}}.$$

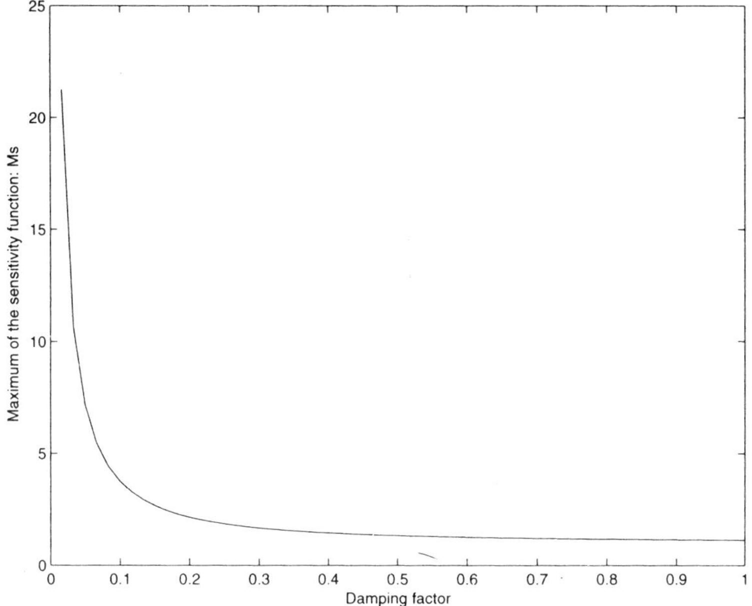

Figure 3.32. *Maximum of the sensitivity function M_s versus the damping factor*

2. Observe that $M_s = \infty$ for $\zeta = 0$. Its evolution as a function of ζ is depicted in Figure 3.32. M_s decreases when ζ increases. The derivative of M_s with respect to ζ is equal to:

$$\frac{dM_s}{d\zeta} = -\frac{8\zeta\left(3 + \sqrt{1 + 24\zeta^2} + 72\zeta^2 + 24\zeta^2\sqrt{1 + 24\zeta^2}\right)}{\dfrac{\left(1 + 24\zeta^2 + \left(4\zeta^2 + 1\right)\sqrt{1 + 24\zeta^2}\right)^{1/2}}{\left(1 + 24\zeta^2 + \left(4\zeta^2 - 1\right)\sqrt{1 + 24\zeta^2}\right)^{-3/2}}} < 0.$$

Chapter 4

Stability and the Root Locus

4.1. Stability

Definition 1 *A system is stable if all its poles are located in the left half-plane (on the left-hand side of the imaginary axis), i.e., the real parts of all the poles are negative.*

Observe that this definition is valid only for causal systems. One general condition of stability is the following.

Definition 2 *If the impulse response of a continuous system has a finite number of local extrema (maxima or minima) and a finite number of discontinuities in any finite interval, then the system is stable if and only if:*

$$\int_{-\infty}^{+\infty} |g(t)|\, dt$$

exists and is finite, where $g(t)$ represents the impulse response, which is equal to zero when t is negative for causal system.

The conditions on extrema and discontinuities are generally fulfilled by real-life systems.

In practice, the parameters of the model are obtained (i.e., estimated) from the available data, i.e., input–output measurements. These measurements are always affected by measurement errors, which depend mainly on the precision of the sensors.

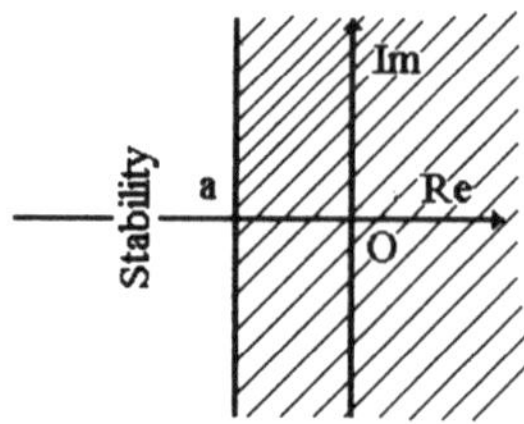

Figure 4.1. *A restricted stability domain*

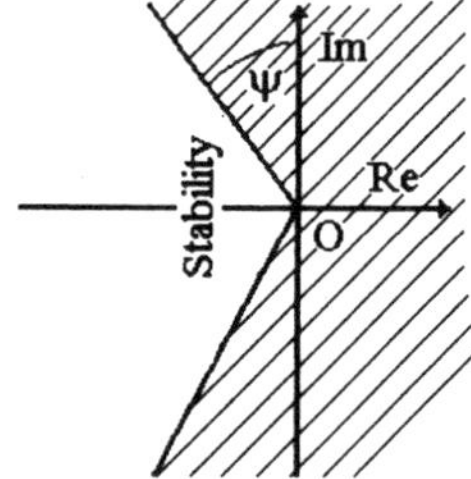

Figure 4.2. *A restricted stability domain with restriction on the damping factor*

In order to take into account this uncertainty about the model parameters, we can consider the domain depicted in Figure 4.1 as the stability domain[1] ($Re\,(poles) < a$). Sometimes, restrictions on the damping are also imposed. These restrictions lead to the stability domain depicted in Figure 4.2 (see Problem 3.20). We may also impose restrictions on $\zeta\omega_n$. These restrictions lead to the domain depicted in Figure 4.3 (see Problem 3.20 again).

In what follows, we briefly present two stability criteria.

---

1. For a second-order system, Descartes' theorem may be used to check if the poles are greater or smaller than a given value.

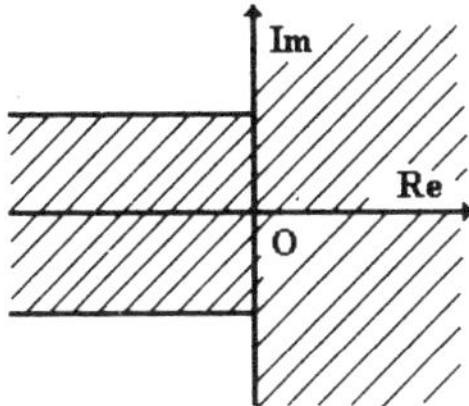

Figure 4.3. *A restricted stability domain for $\zeta \omega_n$ greater than a given constant*

4.1.1. *The Routh–Hurwitz criterion*

We want to know if all the roots of the following polynomial have a negative real part:

$$D\left(s\right) = a_0 s^n + a_1 s^{n-1} + \cdots + a_n, \qquad a_0 > 0. \tag{4.1}$$

where $a_i \in \mathcal{R}, i = 0, 1, \cdots, n$.

Let us construct the determinant:

$$\Delta = \begin{vmatrix} a_1 & a_0 & 0 & \cdot & \cdot & \cdot & \cdot & \cdot & 0 \\ a_3 & a_2 & a_1 & a_0 & 0 & \cdot & \cdot & \cdot & 0 \\ a_5 & a_4 & a_3 & a_2 & a_1 & a_0 & 0 & \cdot & 0 \\ a_7 & \cdots & & & & & & & \\ & \cdots & & & & & & & \\ & \cdots & & & & & & & \end{vmatrix}$$

and compute the determinants Δ_i, where:

$$\Delta_1 = |a_1|, \quad \Delta_2 = \begin{vmatrix} a_1 & a_0 \\ a_3 & a_2 \end{vmatrix}, \quad \Delta_3 = \begin{vmatrix} a_1 & a_0 & 0 \\ a_3 & a_2 & a_1 \\ a_5 & a_4 & a_3 \end{vmatrix}, \quad \text{etc.} \tag{4.2}$$

The roots of equation (4.1) are located in the left half-plane [GRA 00, HUR 64] and [LAV 72] if and only if all the determinants Δ_i are positive, $i = 1, \cdots, n$.

A necessary condition[2] for stability is that all the coefficients $a_i, i = \overline{1, n}$, have the same sign. A proof of this criterion [LAV 72] is given in Appendix A. This criterion

---

2.
$$D\left(s\right) = a_0 \prod_{i=1}^{n} \left(s - p_i\right) = a_0 \left(s^n + \sum_{i=1}^{n} (-1)^n \beta_i s^{n-i}\right),$$

necessitates simple algebraic computations. It is very useful because it does not require the denominator polynomial in the transfer function to be factored.

REMARK 4.1. This criterion can also be used to check the stability of a system given in state-space representation, which can be viewed as one way of parameterizing the transfer function. The characteristic equation of the system is given by

$$|\lambda \mathbf{I} - \mathbf{A}| = \lambda^n + a_1 \lambda^{n-1} + \cdots + a_{n-1}\lambda + a_n, \quad a_0 = 1, \qquad (4.3)$$

where $\mathbf{A}$ and $\mathbf{I}$ represent the state-space matrix and the identity matrix, respectively. The eigenvalues λ all have negative real parts if the determinants Δ_i defined by Equation (4.2) are positive [NAJ 99].

4.1.2. *Revers's criterion*

Revers's criterion is a particular case of the Nyquist criterion (see Appendix A). It is valid for minimum-phase systems and stable open-loop systems (with poles located to the left of the imaginary axis). Revers's criterion can be stated as follows: *if one runs along the Nyquist diagram of the open-loop system in the direction of increasing frequency and the critical point A, with co-ordinates $(-1, 0)$, is always on the left, then the closed-loop system is stable. If the intersection of the Nyquist diagram with the real axis coincides with the critical point, the closed-loop system is oscillatory; otherwise, the closed-loop system is unstable. In the latter case, the complete ($\omega = 0$ to $-\infty$) Nyquist diagram encircles the critical point.* Figure 4.4 shows diagrams for a stable, an oscillatory and an unstable system.

In order to give a statement of Revers's criterion in the context of the Bode and Nichols (or Black) diagrams, let us express the above stability condition as follows. Consider the intersection B of the diagram of the open-loop transfer function with the unit circle. We denote the corresponding frequency by ω_b.

From Figure 4.5, we derive the result that the closed-loop system is:

$$\text{stable if } \psi = \arg \overrightarrow{OB} = \arg T\left(\omega_b\right) > -\pi,$$
$$\text{unstable if } \psi = \arg \overrightarrow{OB} = \arg T\left(\omega_b\right) < -\pi, \qquad (4.4)$$
$$\text{oscillatory if } \psi = \arg \overrightarrow{OB} = \arg T\left(\omega_b\right) = -\pi.$$

---

where:

$$\beta_1 = \sum_{i=1}^{n} p_i, \quad \beta_2 = \sum_{\le i \le j \le n} p_i p_j, \cdots, \quad \beta_n = \prod_{i=1}^{n} p_i .$$

It is clear that the coefficients of the polynomial $\prod_{i=1}^{n}(s - p_i)$ are products and sums of positive numbers p_i (the roots of $D(s)$).

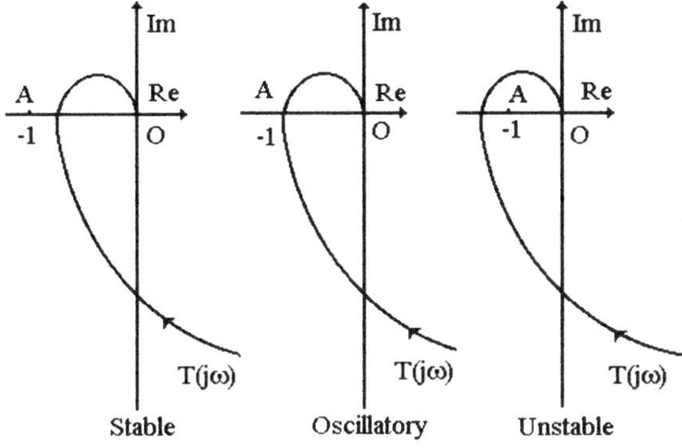

Figure 4.4. *Revers's criterion*

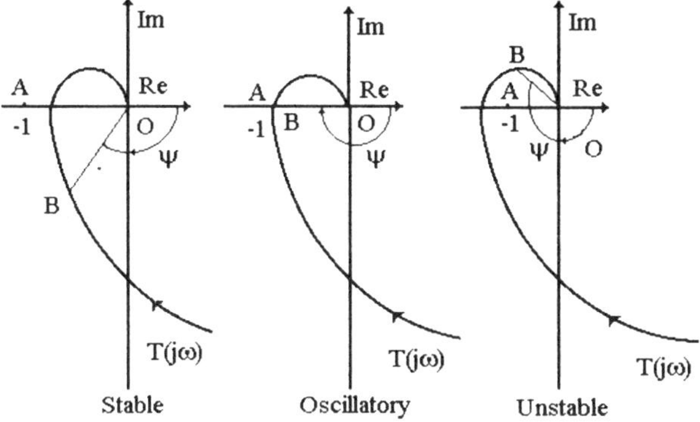

Figure 4.5. *Interpretation of Revers's criterion*

A statement of Revers's criterion for the Bode and Nichols diagrams can be derived easily from Equation (4.4). This criterion is not rigorous, but it is valid for real-life systems.

PROBLEM 4.1. Show that the sum of the real part of the poles of a system is constant when the characteristic equation is given by:

$$1 + kT\left(s\right) = 1 + k\frac{P\left(s\right)}{Q\left(s\right)} = 0,$$

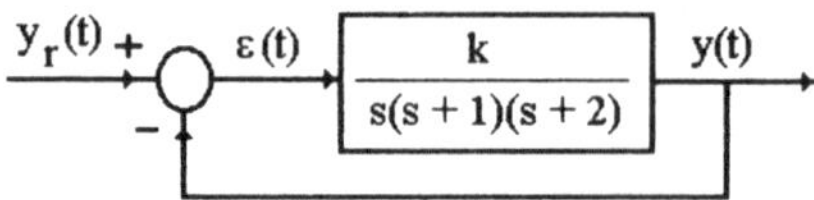

Figure 4.6. *Closed-loop system*

$$\deg P = n, \quad \deg Q = m, \quad n - m \geq 2,$$

$$Q(s) = 1 + a_1 s + a_2 s^2 + \cdots + a_n s^n,$$

$$P(s) = 1 + b_1 s + b_2 s^2 + \cdots + b_m s^m.$$

SOLUTION 4.1. Let us rewrite the characteristic equation in the following form:

$$kP(s) + Q(s) = 0,$$

$$a_n s^n + a_{n-1} s^{n-1} + a_{n-2} s^{n-2} + \cdots + 1 + k\left(b_m s^m + b_{m-1} s^{m-2} + \cdots + 1\right) = 0. \tag{4.5}$$

Under the condition $n - m \geq 2$, the sum of the zeros of the polynomial (4.5) is equal to a_{n-1}, the coefficient of s^{n-1}, which corresponds to the desired result, i.e.:

$$\sum_{i=1}^{n} p_i = a_{n-1} = const.$$

The sum of the zeros of the characteristic equation is independent of the gain k. In other words, if some roots of the characteristic equation move to the left when the gain increases, the other roots must move to the right.

PROBLEM 4.2. Consider the system depicted in Figure 4.6.

1. Determine the values of the static gain for which the asymptotic error is less than 0.1 for a ramp input with a slope equal to 1.

2. Analyze the stability of this system.

3. A regulator with transfer function:

$$R(s) = s + z$$

is introduced. Answer questions 1 and 2 above for this case. For what value of z is the system is stable $\forall k$?

SOLUTION 4.2. 1. The Laplace transform of the error induced by a ramp input of slope 1 is given by:

$$\Xi\left(s\right) = \frac{1}{1 + k/\left[s\left(s+1\right)\left(s+2\right)\right]} Y_r\left(s\right),$$

$$Y_r\left(s\right) = \frac{1}{s^2},$$

$$\Xi\left(s\right) = \frac{s\left(s+1\right)\left(s+2\right)}{s\left(s+1\right)\left(s+2\right) + k} \frac{1}{s^2}.$$

Using Tauber's theorem (the initial- and final-value theorem), we obtain:

$$\varepsilon\left(\infty\right) = \lim_{s\to 0} s \frac{s\left(s+1\right)\left(s+2\right)}{s\left(s+1\right)\left(s+2\right) + k\,s^2} \frac{1}{s^2} = \frac{2}{k} < 0.1,$$

which leads to:

$$k > 20.$$

2. The characteristic equation is given by:

$$s^3 + 3s^2 + 2s + k = 0$$

$$a_0 = 1, \quad a_1 = 3, \quad a_2 = 2, \quad a_3 = k.$$

Using the Routh–Hurwitz criterion, we obtain the following conditions:

$$\Delta_1 = |3| > 0, \quad \Delta_2 = \begin{vmatrix} 3 & 1 \\ k & 2 \end{vmatrix} = 6 - k > 0 \Longrightarrow k < 6.$$

The system is stable for $k < 6$.

Notice that the previous condition ($\varepsilon\left(\infty\right) < 0.1$) cannot be fulfilled (see also the comment at the end of this solution).

3. The steady-state error is given by:

$$\varepsilon\left(\infty\right) = \lim_{s\to 0} s \frac{1}{1 + /\left[s\left(s+1\right)\left(s+2\right)\right]} \frac{1}{s^2}$$

$$= \lim_{s\to 0} s \frac{s\left(s+1\right)\left(s+2\right)}{s\left(s+1\right)\left(s+2\right) + k\left(s+z\right)} \frac{1}{s^2} = \frac{2}{zk} < 0.1,$$

and the gain k must be greater than $20/z$. We observe that the desired value of the gain has been divided by z. This result is general. The introduction of a zero reduces the steady-state error due to a ramp input.

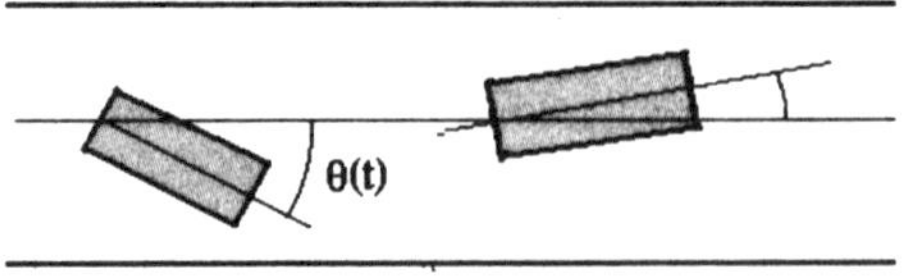

Figure 4.7. *Schematic representation of a car*

The characteristic equation becomes:

$$s^3 + 3s^2 + (k+2)\,s + zk = 0.$$

Using the Routh–Hurwitz criterion, we obtain:

$$\Delta_1 = |3| > 0, \quad \Delta_2 = \begin{vmatrix} 3 & 1 \\ zk & k+2 \end{vmatrix} = (3-z)\,k + 6 > 0.$$

For $z = 3$, the system is stable $\forall k$.

Comment. Many systems become unstable when the gain exceeds a certain bound. Let us illustrate this fact by a simple example dealing with the driving of a car. A system which consists of a car is schematically represented in Figure 4.7.

The control objective is to keep the car on the centre line of the road. The position of the car with respect to this line is denoted by $\theta(t)$. The angle $\varphi(t)$ formed by the steering wheel and a fixed point on the dashboard represents the control variable of this system. In other words, the control objective is to act on the angle $\varphi(t)$ in order to maintain $\theta(t) = 0$. If the driver acts as a proportional controller, the control action will be $k_p\theta(t)$. We may easily imagine that if the value of the design parameter k_p is relatively high, the car will zigzag around the centre line of the road, and will eventually leave the road (possibly into a ravine).

PROBLEM 4.3. Show that the system with the characteristic equation:

$$s^4 + 2s^3 + s^2 + 5s + 7 = 0$$

is unstable.

SOLUTION 4.3. Let us use the Routh–Hurwitz criterion. The coefficients a_i, $i = 0, 1, \cdots, 4$ are:

$$a_0 = 1, \quad a_1 = 2, \quad a_2 = 1, \quad a_3 = 5, \quad a_4 = 7$$

and the determinants Δ_i are given by:

$$\Delta_1 = 1 > 0, \quad \Delta_2 = \begin{vmatrix} a_1 & a_0 \\ a_3 & a_2 \end{vmatrix} = \begin{vmatrix} 2 & 1 \\ 5 & 2 \end{vmatrix} = -1 < 0.$$

The system is unstable.

PROBLEM 4.4. Show that the real parts of the zeros z_i, $i = 1, \cdots, 5$ of the characteristic function:

$$s^5 + 37.0s^4 + 531.75s^3 + 3706.3s^2 + 12529s + 16459 = 0 \tag{4.6}$$

are all less than -3.

SOLUTION 4.4. Observe that in order to fulfill the conditions $Re z_i < -3$, the zeros of this characteristic function have to belong to the restricted stability domain depicted in Figure 4.1, where $a = -3$. Note also that the co-ordinates (x', y'), where $s' = x' + jy'$, of a given point M in the restricted stability domain can be derived easily from its co-ordinates in the stability domain (to the left of the imaginary axis) as follows: $x' = x + 3$, $y' = y$. Let us introduce the variable change $s' = s + 3$; we obtain:

$$s'^5 + 22.0s'^4 + 177.75s'^3 + 648.5s'^2 + 1058s' + 624 = 0. \tag{4.7}$$

Now, we can directly use the Routh–Hurwitz criterion to study the stability of the system associated with the characteristic function (4.7):

$$\Delta_1 = |22| > 0, \quad \Delta_2 = \begin{vmatrix} 22 & 1 \\ 648.5 & 177.75 \end{vmatrix} = 3262 > 0,$$

$$\Delta_3 = \begin{vmatrix} 22 & 1 & 0 \\ 648.5 & 177.75 & 22 \\ 624 & 1058.0 & 648.5 \end{vmatrix} = 1.6171 \times 10^6 > 0,$$

$$\Delta_4 = \begin{vmatrix} 22 & 1 & 0 & 0 \\ 648.5 & 177.75 & 22 & 1 \\ 624 & 1058 & 648.5 & 177.75 \\ 0 & 0 & 624.0 & 1058 \end{vmatrix} = 1.3632 \times 10^9 > 0.$$

We conclude that the real parts of the zeros z_i of the characteristic Equation (4.6) are less than -3.

PROBLEM 4.5. Consider the system:

$$G(s) = \frac{k(1 + s/\alpha)(1 + s/3.5)}{s(1 + 0.25s)(0.2s^2 + 0.8s + 1)}, \quad k > 0,$$

closed under negative unity feedback. Determine the values of the parameter α such that the closed-loop system will be stable $\forall k$.

SOLUTION 4.5. The characteristic equation is:

$$0.05s^4 + 0.4s^3 + \left(1.05 + \frac{k}{3.5\alpha}\right)s^2 + \left[1 + k\left(\frac{1}{3.5} + \frac{1}{\alpha}\right)\right]s + k = 0.$$

Using the Routh–Hurwitz criterion, we obtain the following conditions:

$$\Delta_1 = |a_1| = 0.4 > 0,$$

$$\Delta_2 = \begin{vmatrix} 0.4 & 0.05 \\ 1 + k\left(\dfrac{1}{3.5} + \dfrac{1}{\alpha}\right) & 1.05 + \dfrac{k}{3.5\alpha} \end{vmatrix}$$

$$= 0.4\left(1.05 + \frac{k}{3.5\alpha}\right) - 0.05\left[1 + k\left(\frac{1}{3.5} + \frac{1}{\alpha}\right)\right]$$

$$= 0.4 \times 1.05 + k\left[\frac{0.4}{3.5\alpha} - 0.05\left(\frac{1}{3.5} + \frac{1}{\alpha}\right)\right] - 0.05 > 0\,\forall k.$$

Since Δ_2 must be independent of k:

$$\frac{0.4}{3.5\alpha} - 0.05\left(\frac{1}{3.5} + \frac{1}{\alpha}\right) = 0 \Longrightarrow \alpha = 4.5$$

Now, let us consider the determinant Δ_3:

$$\Delta_3 = \begin{vmatrix} 0.4 & 0.05 & 0 \\ 1 + k\left(\dfrac{1}{3.5} + \dfrac{1}{4.5}\right) & 1.05 + \dfrac{k}{3.5 \times 4.5} & 0.4 \\ 0 & k & 1 + k\left(\dfrac{1}{3.5} + \dfrac{1}{4.5}\right) \end{vmatrix}$$

$$= 0.37 + 2.7937 \times 10^{-2}k - 1.0 \times 10^{-11}k^2$$

$$= -1.0 \times 10^{-11}\left(k + 13.244\right)\left(k - 2.7937 \times 10^9\right) > 0, \quad k > 0,$$

$$k - 2.7937 \times 10^9 < 0 \Longrightarrow k < k_0 = 2.7937 \times 10^9.$$

Taking into account the fact that the upper bound k_0 has a very high value, in practice we can consider that $\Delta_3 > 0$ independently of the gain k. Now, let us consider Δ_4:

$$\Delta_4 = \begin{vmatrix} 0.4 & 0.05 & 0 & 0 \\ 1 + k\left(\dfrac{1}{3.5} + 1/\alpha\right) & 1.05 + \dfrac{k}{3.5\alpha} & 0.4 & 0.05 \\ 0 & k & 1 + k\left(\dfrac{1}{3.5} + 1/\alpha\right) & 1.05 + \dfrac{k}{3.5\alpha} \\ 0 & 0 & 0 & k \end{vmatrix}.$$

By expanding this determinant with respect to the fourth row, we obtain $\Delta_4 = k\Delta_3$. The stability of this system is independent of the gain k for all values $k < 2.7937 \times 10^9$.

PROBLEM 4.6. Calculate the impulse response of the following unity-feedback system:

$$G(s) = \frac{2}{s^2 + 8s + 13}, \quad H(s) = 1.$$

Analyze its stability.

SOLUTION 4.6. The closed-loop transfer function is given by:

$$F(s) = \frac{2}{s^2 + 8s + 15} = \frac{1}{s+3} - \frac{1}{s+5}.$$

The impulse response is equal to:

$$g(t) = \exp(-3t) - \exp(-5t).$$

Let us now calculate the following integral:

$$\int_{-\infty}^{+\infty} |g(t)|\, dt = \int_{0}^{+\infty} |\exp(-3t) - \exp(-5t)|\, dt.$$

As $\exp(5t) > \exp(3t)$ for $t > 0$, we obtain:

$$\int_{-\infty}^{+\infty} |g(t)|\, dt = \int_{0}^{+\infty} |\exp(-3t) - \exp(-5t)|\, dt = \int_{0}^{+\infty} [\exp(-3t) - \exp(-5t)]\, dt$$

$$= \left[-\frac{1}{3}\exp(-3t) + \frac{1}{5}\exp(-5t) \right]_{0}^{+\infty} = \frac{2}{15}.$$

Therefore, the system is stable.

PROBLEM 4.7. On the basis of Revers's criterion, check the stability of the closed-loop system defined by:

$$G(s) = \frac{2}{s^3 + 3s^2 + 2s + 3}, \quad H(s) = 1.$$

SOLUTION 4.7. From the Nyquist diagram depicted in Figure 4.8 and Revers's criterion, we conclude that the closed-loop system is stable.

This result can be verified by application of the Routh–Hurwitz criterion. The closed-loop transfer function is given by:

$$F(s) = \frac{2}{s^3 + 3s^2 + 2s + 5},$$

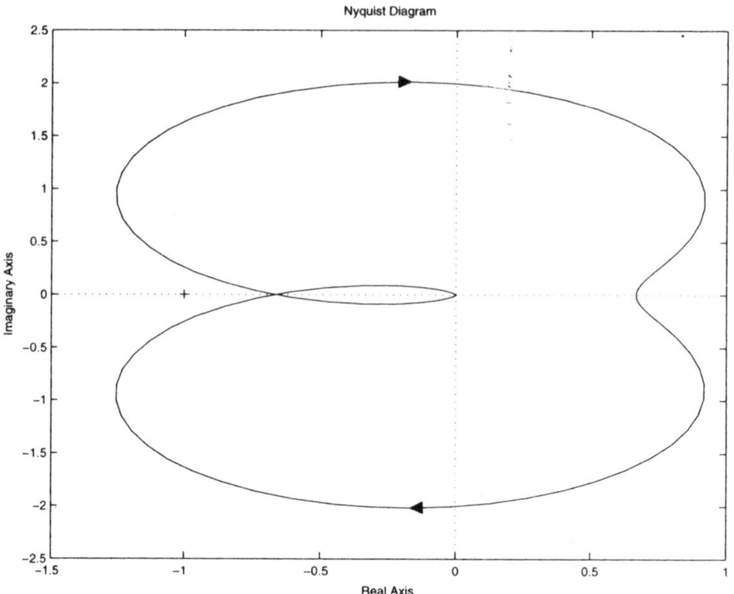

Figure 4.8. *Nyquist diagram of the system* $G(s) = 2/\left(s^3 + 3s^2 + 2s + 3\right)$

and the stability conditions:

$$\Delta_1 = |3|, \quad \Delta_2 = \begin{vmatrix} 3 & 1 \\ 5 & 2 \end{vmatrix} = 1 > 0, \quad \Delta_3 = \begin{vmatrix} 3 & 1 & 0 \\ 5 & 2 & 3 \\ 0 & 0 & 5 \end{vmatrix} = 5 > 0$$

are fulfilled.

PROBLEM 4.8. The characteristic function of a system is given by:

$$D(s) = \alpha_0 s^n + \alpha_1 s^{n-1} + \cdots + \alpha_n = \alpha_0 \prod_{i=1}^{n} (s - z_i). \qquad (4.8)$$

Calculate the variations δz_i, $i = 1, \cdots, n$ of the roots z_i induced by a variation $\delta \alpha_k$ of the coefficient α_k.

SOLUTION 4.8. The characteristic equation may be written as a function of the roots z_i and the coefficients α_i as follows:

$$D(z_i, \alpha_k) = 0.$$

A variation $\delta\alpha_k$ of the coefficient α_k will induce a variation δz_i in the root z_i. Therefore, the characteristic equation becomes:

$$D\left(z_i + \delta z_i, \alpha_k + \delta\alpha_k\right) = 0.$$

Let us now expand the left-hand side of this equation around the point defined by the "nominal" roots z_i and the "nominal" coefficients α_k:

$$D\left(z_i + \delta z_i, \alpha_k + \delta\alpha_k\right) = D\left(z_i, \alpha_k\right) + \left.\frac{\partial D}{\partial s}\right|_{s=z_k} \delta z_i + \left.\frac{\partial D}{\partial \alpha_k}\right|_{s=z_i} \delta\alpha_k + \cdots .$$

By neglecting the terms of order greater than 1, we derive:

$$\delta z_i = -\frac{\left.\partial D/\partial\alpha_k\right|_{s=z_i}}{\left.\partial D/\partial s\right|_{s=z_k}} \delta\alpha_k.$$

Observe that from Equation (4.8), we can easily obtain:

$$\left.\frac{\partial D}{\partial\alpha_k}\right|_{s=z_i} = z_i^{n-k} \quad \text{and} \quad \left.\frac{\partial D}{\partial s}\right|_{s=z_k} = \alpha_0 \prod_{j=1,j\neq i}^{n} \left(z_i - z_j\right).$$

Finally, we obtain:

$$\delta z_i = \frac{z_i^{n-k}}{\alpha_0 \prod\limits_{j=1,j\neq i}^{n} \left(z_i - z_j\right)} \delta\alpha_k.$$

Let us consider the following numerical example:

$$D\left(s\right) = s^3 + 6s^2 + 11s + 59, n = 3,$$

$$\begin{cases} z_1 = -5.845, \\ z_2 = -7.7495 \times 10^{-2} - j3.1762, \\ z_3 = -7.7495 \times 10^{-2} + j3.1762. \end{cases}$$

The effect of a variation $\delta\alpha_3$ on the root r_3 is given by:

$$\begin{aligned} \delta z_3 &= \frac{1}{\left(z_3 - z_1\right)\left(z_3 - z_2\right)} \delta\alpha_k \\ &= \frac{1}{\left(5.7675 + j3.1762\right)\left(j6.3524\right)} \delta\alpha_k \\ &= -\left(1.1533 \times 10^{-2} + j2.0943 \times 10^{-2}\right) \delta\alpha_k. \end{aligned}$$

For $\delta\alpha_k = 0.05$, the variations of the real and imaginary parts of the root z_3 are:

$$\begin{aligned} Re\,\delta z_3 &= -1.1533 \times 10^{-2}\,(0.05) = 0.00057, \\ Im\,\delta z_3 &= 2.0943 \times 10^{-2}\,(0.05) = 0.00104. \end{aligned}$$

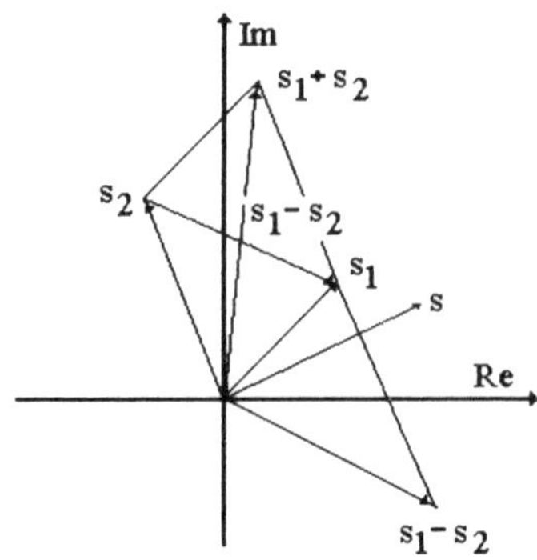

Figure 4.9. *Graphical representation of some complex numbers*

The system remains stable.

PROBLEM 4.9. Let us denote by $z_1, z_2, \cdots, z_k$ the set of distinct zeros of the following polynomial:

$$D(s) = s^n + \alpha_1 s^{n-1} + \cdots + \alpha_n. \tag{4.9}$$

Assume that z_j is a zero of order m_j (multiplicity m_j), and let:

$$\rho = \min_{i \neq j} \left\{ |z_i - z_j| \right\}. \tag{4.10}$$

For all $\varepsilon \in \,]0, \rho[$, show that there exists $\delta > 0$ such that any polynomial:

$$Q(s) = s^n + \beta_1 s^{n-1} + \cdots + \beta_n \tag{4.11}$$

for which:

$$\max_{1 \leq i \leq n} \left\{ \beta_i - \alpha_i \right\} < \delta \tag{4.12}$$

has exactly m_j zeros in the disc $B(z_j; \varepsilon)$, $j = 1, \cdots, k$.

SOLUTION 4.9. The proof of this result is based on Rouché's theorem [CHA 97, LAV 72]. Let us first recall some simple results concerning complex numbers. Figure 4.9 shows a graphical representation of a set of complex numbers.

From this figure, we derive[3]:

$$|s_1 + s_2| \le |s_1| + |s_2|, |s_1 - s_2| \ge |s_1| - |s_2|. \qquad (4.13)$$

The equalities in Equation (4.13) are achieved only when $\arg s_1 = \arg s_2$. The polynomial (4.9) may be written in the form:

$$D(s) = (s - z_1)^{m_1} \cdots (s - z_k)^{m_k}.$$

On the circle Γ_j defined by:

$$|s - z_j| = \varepsilon \qquad (4.14)$$

and for $i \ne j$, we derive:

$$|s - z_i| = |z_j - z_i - (z_j - s)| \ge |z_j - z_i| - |z_j - s| = |z_j - z_i| - |s - z_j|,$$

which, in view of (4.10), implies:

$$|s - z_i| \ge |z_j - z_i| - |s - z_j| \ge \rho - \varepsilon.$$

It follows that:

$$|D(s)| = |(s - z_1)^{m_1} \cdots (s - z_1)^{m_k}| = |(s - z_1)^{m_1}| \cdots |(s - z_1)^{m_k}|$$

$$\ge \varepsilon^{m_j} (\rho - \varepsilon)^{n - m_j},$$

and from Equation (4.12), we obtain:

$$|Q(s) - D(s)| \le \delta \sum_{t=0}^{n-1} |s|^t. \qquad (4.15)$$

From Equation (4.13), we obtain:

$$|s - z_j| \ge |s| - |z_j| \Rightarrow |s| \le |s - z_j| + |z_j|.$$

---

3. The triangle inequality may proved as follows:

$$|s + s_1|^2 = (s + s_1)(s^* + s_1^*) = ss^* + ss_1^* + s_1 s^* + s_1 s_1^*$$

$$= |s|^2 + 2Re(ss_1^*) + |s_1|^2 \le |s|^2 + 2|s||s_1| + |s_1|^2 = (|s| + |s_1|)^2,$$

where s^* represents the conjugate of s. The general result may be proved by induction. Let us assume that:

$$|s_1 + s_2 + \cdots + s_n| \le |s_1| + \cdots + |s_n|$$
$$|S_n| \le |s_1| + \cdots + |s_n|, \quad S_n = s_1 + s_2 + \cdots + s_n,$$

and consider:

$$|S_n + s_{n+1}|.$$

we obtain:

$$|S_n + s_{n+1}| \le |S_n| + |s_{n+1}| \le |s_1| + \cdots + |s_n| + |s_{n+1}|.$$

This, together with Equation (4.14), leads to:

$$|s| \geq \varepsilon + |z_j|. \tag{4.16}$$

Combining Equations (4.15) and (4.16) yields:

$$|Q(s) - D(s)| \leq \delta \sum_{l=0}^{n-1} (|z_j| + \varepsilon)^l. \tag{4.17}$$

If:

$$0 < \delta < \min_{1 \leq j \leq k} \left\{ \frac{\varepsilon^{m_j} (\rho - \varepsilon)^{n-m_j}}{\sum\limits_{l=0}^{n-1} (|z| + \varepsilon)^l} \right\},$$

then:

$$|Q(s) - D(s)| \leq |D(s)|$$

on the circle Γ_j, for $|s - z_j| = \varepsilon, j = 1, \cdots, k$. Using Rouché's theorem[4] [CHA 97, LAV 72], we derive that the polynomial $Q(s)$ has exactly m_j zeros (taking into account their respective orders) in the disc[5] $B(z_j; \varepsilon), j = 1, \cdots, k$.

PROBLEM 4.10. Determine the gain and phase margins of the following system (see [CEN 02]):

$$F(s) = \frac{0.38 \left(s^2 + 0.1s + 0.55\right)}{s(s+1)\left(s^2 + 0.06s + 0.5\right)}.$$

SOLUTION 4.10. The Nichols diagram of this system is plotted in Figure 4.10. From this figure, we deduce that:

$$\varphi_m = 70^o \quad \text{and} \quad k_m = \infty.$$

The Nichols diagram does not cross the vertical axis.

---

4. Rouché's theorem can be stated as follows. Let Γ be a closed curve in a domain D. Assume that the functions f and g are analytic in D and have a finite number of zeros located in D. If:

$$|f(s) - g(s)| < |g(s)|, \quad s \in \Gamma,$$

then:

$$N(f, D) = N(g, D)$$

where $N(f, D)$ represents the numbers of zeros of f located in the domain D.
This theorem finds applications in many areas, such as in queueing theory.
5. In $\mathcal{R}^2$, a disc $B(s_j; \varepsilon)$ corresponds to the inside of a circle centred on s_j with radius ε.

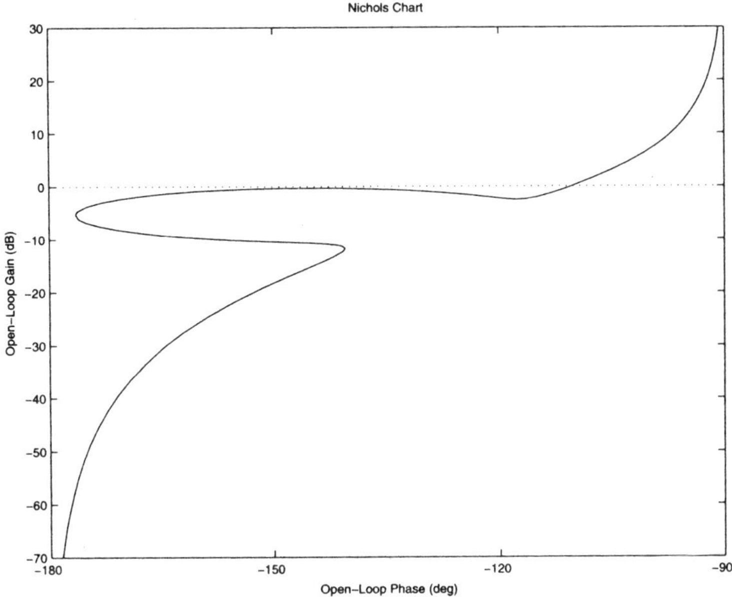

Figure 4.10. *Nichols diagram of the system*
$$F\left(s\right) = 0.38\left(s^2 + 0.1s + 0.55\right)/\left[s\left(s+1\right)\left(s^2 + 0.06s + 0.5\right)\right]$$

PROBLEM 4.11. The open-loop transfer function of a unity-feedback system is given by:

$$T\left(s\right) = \frac{1}{s^2 + 2s + 2}.$$

Draw the Nyquist diagram and calculate the shortest distance from this diagram to the critical point $A\left(-1, 0\right)$.

SOLUTION 4.11. The Nyquist diagram is depicted in Figure 4.11. This diagram does not encircle the critical point. The closed-loop system is stable. Let us consider a point $M\left(x, y\right)$ on this diagram. We have:

$$T\left(j\omega\right) = x + jy = \frac{2 - \omega^2}{\left(2 - \omega^2\right)^2 + 4\omega^2} + \frac{-j2\omega}{\left(2 - \omega^2\right)^2 + 4\omega^2}$$

and

$$AM^2 = \left(x + 1\right)^2 + y^2 = \frac{\omega^8 - 2\omega^6 + 13\omega^4 - 8\omega^2 + 36}{\left(\left(2 - \omega^2\right)^2 + 4\omega^2\right)^2}.$$

The shortest distance from the Nyquist diagram to the critical point is given by:

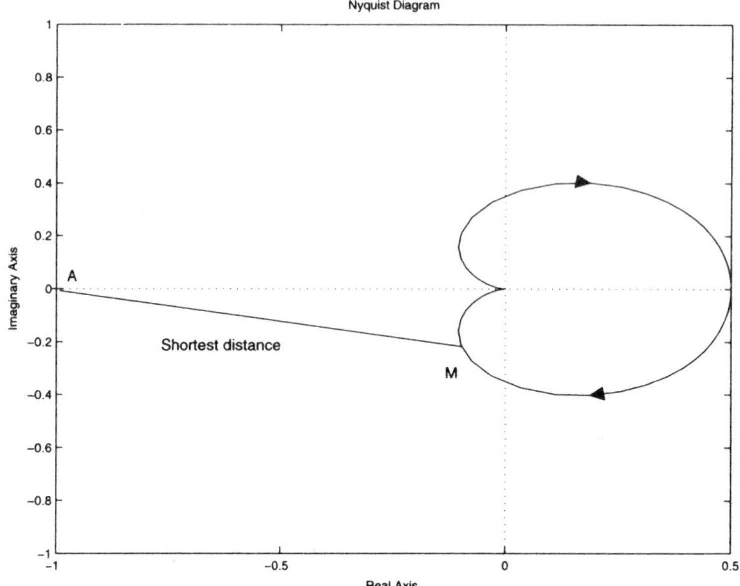

Figure 4.11. *Nyquist diagram of the system* $T\left(s\right) = 1/\left[s^2 + 2s + 2\right]$

$$\frac{d\left(AM^2\right)}{d\omega} = 4\omega\frac{\omega^4 - 5\omega^2 - 4}{\left(4 + \omega^4\right)^2} = 0,$$

which leads to $\omega = 0$ and:

$$\omega^4 - 5\omega^2 - 4 = 0.$$

The real roots are:

$$\omega = \pm 2.3878$$

Therefore, the shortest distance of the Nyquist diagram from the critical point is:

$$\min_{\omega} AM = d = 0.90808, \quad \omega = 2.3878.$$

This shortest distance is usually denoted by $d = 1/M_s$. $M_s = 1/d = 1.101$ is the maximum sensitivity. Observe that:

$$\overrightarrow{AM} \to T\left(j\omega\right) - \left(-1\right).$$

The inverse of the shortest distance from the diagram of the open-loop transfer function to the critical point $A\left(-1, 0\right)$ is equal to:

$$\max_{\omega}\frac{1}{T\left(j\omega\right) + 1} = \max_{\omega} S\left(j\omega\right) = M_s.$$

where $S(j\omega)$ is the sensitivity function. Acceptable values of the maximum sensitivity M_s are in the range 1 to 2.

PROBLEM 4.12. Consider a unity-feedback system with the following open-loop transfer function:

$$T(s) = \frac{k}{s(s-\alpha)}.$$

Study the stability of this system.

SOLUTION 4.12. The closed-loop poles of this system are given by the characteristic equation:

$$1 + T(s) = 0,$$

$$1 + \frac{k}{s(s-\alpha)} = 0,$$

$$s(s-\alpha) + k = s^2 - \alpha s + k = 0.$$

Using the Routh–Hurwitz criterion, we obtain the result that for:

$$\alpha < 0 \quad \text{with } k \in [0, +\infty[,$$

the closed-loop system is stable. Observe that we are concerned with the roots of a second-order algebraic equation. They can be directly calculated:

$$s_1, s_2 = \frac{\alpha \pm \sqrt{\alpha^2 - 4k}}{2} \quad \text{for } \alpha^2 > 4k,$$

$$s_1, s_2 = \frac{\alpha \pm j\sqrt{4k - \alpha^2}}{2} \quad \text{for } \alpha^2 < 4k.$$

Alternatively, we can observe that the sum and the product of the zeros of the characteristic equation are equal to α and k, respectively.

PROBLEM 4.13. Consider a feedback control system with:

$$G(s) = \frac{10}{s^2 - 2s - 3} \quad \text{and } H(s) = 1.$$

Analyze its stability on the basis of the Nyquist criterion.

SOLUTION 4.13. The open-loop transfer function is equal to:

$$T(s) = G(s)H(s) = \frac{10}{s^2 - 2s - 3}.$$

The Nyquist diagram is depicted in Figure 4.12.

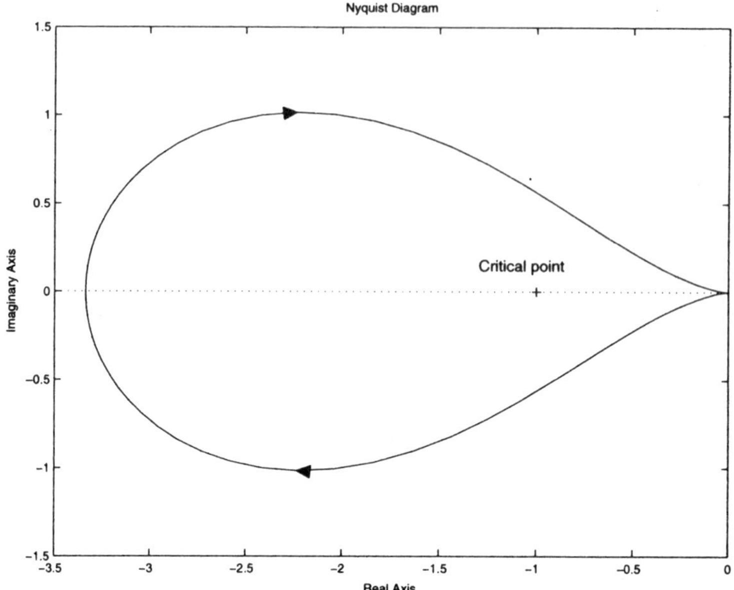

Figure 4.12. *Nyquist diagram of the system* $T(s) = 10/\left(s^2 - 2s - 3\right)$

This figure shows that the Nyquist plot encircles the critical point $A(-1,0)$ once. Consequently, the closed-loop system has a pole located to the right of the imaginary axis.

PROBLEM 4.14. Consider the following unity-feedback system:

$$G(s) = \frac{70}{s^3 + 6s^2 + 11s + 6} \text{ and } H(s) = 1.$$

1. Analyze its stability on the basis of the Nyquist diagram.

2. From the Bode diagram of the transfer function $V(s) = 1 + G(s)$, calculate the number N defined by:

$$N = \frac{2\,\Delta\varphi}{2\pi}, \quad \Delta\varphi = \arg V(j\omega)_{\omega=0} - \arg V(j\omega)_{\omega\to+\infty}. \tag{4.18}$$

SOLUTION 4.14. 1. The open-loop transfer function is given by:

$$T(s) = \frac{70}{s^3 + 6s^2 + 11s + 6}.$$

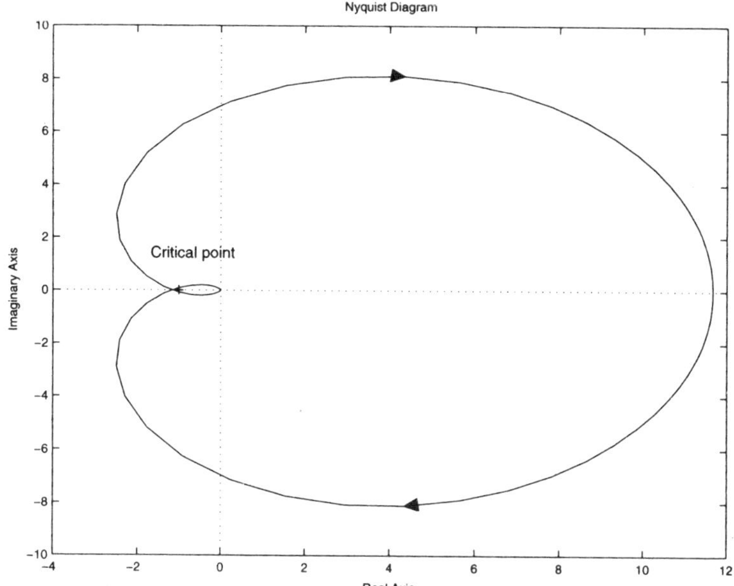

Figure 4.13. *Nyquist diagram of the system*
$$T(s) = 70/\left(s^3 + 6s^2 + 11s + 6\right)$$

The Nyquist plot depicted in Figure 4.13 shows that the closed-loop system is unstable. The encirclement number is equal to 2.

2. A plot of the Bode diagram associated with the transfer function $V(s)$ is shown in Figure 4.14.

From this figure, we can derive the encirclement number N:

$$\Delta\varphi = \arg V(j\omega)_{\omega=0} - \arg V(j\omega)_{\omega\to+\infty} = 2\pi - 0, \quad N = \frac{2\Delta\varphi}{2\pi} = 2.$$

To check the stability of a feedback system with a stable open-loop transfer function on the basis of the Nyquist criterion, it is not necessary to draw the Nyquist diagram. Instead, the encirclement number of the critical point can be calculated from equation (4.18). This technique was proposed by [CEN 02]. It has been implemented in commercial microwave CAD tools.

PROBLEM 4.15. The open-loop transfer function of a unity-feedback system is given by:

$$T(s) = \frac{50}{s^3 + 6s^2 + 11s + 6}.$$

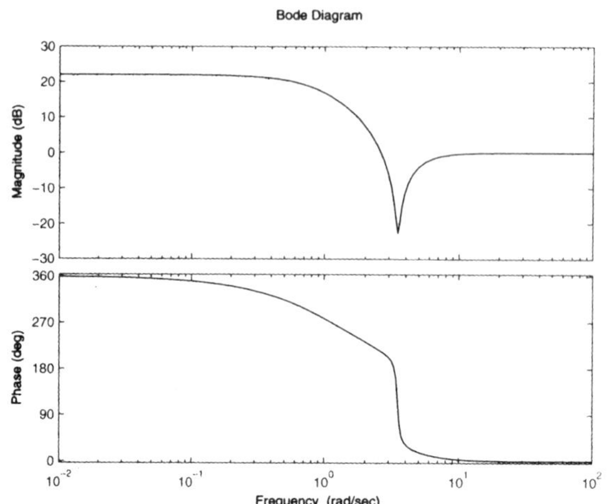

Figure 4.14. *Bode diagram of the system*
$$V(s) = 1 + 70/\left(s^3 + 6s^2 + 11s + 6\right)$$

Calculate the phase variation:

$$\Delta\varphi = \arg V(j\omega)_{\omega=0} - \arg V(j\omega)_{\omega\to+\infty}, \quad V(j\omega) = 1 + T(s),$$

and check the stability of the closed-loop system.

SOLUTION 4.15. The poles of the open-loop system are:

$$p_1 = -1, \quad p_2 = -2, \quad p_3 = -3.$$

From the Bode diagram depicted in Figure 4.15, we derive:

$$\Delta\varphi = \arg V(j\omega)_{\omega=0} - \arg V(j\omega)_{\omega\to+\infty} = 0$$

and

$$\frac{2\,\Delta\varphi}{2\pi} = 0 \implies N = 0.$$

Therefore, the closed-loop system is stable. The closed-loop poles are:

$$p_1 = -5.77, \quad p_2 = -0.11 + j3.11, \quad p_3 = p_2^*.$$

We can easily verify that the ultimate gain is equal to 60.

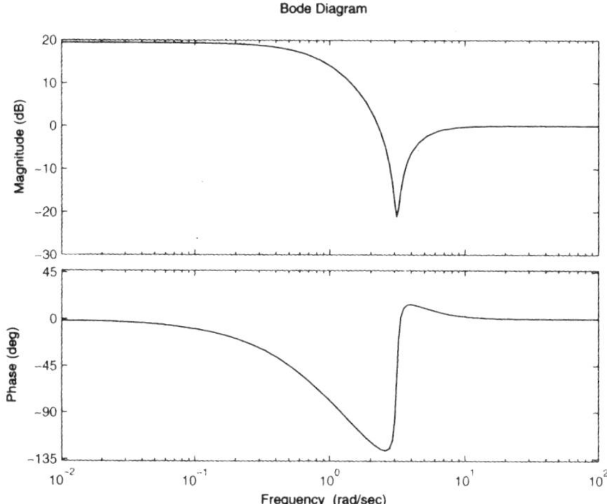

Figure 4.15. *Bode diagram of the system*
$$V\left(s\right) = 1 + 50/\left(s^3 + 6s^2 + 11s + 6\right)$$

PROBLEM 4.16. Consider a feedback control system with:

$$G\left(s\right) = \frac{2s - 4}{s^2 - 4s - 3} \quad \text{and } H\left(s\right) = 1.$$

Analyze its stability on the basis of the Nyquist criterion.

SOLUTION 4.16. The open-loop transfer function is given by:

$$T\left(s\right) = \frac{2s - 4}{s^2 - 4s + 3}.$$

The zero of this transfer function is located to the right of the imaginary axis. Consequently, the open-loop system is a non-minimum-phase system. The Nyquist diagram is plotted in Figure 4.16. This plot encircles the critical point once. Therefore, the system is unstable.

PROBLEM 4.17. Consider a feedback control system with:

$$G\left(s\right) = \frac{2s - 4}{s^2 - 2s - 3} \quad \text{and } H\left(s\right) = 1.$$

Analyze its stability on the basis of the Nyquist criterion.

SOLUTION 4.17. The open-loop system is non-minimum-phase. The zero $s = 2$ is located to the right of the imaginary axis. The Nyquist diagram is plotted in Figure 4.17. It encircles the critical point. The closed-loop system is unstable.

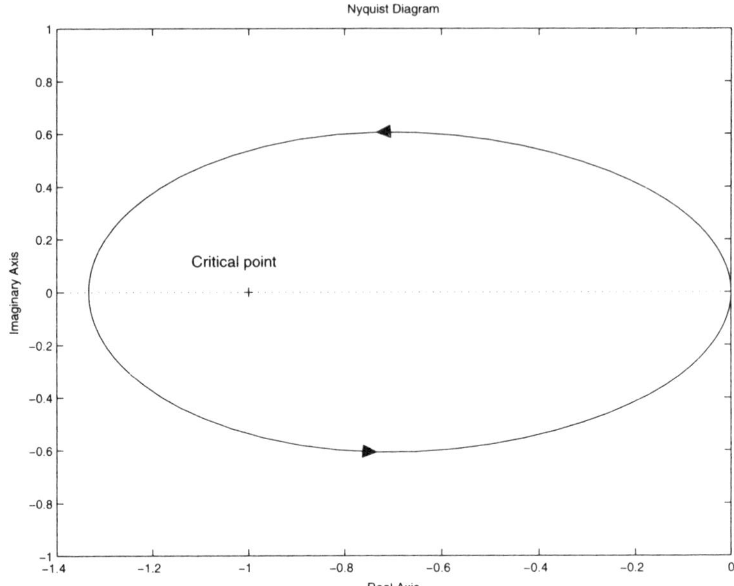

Figure 4.16. *Nyquist diagram of the system* $T\left(s\right) = \left(2s - 4\right) / \left(s^2 - 4s + 3\right)$

PROBLEM 4.18. Analyze the stability of the unity-feedback system:

$$G\left(s\right) = \frac{k}{1 + Ts}\left(k_c + \frac{k_i}{s} + k_d s\right), \quad k > 0,$$

as a function of the parameters k_c, k_i and k_d.

SOLUTION 4.18. Observe that this transfer function corresponds to a PID regulator in series with a system $k/\left(1 + Ts\right)$. The characteristic equation is given by:

$$1 + G\left(s\right) = 0,$$

$$\left(T + kk_d\right)s^2 + \left(1 + kk_c\right)s + kk_i = 0.$$

From the Routh–Hurwitz criterion, this closed-loop system is stable if:

$$a_0 = T + kk_d > 0, \quad \Delta_1 = \left|1 + kk_c\right| > 0,$$

$$\Delta_2 = \begin{vmatrix} 1 + kk_c & T + kk_d \\ 0 & kk_i \end{vmatrix} = kk_i\left(1 + kk_c\right) > 0.$$

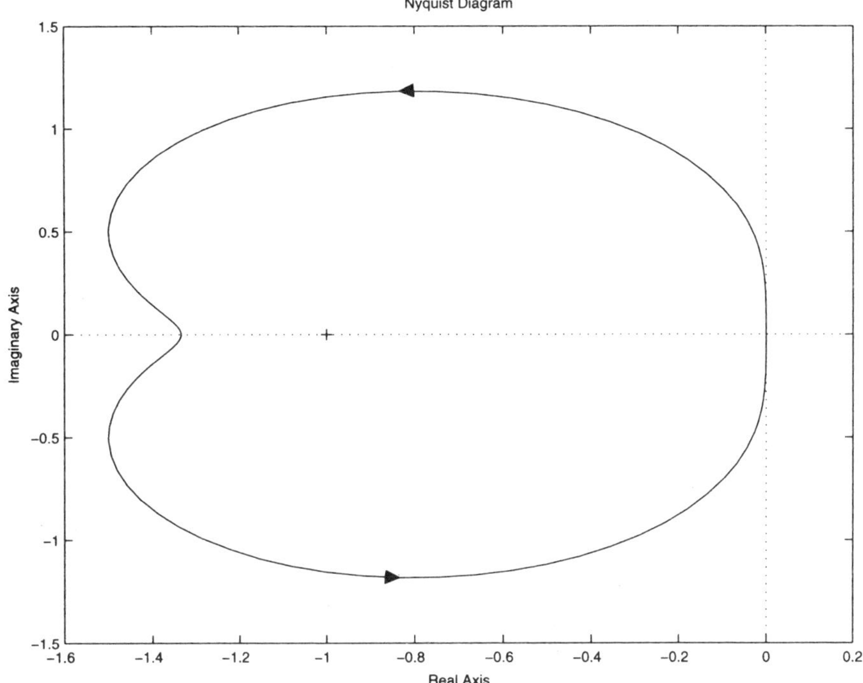

Figure 4.17. *Nyquist diagram of the system* $T\left(s\right) = \left(2s - 4\right) / \left(s^2 - 2s - 3\right)$

For $T + kk_d > 0$, it follows that:

$$k_c > -\frac{1}{k} \quad \text{and } k_i > 0.$$

For $T + kk_d < 0$, we multiply the characteristic equation by -1, and we derive:

$$k_c < -\frac{1}{k} \quad \text{and } k_i < 0.$$

PROBLEM 4.19. Consider the following stable, proper transfer functions:

$$F_1\left(s\right) = \frac{P_1\left(s\right)}{Q_1\left(s\right)}, \quad F_2\left(s\right) = \frac{P_2\left(s\right)}{Q_2\left(s\right)}$$

such that:

$$\left|\frac{P_1\left(j\omega\right)}{Q_1\left(j\omega\right)}\right| + \left|\frac{P_2\left(j\omega\right)}{Q_2\left(j\omega\right)}\right| < 1. \tag{4.19}$$

For all θ and $\varphi \in [0, \pi[$, show that the zeros of:

$$Q_1\left(j\omega\right) Q_2\left(j\omega\right) + \exp\left(j\theta\right) P_1\left(j\omega\right) Q_2\left(j\omega\right) + \exp\left(j\varphi\right) P_2\left(j\omega\right) Q_1\left(j\omega\right)$$

are located to the left of the imaginary axis [MIN 03].

SOLUTION 4.19. From the condition (4.19), we derive:

$$\left|\frac{P_1\left(j\omega\right)}{Q_1\left(j\omega\right)}\right| + \left|\frac{P_2\left(j\omega\right)}{Q_2\left(j\omega\right)}\right| = \frac{\left|P_1\left(j\omega\right)\right|}{\left|Q_1\left(j\omega\right)\right|} + \frac{\left|P_2\left(j\omega\right)\right|}{\left|Q_2\left(j\omega\right)\right|}$$

$$= \frac{\left|P_1\left(j\omega\right)\right|\left|Q_2\left(j\omega\right)\right| + \left|P_2\left(j\omega\right)\right|\left|Q_1\left(j\omega\right)\right|}{\left|Q_1\left(j\omega\right)\right|\left|Q_2\left(j\omega\right)\right|}$$

$$= \frac{\left|P_1\left(j\omega\right) Q_2\left(j\omega\right)\right| + \left|Q_1\left(j\omega\right) P_2\left(j\omega\right)\right|}{\left|Q_1\left(j\omega\right) Q_2\left(j\omega\right)\right|}$$

$$= \frac{\left|\exp\left(j\theta\right) P_1\left(j\omega\right) Q_2\left(j\omega\right)\right| + \left|\exp\left(j\varphi\right) Q_1\left(j\omega\right) P_2\left(j\omega\right)\right|}{\left|Q_1\left(j\omega\right) Q_2\left(j\omega\right)\right|} < 1,$$

$$\left|Q_1\left(j\omega\right) Q_2\left(j\omega\right)\right| > \left|\exp\left(j\theta\right) P_1\left(j\omega\right) Q_2\left(j\omega\right)\right| + \left|\exp\left(j\varphi\right) Q_1\left(j\omega\right) P_2\left(j\omega\right)\right|.$$

As $\left|s_1 + s_2\right| \leq \left|s_1\right| + \left|s_2\right|$, we obtain:

$$\left|Q_1\left(j\omega\right) Q_2\left(j\omega\right)\right| > \left|\exp\left(j\theta\right) P_1\left(j\omega\right) Q_2\left(j\omega\right) + \exp\left(j\varphi\right) Q_1\left(j\omega\right) P_2\left(j\omega\right)\right|.$$

Using Rouché's Theorem,[6] and taking into account the fact that the systems considered are stable, we conclude that the zeros of:

$$Q_1\left(j\omega\right) Q_2\left(j\omega\right) + \exp\left(j\theta\right) P_1\left(j\omega\right) Q_2\left(j\omega\right) + \exp\left(j\varphi\right) Q_1\left(j\omega\right) P_2\left(j\omega\right)$$

are located to the left of the imaginary axis.

Notice that if we change the condition (4.19) to:

$$\left|\frac{P_1\left(j\omega\right)}{Q_1\left(j\omega\right)}\right|_{\omega=\infty} + \left|\frac{P_2\left(j\omega\right)}{Q_2\left(j\omega\right)}\right|_{\omega=\infty} < 1,$$

we obtain:

$$\left|\frac{a_n}{b_n}\right| + \left|\frac{c_m}{d_m}\right| < 1$$

---

6. We shall give another statement of Rouché's theorem here. If $f\left(s\right)$ and $g\left(s\right)$ are analytic functions inside a contour C that are continuous on C and satisfy the condition:

$$\left|f\left(s\right)\right| > \left|g\left(s\right)\right|,$$

then the functions $f\left(s\right)$ and $f\left(s\right) + g\left(s\right)$ have the same number of zeros inside C.

where:

$$P_1\left(s\right) \;=\; \sum_{i=0}^{n} a_i s^i, \quad Q_1\left(s\right) = \sum_{i=0}^{n} b_i s^i,$$

$$P_2\left(s\right) \;=\; \sum_{i=0}^{m} c_i s^i, \quad Q_2\left(s\right) = \sum_{i=0}^{m} d_i s^i.$$

4.2. The root locus

The root locus consists of a graphical representation of the roots of the characteristic equation when the gain (or some other variable design parameter) varies from zero to infinity. The characteristic equation is given by:

$$1 + kT\left(s\right) \;=\; 1 + k\,\frac{\displaystyle\prod_{i=1}^{m}\left(s - z_i\right)}{\displaystyle\prod_{i=1}^{n}\left(s - p_i\right)} = 0,$$

$$\frac{\displaystyle\prod_{i=1}^{m}\left(s - z_i\right)}{\displaystyle\prod_{i=1}^{n}\left(s - p_i\right)} \;=\; -\frac{1}{k}, \quad k \in [0, +\infty[$$

or

$$\left|\frac{\displaystyle\prod_{i=1}^{m}\left(s - z_i\right)}{\displaystyle\prod_{i=1}^{n}\left(s - p_i\right)}\right| = \frac{1}{k} \quad \text{and} \quad \arg\frac{\displaystyle\prod_{i=1}^{m}\left(s - z_i\right)}{\displaystyle\prod_{i=1}^{n}\left(s - p_i\right)} = -\pi\left(2\pi\right), \tag{4.20}$$

where z_i and p_i represent the zeros and the poles of the open-loop transfer function. The latter two equalities correspond to the magnitude and angle requirements. In what follows, we shall present a set of useful rules for drawing the root locus of a given system. Proofs of the rules presented below are given in Appendix A.

Notice first that if $s = \alpha + j\beta$ is a solution of Equation (4.20), then $s^* = \alpha - j\beta$ is also a solution. As a consequence, the root locus is *symmetric* about the real axis. Usually, the poles and the zeros of the open-loop transfer function are represented by the symbols "×" and "o", respectively. The rules for drawing the root locus are as follows:

1) *Starting points* ($k = 0$). The poles (p_i) of the transfer function of the open-loop system.

2) *Ending points* ($k = \infty$). The zeros (z_i) of the transfer function of the open-loop system.

3) *Real-axis portions of the root locus.* A point M on the real axis belongs to the root locus if the number of zeros and poles of the open-loop transfer function located to the right of M is odd[7].

4) *Number of asymptotic directions.* $n - m$.

5) *Positions of asymptotic directions with respect to the real axis.* The angles between the the asymptotic directions and the real axis are equal to:

$$\beta = \frac{2\lambda + 1}{n - m}\pi, \quad \lambda = 0, 1, 2, \cdots .$$

6) *Intersections of the asymptotic directions with the real axis.* The abscissa of these intersections are given by:

$$\delta = \frac{\displaystyle\sum_{i=1}^{n} p_i - \sum_{i=1}^{m} z_i}{n - m}.$$

The asymptotes start from the point defined by the co-ordinates $(\delta, 0)$. This point is called the *centroid.*

7) *Intersections of the root locus with the real axis (breakaway points).* The abscissa x of these intersections are given by:

$$\sum_{i=1}^{n} \frac{1}{x - p_i} = \begin{cases} \displaystyle\sum_{i=1}^{m} \frac{1}{x - z_i} & \text{if } m \neq 0, \\ 0 & \text{if } m = 0. \end{cases}$$

8) *Intersections of the root locus with the imaginary axis (crossover points, or imaginary intercepts).* These intersections are given by the solution of the characteristic equation for $s = j\omega$,

$$1 + kT(j\omega) = 0,$$

which leads to two algebraic equations and gives also the ultimate (critical) gain and frequency.

9) *Positions of the tangents to the root locus at the complex poles p_i (angles of departure) and zeros z_i (angles of arrival) with respect to the real axis.* These tangents form angles θ with the real axis equal to:

$$\theta \;=\; \arg T'(p_i) + 180 \text{ for complex poles,}$$

$$\theta \;=\; 180 - \arg T'(z_i) \text{ for complex zeros,}$$

---

7. Observe that if a complex number is a pole (or zero) of the open-loop transfer function, its conjugate is also a pole (or zero).

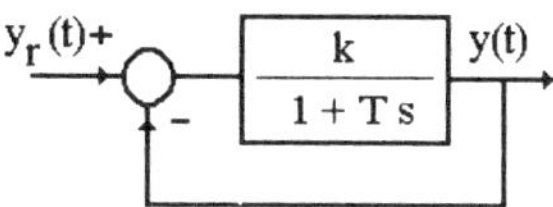

Figure 4.18. *System considered in Problem 4.20*

where:
$$T'(s) = \begin{cases} T(s)(s - p_i) & \text{for a complex pole,} \\ \dfrac{T(s)}{s - z_i} & \text{for a complex zero.} \end{cases}$$

These angular positions are called *angles of departure* (for complex poles p_i) and *angles of arrival* (complex zeros z_i). The poles and zeros of the open-loop transfer function represent the starting and ending points, respectively, of the root locus.

Finally, observe that, in order to refine the drawing of the root locus, the intersections of the asymptotic directions with the imaginary axis can be calculated and compared with the intersections of the root locus with the imaginary axis in order to know whether the latter intersections are located above or below the former intersections. The equations of the asymptotic directions can easily be calculated from a knowledge of δ and β. Their parametric representations are given in Appendix A.

PROBLEM 4.20. Draw the root locus of the system depicted in Figure 4.18.

SOLUTION 4.20. The characteristic equation is given by:
$$1 + k + Ts = 0, \ T > 0,$$

and leads to:
$$s = -\frac{1 + k}{T}.$$

For $k \in [0, \infty[$, the corresponding root locus is drawn in Figure 4.19.

Observe that from rule 3 we can immediately deduce that the semi-axis $]-\infty, -1/T]$ belongs to the root locus. In general, an algebraic approach is not feasible for more complex systems. Under these conditions, it is preferable to use a geometric (graphical) approach based on the rules presented above.

PROBLEM 4.21. Draw the root locus for a system having the following open-loop transfer function:
$$T(s) = \frac{k(1 + 2.5s)}{s(2s^2 + 3s + 1)}.$$

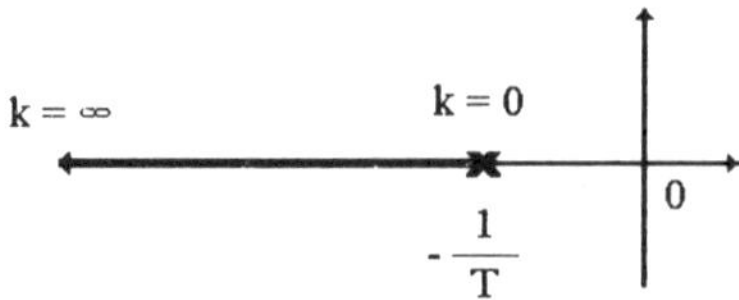

Figure 4.19. *Root locus of the system considered in Problem 4.20*

SOLUTION 4.21. The poles and zeros of the open-loop transfer function are:

$$z_1 = -0.4, \quad p_1 = 0, \quad p_2 = -1 \quad \text{and } p_3 = -0.5.$$

The segments $[-0.4, 0]$ and $[-1, -0.5]$ belong to the root locus, from rule 3. There certainly exists an intersection between the real axis and the root locus located in the segment $[-1, -1/2]$. The number of asymptotic directions is equal to 2, from rule 4, and their positions with respect to the real axis are $\beta = \pi/2$ and $-\pi/2$ (or $+\frac{3\pi}{2}$),[8] from rule 5. The abscissa of the intersections of the asymptotic directions with the real axis is, according to 6:

$$\delta = \frac{\sum\limits_{i=1}^{3} p_i - \sum\limits_{i=1}^{1} z_i}{2} = \frac{-1 - 0.5 + 1/2.5}{2} = -\frac{11}{20},$$

and the abscissa of the intersection of the root locus with the real axis, from rule 7, is given by:

$$\sum_{i=1}^{3} \frac{1}{x - p_i} = \sum_{i=1}^{1} \frac{1}{x - z_i},$$

$$\frac{1}{x} + \frac{1}{x+1} + \frac{1}{x+0.5} - \frac{1}{x+1/2.5} = 0 \Longrightarrow 2x^3 + 2.7x^2 + 1.2x + 0.2 = 0.$$

Let us write this equation in the form[9]:

$$f(x) = 0,$$

---

8. The directions $3\pi/2$ and $-\pi/2$ are the same.

9. We can find an approximate solution of:

$$f(x) = 0$$

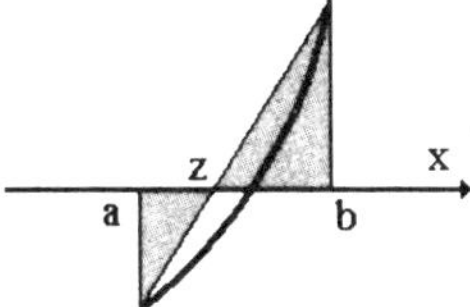

Figure 4.20. *Localization of a root of $f(x)$ in the interval $[a, b]$*

and calculate $f(-11/20)$, $f(-1/2)$ and $f(-1)$. $f\left(-\dfrac{1}{2}\right)$ and $f(-11/20)$ are positive, and $f(-1)$ is negative, and therefore the intersection of the root locus with the real axis is in the segment $[-1, -11/20]$ (see Figure 4.20).

The roots of the polynomial:

$$2x^3 + 2.7x^2 + 1.2x + 0.2$$

are:

$$x_1 = -0.69136, \quad x_2 = -0.32932 - j0.19024, \quad \text{and } x_3 = -0.32932 + j0.19024.$$

The complex solutions have to be rejected. The intersection of the root locus with the imaginary axis is given by:

$$-j2\omega^3 - 3\omega^2 + j(1 + 2.5k)\omega + k = 0,$$
$$2\omega^2 = 1 + 2.5k, \quad k = 3\omega^2 \implies 2\omega^2 = 1 + 7.5\omega^2.$$

The root locus does not cross the imaginary axis.

The root locus is drawn in Figure 4.21. From this figure, we deduce that the system is stable $\forall k$.

---

in the interval $[a, b]$ in the following way. First, observe that if $f(a) f(b) < 0$, there exists a root of $f(x)$ in the segment $[a, b]$. If $f(a) f(b) < 0$, we may divide the interval $[a, b]$ and calculate the sign of $f(a) f((a + b)/2)$, and so on until we obtain the desired accuracy. Observe, however, that instead of dividing the successive intervals by 2, it is preferable to consider an interval $[a, z]$ which corresponds to a division by $f(a)/f(b)$. From the shaded triangles in Figure 4.20, we obtain:

$$\frac{z - a}{b - z} = \frac{f(a)}{f(b)}.$$

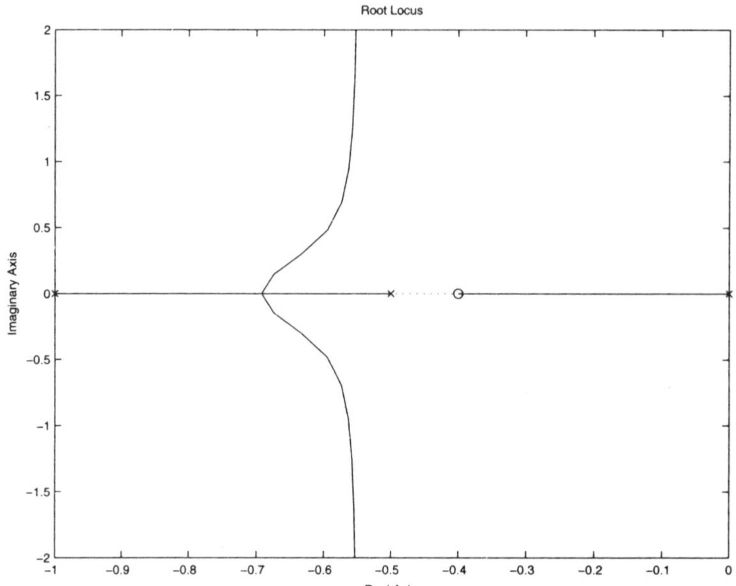

Figure 4.21. *Root locus of the system*
$$T\left(s\right) = k\left(1 + 2.5s\right) / \left[s\left(2s^2 + 3s + 1\right)\right]$$

PROBLEM 4.22. Draw the root locus for a system having the following open-loop transfer function:
$$T\left(s\right) = \frac{k\left(1 + s\right)}{s\left(1 + 2s\right)}.$$

SOLUTION 4.22. We have $m = 1$, $n = 2$, $z_1 = -1$, $p_1 = 0$ and $p_2 = -1/2$. On the basis of rule 3, the segment $[-1/2, 0]$ and the semi axis $]-\infty, -1]$ belong to the root locus. The root locus contains one asymptotic direction, represented by the negative real axis, from rule 4, and its intersections with the real axis are given by 7:

$$\frac{1}{x} + \frac{1}{x + 1/2} = \frac{1}{x + 1},$$

$$\left(x + \frac{1}{2}\right)\left(x + 1\right) + x\left(x + 1\right) = x\left(x + \frac{1}{2}\right),$$

$$x^2 + 2x + \frac{1}{2} = 0,$$

which leads to:

$$x_1 = -1 + \frac{\sqrt{2}}{2} \quad \text{and} \quad x_2 = -1 - \frac{\sqrt{2}}{2}.$$

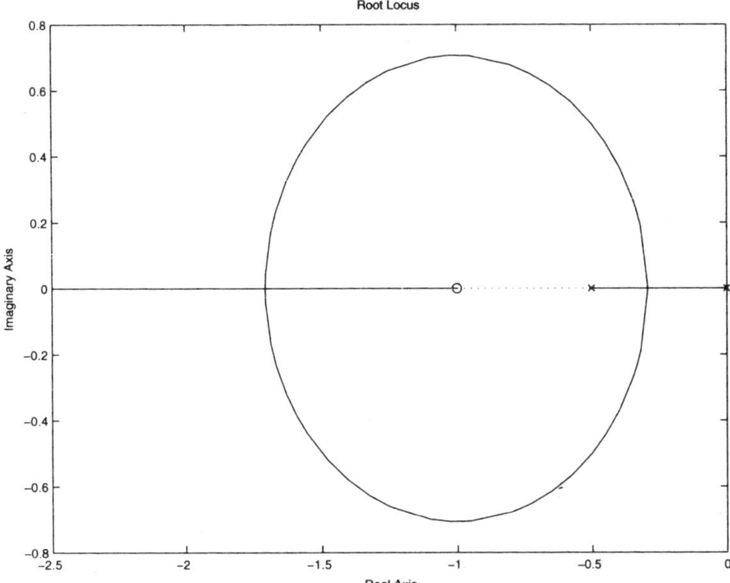

Figure 4.22. *Root locus for the system* $T(s) = k(1+s)/(2s^2 + s)$

Since the segment $[-1/2, 0]$, whose extremities are the poles 0 and $-1/2$, belongs to the root locus, there exists an intersection between the root locus and this segment at x_1, and the branch of the root locus which passes through this intersection has to go to an ending point. In this case, we have only one ending point, which corresponds to the zero -1. As a consequence, the root locus consists of a closed curve which intersects the real axis at the points x_1 (located on the segment $[-1/2, 0]$) and x_2 (located on the negative real axis $]-\infty, -1/2]$). The root locus is plotted in Figure 4.22.

From this plot, we deduce that this system is stable $\forall k$.

PROBLEM 4.23. 1. Draw the root locus for a system having the following open-loop transfer function:

$$T(s) = \frac{k}{s(1+0.4s)(1+0.1s)}.$$

2. Study the effect of the additional complex zeros:

$$s^2 + s + 1, \quad s = -\frac{1}{2} \pm j\frac{1}{2}\sqrt{3}.$$

SOLUTION 4.23. 1. The segment $[-1/0.4, 0]$ and the semi-axis $]-\infty, -1/0.1]$ belong to the root locus. Recall that the poles of the open-loop transfer function correspond to the starting points ($k = 0$) of the root locus. The number of asymptotic directions is equal to 3; $\beta = \pi/3$, π (real axis) and $-\pi/3$; and $\delta = (0 - 2.5 - 10)/3 = -4.166$. The intersection of the root locus with the real axis is given by:

$$\sum_{i=1}^{3} \frac{1}{x - p_i} = \frac{1}{x} + \frac{1}{x + 2.5} + \frac{1}{x + 10} = \frac{3x^2 + 25x + 25}{x(x + 2.5)(x + 10)} = 0,$$

$$\implies x_1 = -25/6 + 5\sqrt{13}/6 = -1.162,$$

$$x_2 = -25/6 - 5\sqrt{13}/6 = -7.1713.$$

The intersection of the root locus is located in the segment $[-2.5, 0]$ because the poles 0 and -2.5 are departure points of the root locus. Consequently, the solution x_2 must be rejected, and the abscissa of the intersection corresponds to $x_1 = -1.162$. Intersections of the root locus with the imaginary axis occur for:

$$0.04s^3 + 0.5s^2 + s + k = 0, \quad s = j\omega,$$

$$-0.04j\omega^3 - 0.5\omega^2 + j\omega + k = 0,$$

$$\implies k - 0.5\omega^2 = 0, \quad \omega(1 - 0.04\omega^2) = 0, \quad \omega = 5 \quad \text{and } k = 12.5.$$

This system is stable for $k < k_{\lim} = 12.5$. A plot of the root locus is given in Figure 4.23.

2. The introduction of a complex pair of zeros has no effect on the segments and semi-segments of the root locus on the real axis. The number of asymptotes is equal to 1. The asymptotic direction corresponds to the semi-segment $]-\infty, -10]$. The poles are departure points. Therefore, there exists an intersection of the root locus with the real axis located on the segment $[-1/0.4, 0]$. The abscissa of this intersection is given by:

$$\sum_{i=1}^{3} \frac{1}{x - p_i} = \sum_{i=1}^{2} \frac{1}{x - z_i},$$

$$f(x) = \frac{1}{x} + \frac{1}{x + 2.5} + \frac{1}{x + 10} - \frac{1}{x + 1/2 + j\sqrt{3}/2} - \frac{1}{x + 1/2 - j\sqrt{3}/2}.$$

We find that $f(-0.8) < 0$ and $f(-0.7) > 0$. Therefore, the intersection of the root locus with the real axis is located on the segment $[-0.8, -0.7]$. The root locus is depicted in Figure 4.24. The angles of arrival are given by:

$$\theta_1 = 180 - \arg \left. \frac{k(s + 1/2 + j\sqrt{3}/2)}{s(1 + 0.4s)(1 + 0.1s)} \right|_{s = -1/2 + j\sqrt{3}/2}$$

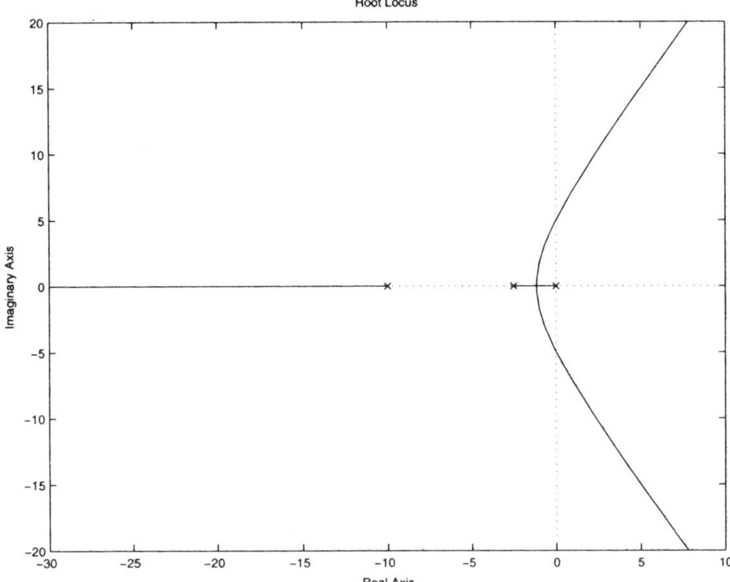

Figure 4.23. *Root locus of the system $T(s) = k\left[s\left(1 + 0.4s\right)\left(1 + 0.1s\right)\right]$*

$$= 180 - \arg\left(1.084\,4 - 1.7781j\right)k = 238.63,$$

$$\theta_2 = 180 - \arg \left.\frac{k\left(s + 1/2 - j\sqrt{3}/2\right)}{s\left(1 + 0.4s\right)\left(1 + 0.1s\right)}\right|_{s=-1/2-j\sqrt{3}/2}$$

$$= 180 - \left(1.084\,4 + 1.7781j\right)k = 121.38.$$

The system is stable $\forall k$

In practice, a zero is introduced together with a pole (in a proper system). The introduced pole must be located on the negative real axis far from the origin (corresponding to a very small time constant).

PROBLEM 4.24. Draw the root locus for a system having the following open-loop transfer function:

$$T(s) = \frac{k}{(s+1)\left(s^2 + 5s + 6\right)}.$$

SOLUTION 4.24. The open-loop poles are:

$$s = -1, \quad s = -3 \quad \text{and } s = -2.$$

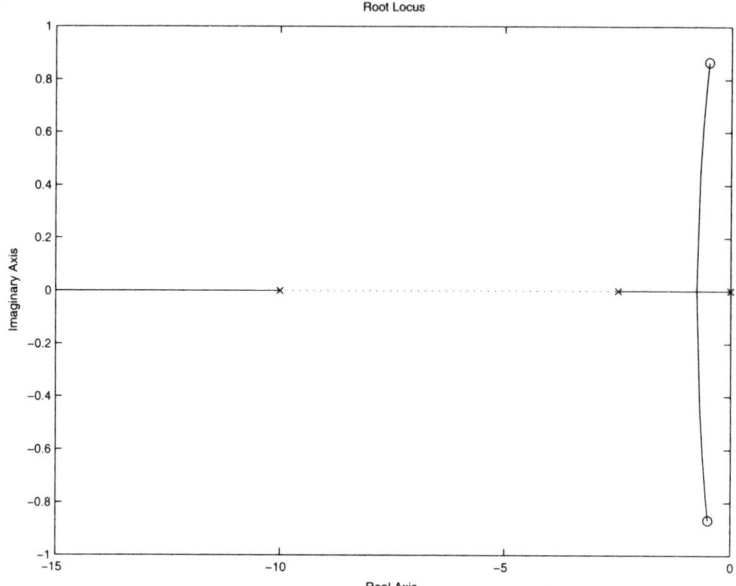

Figure 4.24. *Root locus of the system*
$$T\left(s\right) = k\left(s^2 + s + 1\right) / \left[s\left(1 + 0.4s\right)\left(1 + 0.1s\right)\right]$$

The unit segment $[-2, -1]$ and the semi-axis $]-\infty, -3]$ belong to the root locus. This root locus has three asymptotic directions, the positions of which are $\pi/3, \pi$ and $-\pi/3$. Their intersections with the real axis are given by $\delta = (-1 - 2 - 3)/3 = -2$. There certainly exists an intersection of the root locus with the real axis located in the segment $[-2, -1]$. This intersection is given by:

$$\sum_{i=1}^{3} \frac{1}{x - p_i} = \frac{1}{x+1} + \frac{1}{x+2} + \frac{1}{x+3} = \frac{3x^2 + 12x + 11}{(x+1)(x+2)(x+3)} = 0,$$

$$\implies x_1 = -2 + \frac{1}{3}\sqrt{3} \in [-2, -1], \quad x_2 = -2 - \frac{1}{3}\sqrt{3} \notin [-2, -1].$$

The solution x_2 has to be rejected. The intersection of the root locus with the imaginary axis is given by:

$$-j\omega^3 - 6\omega^2 + 11j\omega + 6 + k = 0,$$

$$k = 6\left(\omega^2 - 1\right), \quad \omega\left(11 - \omega^2\right) = 0 \implies \omega = \sqrt{11}, \quad k = 60.$$

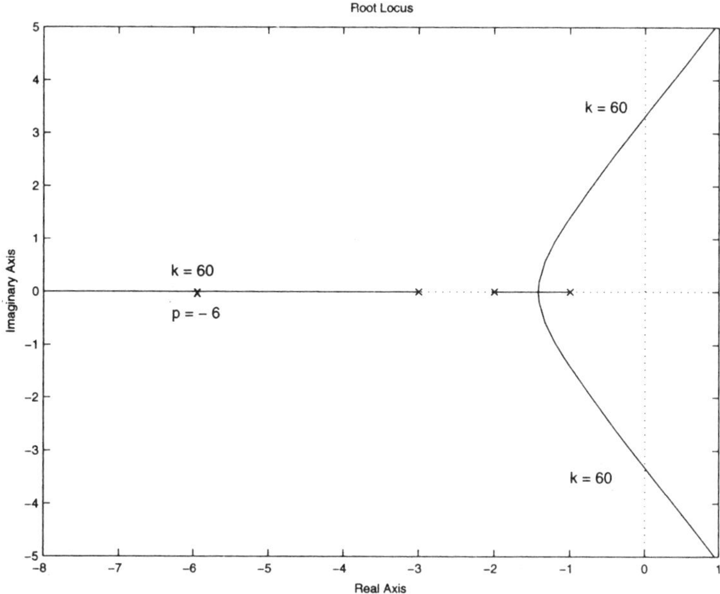

Figure 4.25. *Root locus of the system* $T(s) = k/\left[(s+1)\left(s^2 + 5s + 6\right)\right]$

The system is stable for $k < k_{\text{lim}} = 60$. This result is confirmed by the root locus, which is drawn in Figure 4.25. The third pole, corresponding to $k = 60$, is given by $p_3 = -6$.

PROBLEM 4.25. Draw the root locus for a system having the following open-loop transfer function:

$$\Gamma(s) = \frac{k}{s(1+s)(2+s)^2}.$$

SOLUTION 4.25. We advise the reader to represent N-fold multiple poles and zeros by N crosses and N circles, to avoid errors in the use of rule 3. The segment $[-1, 0]$ belongs to the root locus. There exist necessarily two intersections of the root locus with the real axis. The first of them is located on the segment $[-1, 0]$, and the second coincides with the double pole -2 of the open-loop transfer function. This root locus has four asymptotic directions; their positions are given by $\beta = \pi/4, 3\pi/4, -\pi/4$, and $-3\pi/4$. Their intersections with the real axis are located at $\delta = (0 - 1 - 2 - 2)/4 =$

-1.25. The intersection of the root locus with the real axis occurs at:

$$\sum_{i=1}^{4} \frac{1}{x - p_i} \;=\; \frac{1}{x} + \frac{1}{x+1} + \frac{2}{x+2} = \frac{4x^2 + 7x + 2}{x\,(x+1)\,(x+2)} = 0,$$

$$\Longrightarrow \quad x_1 = -\frac{7}{8} + \frac{1}{8}\sqrt{17}, \quad x_2 = -\frac{7}{8} - \frac{1}{8}\sqrt{17}.$$

The solution $x_2 \notin [-1, 0]$ and does not coincide with the double pole $p = -2$. It has to be rejected. We conclude that the root locus has two intersections with the real axis, located at the points of abscissa x_1 and -2.

The characteristic equation is:

$$s^4 + 5s^3 + 8s^2 + 4s + k = 0.$$

Using the Routh–Hurwitz criterion, we obtain:

$$\Delta_1 = |5|, \quad \Delta_2 = \begin{vmatrix} 5 & 1 \\ 4 & 8 \end{vmatrix} = 36 > 0,$$

$$\Delta_3 = \begin{vmatrix} 5 & 1 & 0 \\ 4 & 8 & 5 \\ 0 & k & 4 \end{vmatrix} = 144 - 25k = 0 \Longrightarrow k = k_{\mathrm{lim}} = 5.76.$$

The intersection of the root locus with the imaginary axis is given by:

$$\omega^4 - 5j\omega^3 - 8\omega^2 + 4j\omega + k = 0 \Rightarrow \omega = \frac{2}{\sqrt{5}},$$

$$k = 8\omega^2 - \omega^4 = k_{\mathrm{lim}} = 5.76.$$

For k_{lim}, the other two poles are equal to $-2.5 \pm j0.974$. The root locus is shown in Figure 4.26.

PROBLEM 4.26. Draw the root locus for a system having the following open-loop transfer function:

$$T\,(s) = \frac{k}{s\,(s^2 + 2s + 2)}.$$

SOLUTION 4.26. The poles of the open-loop transfer function are:

$$p_1 = 0, \quad p_2 = -1 + j \quad \text{and } p_2 = -1 - j.$$

The real semi-axis $\,]-\infty, 0]$ belongs to the root locus. The root locus has three asymptotic directions, the intersection of which with the real axis is located at $\delta =$

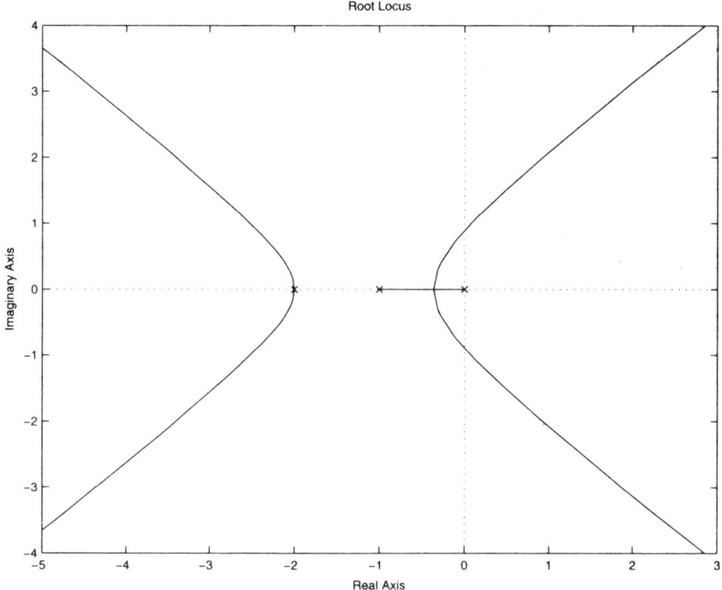

Figure 4.26. *Root locus of the system* $T(s) = k/\left[s(1+s)(2+s)^2\right]$

$(0 + -1 + j - 1 - j)/3 = -\dfrac{2}{3}$. Their positions with respect to the real axis are $\beta = \pi/3, \pi$ and $-\pi/3$. Now we determine the positions of the tangents to the root locus at the complex poles of the open-loop transfer function:

$$\theta_1 = \arg T'(-1+j) + 180 = \arg\left[\frac{k}{s(s+1+j)}\right]_{s=-1+j} + \pi = \frac{7}{4}\pi,$$

$$\theta_2 = \arg T'(-1-j) + 180 = \arg\left[\frac{k}{s(s+1-j)}\right]_{s=-1-j} + \pi = \frac{1}{4}\pi.$$

The intersections of the root locus with the imaginary axis are given by:

$$1 + \frac{k}{j\omega\left((j\omega)^2 + 2j\omega + 2\right)} = 0 \implies k = 2\omega^2, \omega^2 = 2,$$

which leads to:

$$\omega = \sqrt{2} \quad \text{and} \quad k = k_{\lim} = 4.$$

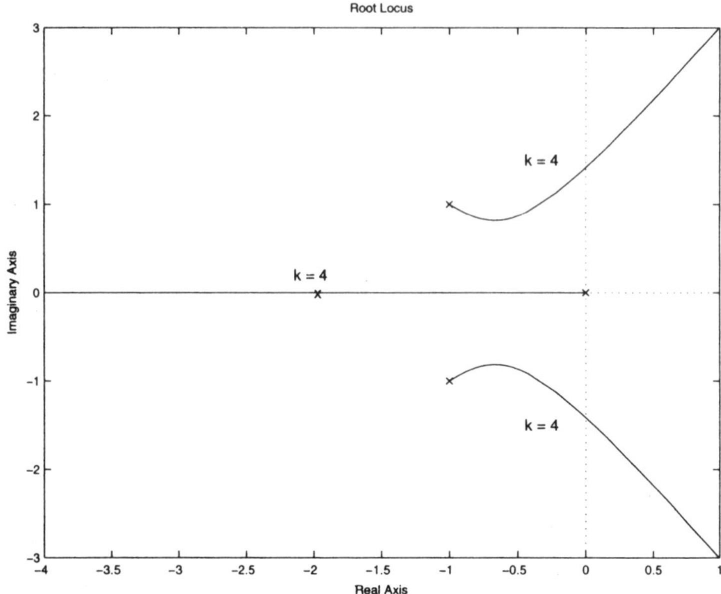

Figure 4.27. *Root locus of the system* $T(s) = k / \left[s \left(s^2 + 2s + 2 \right) \right]$

For $k = k_{\text{lim}}$, the third pole is given by $p = -2$. Figure 4.27 shows the root locus of this system. The system is stable for $k < 4$.

PROBLEM 4.27. 1. Draw the root locus of a unity-feedback system having the following open-loop transfer function:

$$T(s) = \frac{k(s+1)^2}{s^3(s^2 + 5s + 6)}.$$

2. We introduce, in the forward path, a system described by a transfer function $F_1(s) = 4(s + 1.5)/s$. Using the root locus, show the effect of this system on the process.

SOLUTION 4.27. 1. The poles and zeros of the open-loop transfer function are:

$$z_1 = -1, \quad z_2 = -1, \quad p_1 = 0, \quad p_2 = 0, \quad p_3 = 0, \quad p_4 = -2, \quad p_5 = -3.$$

The segments $[-1, 0]$ and $[-2, -1]$ and the semi-axis $]-\infty, -3]$ belong to the root locus. The number of asymptotic directions is 3. Their positions relative to the real axis are $\pi/3, \pi$ and $-\pi/3$. Their intersections with the real axis are $\delta = (-2 - 3 + 1 + 1) / 3 = -1$. Apart the triple pole, there are no intersections of the root locus with the

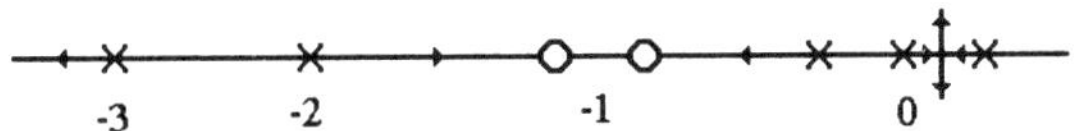

Figure 4.28. *Location of the real poles and zeros, and the segments of the real axis belonging to the root locus*

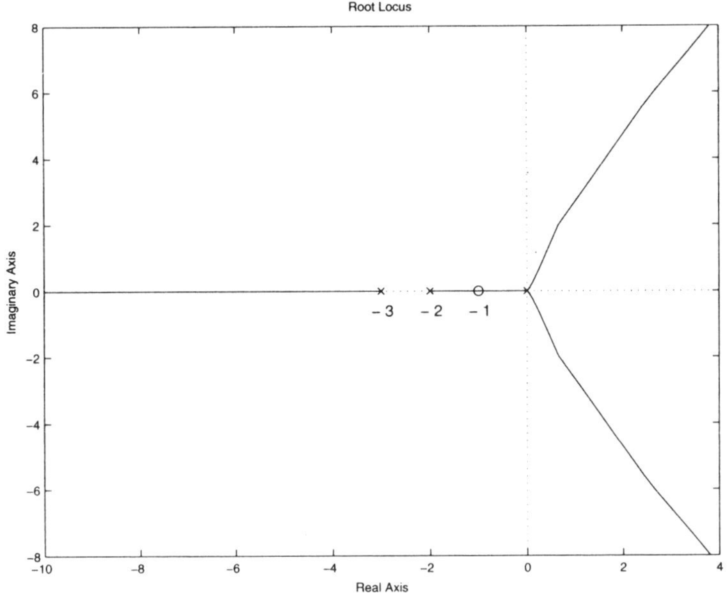

Figure 4.29. *Root locus of the system*
$$T\left(s\right) = k\left(s+1\right)^{2} / \left[s^{3}\left(s^{2}+5s+6\right)\right]$$

real and imaginary axes (see Figure 4.28, where the triple pole and the double zero are represented by three crosses and two circles, respectively). The root locus is plotted in Figure 4.29. We observe that the system is unstable $\forall k > 0$.

2. The open-loop transfer function of the modified system is:

$$T\left(s\right) = \frac{4k\left(s+1\right)^{2}\left(s+1.5\right)}{s^{4}\left(s^{2}+5s+6\right)},$$

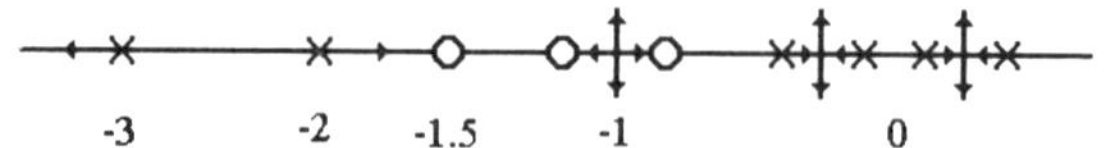

Figure 4.30. *Location of the real poles and zeros in Problem 4.27*

and the zeros and poles are given by:

$$z_1 = -1, \quad z_2 = -1, \quad z_3 = -1.5, \quad p_1 = p_2 = p_3 = p_4 = 0,$$
$$p_5 = -2, \quad p_6 = -3.$$

The number of asymptotic directions is $n - m = 6 - 3 = 3$. Their positions with respect to the real axis are $\beta = \pi/3$, π and $-\pi/3$. The intersection of the asymptotic directions is given by $\delta = (-2 - 3 + 1 + 1 + 1.5)/3 = -0.5$. The segment $[-2, -1.5]$ and the semi-segment $]-\infty, -3]$ belong to the root locus. There exist three intersections of the root locus with the real axis. Two of them are located at the quadruple pole, and the third is located at the double zero -1 (see Figure 4.30). The root locus certainly contains a closed curve. There exist three intersections of the root locus with the real axis. From one of the intersections located at the origin, the root locus tends towards the asymptotes at angles $\pi/3$ and $-\pi/3$, and from the other of these two intersections it goes to the third intersection, located at the double zero -1. The root locus is depicted in Figure 4.31.

PROBLEM 4.28. 1. Draw the root locus for a system having the following open-loop transfer function:

$$T(s) = \frac{k}{s(s^2 + 2s + 1)}, \quad k \in [0, +\infty[.$$

2. Determine the value of the static gain k for which this system is equivalent to a second-order system with a damping factor equal to $\zeta = \sqrt{2}/2$.

3. Analyze briefly the effect of the introduction of a lead compensator on the root locus.

SOLUTION 4.28. 1. The poles of the open-loop transfer function are:

$$s = 0, \quad s = -1 \quad \text{and} \quad s = -1.$$

The unit segment $[-1, 0]$ and the semi-axis $]-\infty, -1]$ belong to the root locus. There exists an intersection of the root locus with the real axis, located on the segment $[-1, 0]$

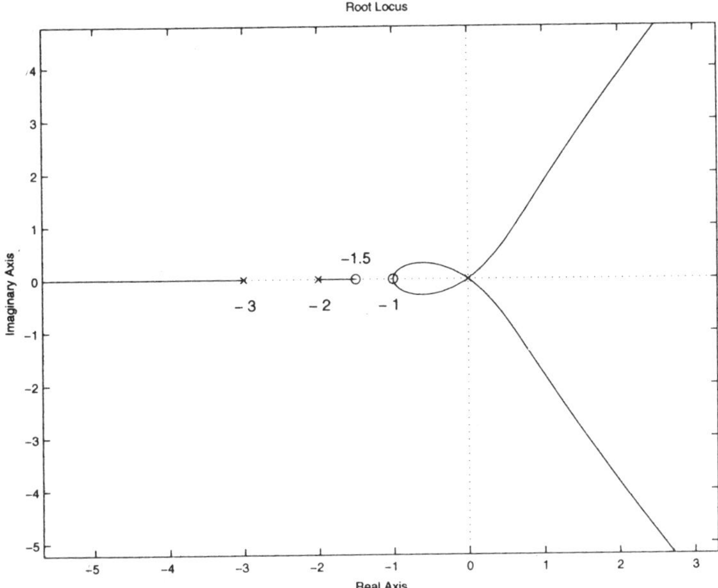

Figure 4.31. *Root locus of the system*
$$T\left(s\right) = 4k\left(s+1\right)^2\left(s+1.5\right)/\left[s^4\left(s^2+5s+6\right)\right]$$

(two departure points). This intersection is given by:

$$\sum_{i=1}^{n}\frac{1}{x-p_i} = 0, \quad \frac{1}{x}+\frac{2}{x+1} = 0 \implies x = -\frac{1}{3}.$$

The number of asymptotic directions is 3. The positions of these directions with respect to the real axis are $\beta = \pi/3, \pi$ and $-\pi/3$. Their intersections with the real axis correspond to $\delta = \left(-1-1\right)/3 = -2/3$. The intersection of the root locus with the imaginary axis is given by:

$$-j\omega^3 - 2\omega^2 + j\omega + k = 0 \implies k = 2\omega^2, \quad \omega\left(1-\omega^2\right) = 0,$$

$$\omega = 1, \quad k = 2.$$

The third pole, corresponding to $k = 2$, is equal to -2. The system is unstable for $k > 2$. The root locus is depicted in Figure 4.32.

2. Recall the canonical characteristic equation of a second-order system:

$$s^2 + 2\zeta\omega_n s + \omega_n^2 = 0,$$

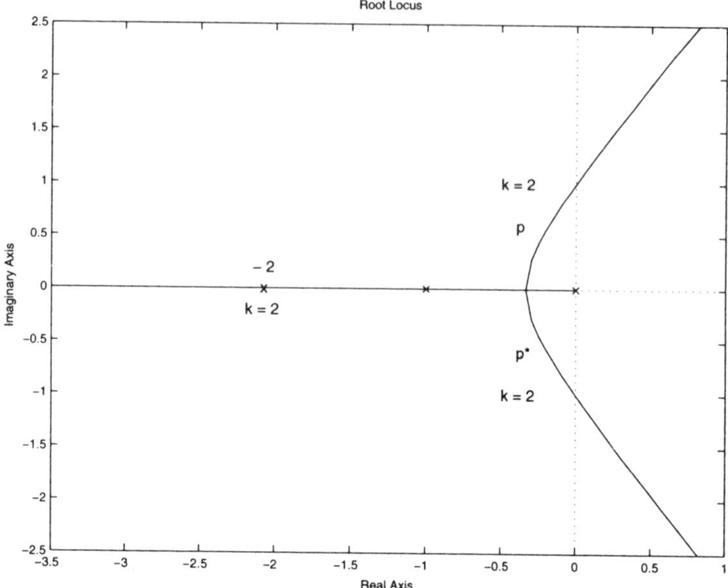

Figure 4.32. *Root locus of the system* $T(s) = k/\left[s\left(s^2 + 2s + 1\right)\right]$

where ζ and ω_n represent the damping factor (coefficient) and the natural frequency, respectively. The roots of this characteristic equation for $\zeta < 1$ are:

$$p_1 = -\zeta\omega_n + j\omega_n\sqrt{1 - \zeta^2}, \quad p_2 = p_1^* = -\zeta\omega_n - j\omega_n\sqrt{1 - \zeta^2}.$$

These roots are shown in Figure 4.33.

From this figure, using Pythagoras' theorem, we deduce that the damping factor is a measure of the angle Ψ, i.e.:

$$\sin\Psi = \frac{\zeta\omega_n}{\sqrt{\zeta^2\omega_n^2 + \omega_n^2\left(1 - \zeta^2\right)}} = \zeta.$$

Taking this result into account, it follows that $\Psi = 45°$. This problem can be solved analytically or geometrically. For the geometric solution, we draw the second bisectors. Their intersections with the root locus give the poles of the equivalent second-order system. Finally, we obtain:

$$p_1 = -0.29 + j0.29, \quad p_2 = p_1^* = -0.29 - j0.29.$$

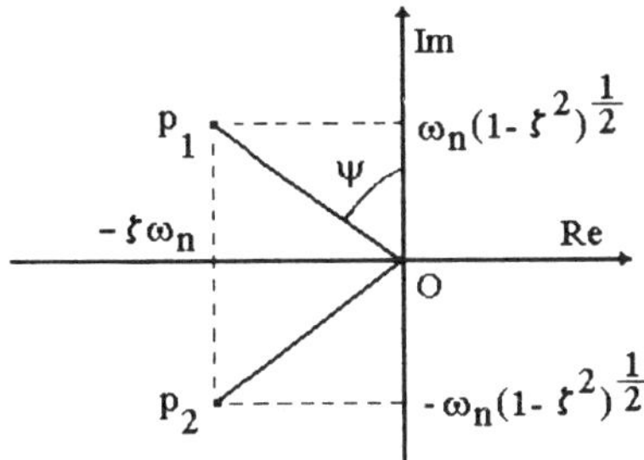

Figure 4.33. *Graphical representation of the poles of a second order system when the damping factor $\zeta > 1$*

Taking into account that $\Psi = 45°$, it follows that for the analytical solution, we have to look for solutions of the characteristic equation of the form:

$$p = \alpha \pm j\alpha, \quad \alpha < 0.$$

For $p = \alpha - j\alpha$, we obtain:

$$\frac{k}{(\alpha - j\alpha)\left((\alpha - j\alpha)^2 + 2(\alpha - j\alpha) + 1\right)} + 1 = 0,$$

$$\left[-2j\alpha^3 - 2\alpha^3 - 4j\alpha^2 + \alpha - j\alpha\right] + k = 0,$$

$$\alpha\left(1 - 2\alpha^2\right) + k = 0, \quad -\alpha\left(1 + 4\alpha + 2\alpha^2\right) = 0,$$

$$2\alpha^2 + 4\alpha + 1 = 0, \quad k = \alpha\left(2\alpha^2 - 1\right),$$

$$\alpha_1 = -1 - \sqrt{1/2} = -1.7071, \quad \alpha_2 = -1 + \sqrt{1/2} = -0.2929,$$

$$\alpha_1 \Longrightarrow k_1 = -8.246, \quad \alpha_2 \Longrightarrow k_2 = 0.2426.$$

Taking into account that $k \in [0, +\infty[$, we derive:

$$\alpha = -1 + \sqrt{\frac{1}{2}} = -0.2929 \quad \text{and} \quad k = 0.2426.$$

The pole of the equivalent second-order system[10] are:

$$p_1 = -0.2929 + j0.2929, \quad p_2 = p_1^* = -0.2929 - j0.2929.$$

We can obtain the same result from a graduated root locus.

3. The difference between the number of poles and the number of zeros remains unchanged. Therefore, the number of asymptotic directions remains unchanged. The intersection of the asymptotes with the real axis moves to the left. We obtain:

$$\delta = \frac{\sum\limits_{i=1}^{3} p_i - (1 - 1/a)/T}{n - m},$$

where $-1/T$ and $-1/aT$ are the pole and the zero, respectively, of lead compensator. $\sum\limits_{i=1}^{3} p_i$ represents the sum of the poles of the uncompensated system. The amplitude of the translation of this intersection increases with a. This translation to the left induces a translation of the root locus to the left. Therefore, the root locus of the compensated system has more portions to the left of the imaginary axis.

PROBLEM 4.29. The open-loop transfer functions of two systems are given by:

$$T_1(s) = (1 + \tau_1 s) Q_1(s), \quad T_2(s) = \frac{1}{(1 + \tau_2 s)} Q_2(s),$$

$$\tau_i \in [0, \infty[, \quad i = 1, 2.$$

Analyze the stability of these systems with respect to the time constants τ_1 and τ_2.

SOLUTION 4.29. For the first system, the closed-loop transfer function is given by:

$$\frac{Y(s)}{Y_r(s)} = F_1(s) = \frac{G_1(s)}{1 + T_1(s)} = \frac{G_1(s)}{1 + (1 + \tau_1 s) Q_1(s)}, \tag{4.21}$$

---

10. The relative error in the impulse response, which is defined as:

$$\frac{\left(\int\limits_{0}^{\infty} g_e^2(t)\, dt \right)^{\frac{1}{2}}}{\left(\int\limits_{0}^{\infty} g^2(t)\, dt \right)^{\frac{1}{2}}}$$

where $g_e(t)$ represents the difference between the impulse response of the model $g(t)$ and the impulse response of the reduced model, is used as a criterion in the model reduction problem. The integrals involved in the relative error in the impulse response can be computed using the algorithm presented in Appendix A (see also Problem 2.5).

$$F_1\left(s\right) = \frac{G_1\left(s\right)/\left[1 + Q_1\left(s\right)\right]}{1 + \tau_1 s Q_1\left(s\right)/\left[1 + Q_1\left(s\right)\right]} = \frac{G_1^*\left(s\right)}{1 + G_1^*\left(s\right) H_1^*\left(s\right)}, \tag{4.22}$$

where:

$$G_1^*\left(s\right) = \frac{G_1\left(s\right)}{1 + Q_1\left(s\right)}, \quad H_1^*\left(s\right) = \tau_1 s \frac{Q_1\left(s\right)}{G_1\left(s\right)},$$

since:

$$G_1^*\left(s\right) H_1^*\left(s\right) = \tau_1 s \frac{Q_1\left(s\right)}{1 + Q_1\left(s\right)}. \tag{4.23}$$

Systems (4.21) and (4.22) are equivalent. The characteristic equation is given by:

$$1 + \tau_1 \left(s \frac{Q_1\left(s\right)}{1 + Q_1\left(s\right)} \right) = 0.$$

On the basis of this expression, we may use the root locus approach to study the stability of this system with respect to the time constant τ_1.

For the second system, we obtain:

$$\frac{Y\left(s\right)}{Y_r\left(s\right)} = F_2\left(s\right) = \frac{G_2\left(s\right)}{1 + T_2\left(s\right)} = \frac{G_2\left(s\right)}{1 + Q_2\left(s\right)/\left(1 + \tau_2 s\right)}, \tag{4.24}$$

$$F_2\left(s\right) = \frac{G_2\left(s\right)\left(1 + \tau_2 s\right)/\left[1 + Q_2\left(s\right)\right]}{1 + \tau_2 s/\left[1 + Q_2\left(s\right)\right]} = \frac{G_2^*\left(s\right)}{1 + G_2^*\left(s\right) H_2^*\left(s\right)}, \tag{4.25}$$

where:

$$G_2^*\left(s\right) = \frac{G_2\left(s\right)\left(1 + \tau_2 s\right)}{1 + Q_2\left(s\right)}, \quad H_2^*\left(s\right) = \frac{\tau_2 s}{G_2\left(s\right)\left(1 + \tau_2 s\right)},$$

$$T_2^*\left(s\right) = \tau_2 \frac{s}{1 + Q_2\left(s\right)}.$$

Systems (4.24) and (4.25) are equivalent. Now, we can use the root locus approach to analyze the stability of this system with respect to the time constant τ_2.

PROBLEM 4.30. Analyze the effect of variation of the parameter τ_1 on the stability of a system for which the open-loop transfer function is given by:

$$T_1\left(s\right) = \left(1 + \tau_1 s\right) Q_1\left(s\right), \quad Q_1\left(s\right) = \frac{2s + 1}{3s^2 + s + 1}, \quad \tau_1 \in \left[0, \infty\right[.$$

SOLUTION 4.30. On the basis of the result obtained in the previous problem, the characteristic equation is given by:

$$1 + \tau_1 s \frac{Q_1\left(s\right)}{1 + Q_1\left(s\right)} = 1 + \tau_1 s \frac{2s + 1}{3s^2 + 3s + 2} = 0.$$

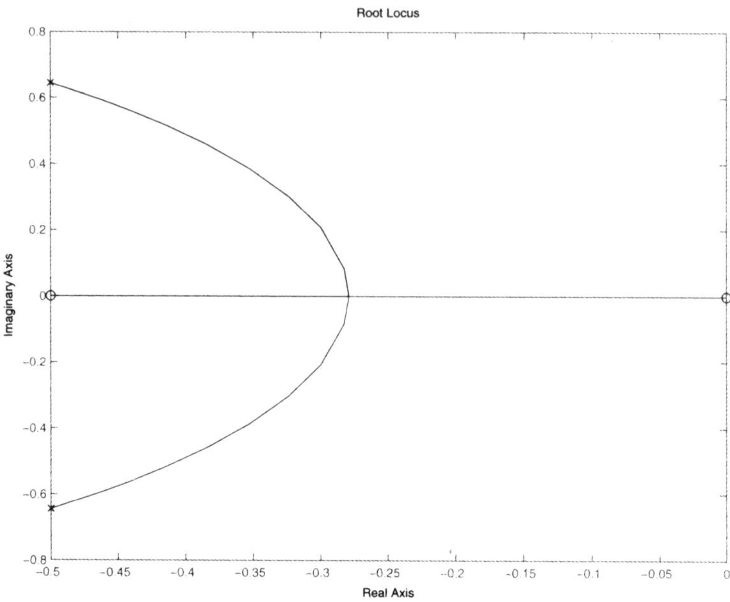

Figure 4.34. *Root locus associated with a varying time constant*

On the basis of this expression, we may use the root locus approach to study the stability of this system with respect to the time constant τ_1. The poles and zeros are given by:

$$z_1 = 0, \quad z_2 = -0.5, \quad p = -\frac{1}{2} + j\frac{1}{6}\sqrt{15} \quad \text{and } p^* = -\frac{1}{2} - j\frac{1}{6}\sqrt{15}.$$

The segment $[-1/2, 0]$ belongs to the root locus. Recall that the poles and zeros of the open-loop transfer function correspond to departure and arrival points, respectively. In this case we have two departure points (p and p^*) and two arrival points (z). Therefore, there exists an intersection of the root locus with the real axis. This intersection is located on the segment $[-1/2, 0]$ and is given by:

$$\sum_{i=1}^{n} \frac{1}{x - p_i} = \sum_{i=1}^{m} \frac{1}{x - z_i},$$

$$\frac{1}{x - \left(-1/2 + j\sqrt{15}/6\right)} + \frac{1}{x - \left(-1/2 - j\sqrt{15}/6\right)} = \frac{1}{x} + \frac{1}{x + 0.5},$$

$$0.5x^2 + \frac{4}{3}x + \frac{1}{3} = 0 \implies x_1 = -0.27925, \quad x_2 = -2.3874.$$

Since the solution is on the segment $[-1/2, 0]$, the solution x_2 has to be rejected. The remainder of the root locus consists of an arc connecting the complex poles. The root locus is plotted in Figure 4.34.

Let us now calculate the positions of the tangents to the complex poles with respect to the real axis (the angles of departure):

$$\theta_1 = \arg T'_p \left(-\frac{1}{2} + j\frac{1}{6}\sqrt{15} \right) + 180,$$

$$\theta_2 = \arg T'_{p^*} \left(-\frac{1}{2} - j\frac{1}{6}\sqrt{15} \right) + 180.$$

where:

$$T'_p(s) = s\frac{2s+1}{3\left(s+1/2+j\sqrt{15}/6\right)}, \quad T'_{p^*}(s) = s\frac{2s+1}{3\left(s+1/2-j\sqrt{15}/6\right)}.$$

It follows that:

$$T'_p(p) = -\frac{1}{6} + j\frac{1}{18}\sqrt{15}, \quad T'_{p^*}(p^*) = -\frac{1}{6} - j\frac{1}{18}\sqrt{15},$$

and

$$\theta_1 = 127.74, \quad \theta_2 = 232.26.$$

This system is stable $\forall \tau_1$.

Chapter 5

Regulation and PID Regulators

5.1. Introduction

Let us first consider a simple problem which illustrates the usefulness of feedback control in the presence of disturbances.

PROBLEM 5.1. Consider a system with a unit transfer function and a disturbed measured output $y(t)$, and the same system compensated with a proportional regulator (see Figure 5.1). For a given input $y_r(t)$, calculate the outputs of the open-loop and closed-loop systems.

SOLUTION 5.1. The output of the open-loop system is equal to:

$$y(t) = y_r(t) + d(t). \tag{5.1}$$

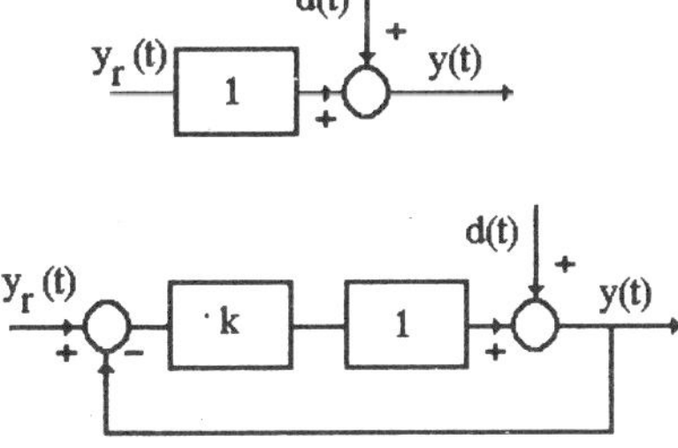

Figure 5.1. *Open-loop and closed-loop control*

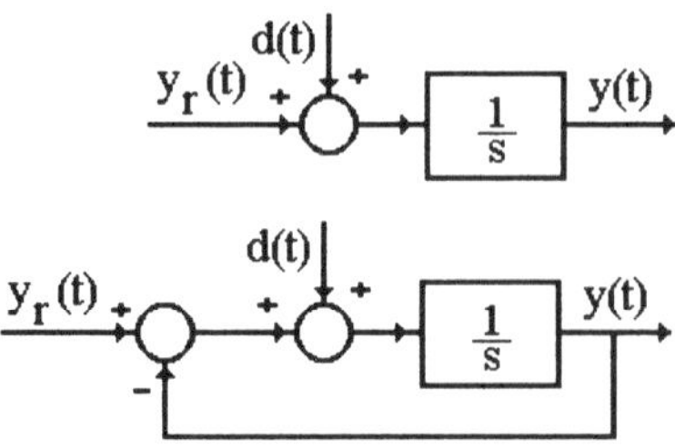

Figure 5.2. *Integrator and feedback*

The output of the closed-loop system is given by:

$$k\left(y_r\left(t\right) - y\left(t\right)\right) + d(t) = y\left(t\right),$$
$$y\left(t\right) = \frac{k}{1+k}y_r\left(t\right) + \frac{1}{1+k}d(t). \tag{5.2}$$

The effect of the disturbance on the system output has been reduced. For large values of the gain k, we derive:

$$\frac{k}{1+k} \simeq 1, \quad \frac{1}{1+k} \simeq 0,$$
$$y\left(t\right) \simeq y_r\left(t\right).$$

The effect of the disturbance on the measured output is now negligible.

PROBLEM 5.2. Consider the systems depicted in Figure 5.2. Analyze their behavior for $y_r\left(t\right) = 0$ and $d\left(t\right) = d_0$.

SOLUTION 5.2. For the open-loop system, we obtain:

$$Y\left(s\right) = \frac{1}{s}\left(Y_r\left(s\right) + D\left(s\right)\right) = \frac{d_0}{s^2}, \quad y\left(t\right) = d_0 t 1\left(t\right),$$

where $1\left(t\right)$ represents the unit step. The output increases with time. Therefore, the integrator will saturate.

For the closed-loop system, we derive:

$$\left[\left(Y_r\left(s\right) - Y\left(s\right)\right) + D\left(s\right)\right]\frac{1}{s} = Y\left(s\right),$$

$$Y(s) = \frac{1}{s+1}\left[Y_r(s) + D(s)\right] = \frac{d_0}{s(s+1)} = d_0\left[\frac{1}{s} - \frac{1}{s+1}\right],$$

$$y(t) = d_0\left(1 - \exp(-t)\right).$$

The closed-loop system prevents itself from driving the output to saturation.

PROBLEM 5.3. Study the effect of a proportional controller on first- and second-order systems.

SOLUTION 5.3. Let us first consider a first order system:

$$G(s) = \frac{k}{1 + Ts}.$$

The closed-loop transfer function is given by:

$$F(s) = \frac{G(s)R(s)}{1 + G(s)R(s)}, \quad R(s) = k_c,$$

$$F(s) = \frac{kk_c}{1 + kk_c + Ts} = \frac{kk_c/(1 + kk_c)}{1 + Ts/(1 + kk_c)}.$$

The static gain of the compensated system becomes equal to $kk_c/(1 + kk_c) < 1$ and tends to 1 when k_c tends to infinity. The time constant becomes equal to $T/(1 + kk_c)$. It decreases as k_c increases.

For a second-order system, we obtain:

$$F(s) = \frac{kk_c\omega_n^2}{s^2 + 2\zeta\omega_n s + \omega_n^2(1 + kk_c)}$$

$$= \frac{kk_c/(1 + kk_c)}{s^2/\left[\omega_n^2(1 + kk_c)\right] + 2\zeta s/\left[\omega_n(1 + kk_c)\right] + 1}.$$

The parameters of this second-order system are:

$$k_m = \frac{kk_c}{1 + kk_c}, \quad \omega_{nm}^2 = \omega_n^2(1 + kk_c), \quad \omega_{nm} = \omega_n\sqrt{1 + kk_c},$$

$$\zeta_m = \frac{\zeta}{\sqrt{1 + kk_c}}.$$

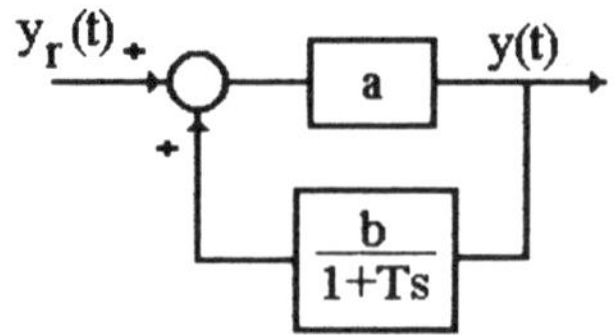

Figure 5.3. *Positive-feedback system*

The natural frequency ω_{nm} is increased and the static gain becomes less than one, i.e., $k_m < 1$. The damping factor ζ_m is decreased. If the initial system is overdamped ($\zeta > 1$), the introduction of a proportional controller can lead to an oscillatory system.

PROBLEM 5.4. Show that for $ab = 1$ and for $ab < 1$, the system depicted in Figure 5.3 is equivalent to a PI controller and to a lag compensator system, respectively.

SOLUTION 5.4. Let us denote by $G(s)$ and $H(s)$ the transfer functions of the forward and the feedback path, respectively. The transfer function of this system is equal to

$$F(s) = \frac{G(s)}{1 - G(s)H(s)} = \frac{a}{1 - ab/(1 + Ts)} = \frac{a(1 + Ts)}{1 - ab + Ts}.$$

For $ab = 1$, we obtain:

$$F(s) = a + \frac{a}{Ts},$$

which corresponds to the transfer function of a PI controller.

For $ab < 1$, we obtain:

$$F(s) = \frac{a}{1 - ab} \frac{1 + Ts}{1 + Ts/(1 - ab)},$$

which represents the transfer function of a lag compensator.

PROBLEM 5.5. Consider the unstable system:

$$G(s) = \frac{k}{(s + 2)(s - 1)},$$

and introduce a gain feedback compensator $H(s) = a$. For what values of a is the feedback system stable?

SOLUTION 5.5. The closed-loop transfer function is given by:

$$F\left(s\right) = \frac{G\left(s\right)}{1 + G\left(s\right)H\left(s\right)} = \frac{k}{\left(s+2\right)\left(s-1\right)+ka}.$$

From the Routh–Hurwitz criterion, the closed-loop system is stable for:

$$ka > 2, \quad a > \frac{2}{k}.$$

In what follows we shall present two problems related to the design of a closed-loop transfer function which fulfils the control specifications.

5.2. Direct design

PROBLEM 5.6. Determine the parameters of a stable second-order system which fulfils the following conditions:

1. $\varepsilon\left(\infty\right) = 0$ for $y_r\left(t\right) = 1\left(t\right)$ (unit step);

2. $\varepsilon\left(\infty\right) < 0.33$ for $y_r\left(t\right) = t$ (unit ramp);

3. overshoot $(OS) < 20\%$;

4. the frequency of free oscillation must satisfy $2.5 < \omega_n\sqrt{1-\zeta^2} < 3$.

SOLUTION 5.6. 1. The transfer function of a second-order system is given by:

$$F\left(s\right) = \frac{k}{1 + \dfrac{2\zeta}{\omega_n}s + \dfrac{1}{\omega_n^2}s^2} = \frac{k^*}{\omega_n^2 + 2\zeta\omega_n s + s^2}, \quad k^* = k\omega_n^2.$$

If the static gain is equal to 1, the steady-state error corresponding to a unit step will be equal to zero ($\varepsilon\left(\infty\right) = y_r\left(\infty\right) - y\left(\infty\right)$). We obtain:

$$\Xi\left(s\right) = Y_r\left(s\right) - Y\left(s\right) = \left[1 - F\left(s\right)\right]Y_r\left(s\right) = \frac{s^2 + 2\zeta\omega_n s + \omega_n^2\left(1-k\right)}{s^2 + 2\zeta\omega_n s + \omega_n^2}\frac{1}{s}$$

and

$$\varepsilon\left(\infty\right) = \lim_{s\to\infty} s\frac{s^2 + 2\zeta\omega_n s + \omega_n^2\left(1-k\right)}{s^2 + 2\zeta\omega_n s + \omega_n^2}\frac{1}{s} = \omega_n^2\left(1-k\right) = 0 \Longrightarrow k = 1.$$

2. For a ramp, the Laplace transform of the system output[1] is:

$$Y(s) \; = \; \frac{k}{s^2\left(1 + 2\zeta s/\omega_n + s^2/\omega_n^2\right)}$$

$$= -k\frac{2\zeta}{\omega_n}\frac{1}{s} + \frac{k}{s^2} + \frac{\left(2\zeta s/\omega_n\right)^2 - 1/\omega_n^2 + 2\zeta s/\omega_n^3}{1 + 2\zeta s/\omega_n + s^2/\omega_n^2}.$$

For a static gain equal to one, we derive:

$$Y(s) = -\frac{2\zeta}{\omega_n}\frac{1}{s} + \frac{1}{s^2} + \frac{\left(2\zeta/\omega_n\right)^2 - 1/\omega_n^2 + 2\zeta s/\omega_n^3}{1 + 2\zeta s/\omega_n + s^2/\omega_n^2}$$

The output is:

$$y(t) = -\frac{2\zeta}{\omega_n}1(t) + t + v(t).$$

As the second-order system is stable, $v(t)$ contains terms of the form $\exp\left(-\delta t\right)$, $\delta > 0$, tending to zero as $t \to \infty$. The error is given[2] by:

$$y_r(t) - y(t) = t - \left(-\frac{2\zeta}{\omega_n}1(t) + t + v(t)\right) = \frac{2\zeta}{\omega_n}1(t) - v(t)$$

and the steady-state error is equal to:

$$\varepsilon(\infty) = \frac{2\zeta}{\omega_n}. \tag{5.3}$$

From the preceding arguments, the first condition yields:

$$k = 1, \quad k^* = \omega_n^2. \tag{5.4}$$

---

1. The partial fraction expansion of:

$$Y(s) = \frac{k}{s^2\left(1 + as + bs^2\right)}$$

is:

$$Y(s) = -k\frac{a}{s} + \frac{k}{s^2} + k\frac{a^2 - b + abs}{1 + as + bs^2}.$$

2. The Laplace transform of the error, and the steady-state error, may be calculated directly as follows:

$$\Xi(s) \; = \; Y_r(s) - Y(s) = \left(1 - F(s)\right)Y_r(s),$$

$$\varepsilon(\infty) \; = \; \lim_{s \to 0} s\left(1 - F(s)\right)Y_r(s).$$

The second condition implies:

$$\frac{2\zeta}{\omega_n} < 0.33. \tag{5.5}$$

3. Recall the expression for the overshoot as a function of the damping factor:

$$OS = \exp\left(-\frac{\pi\zeta}{\sqrt{1 - \zeta^2}}\right).$$

To have $OS < 20\%$, the damping factor must satisfy the condition:

$$\zeta > 0.45. \tag{5.6}$$

4. The above constraint leads to:

$$2.5 < \omega_n \sqrt{1 - \zeta^2} < 3. \tag{5.7}$$

For $\zeta = 0.5$ and $2\zeta/\omega_n = 0.32 < 0.33$, we obtain:

$$\frac{2\zeta}{\omega_n} = \frac{1}{\omega_n} = 0.32 \implies \omega_n = \frac{1}{0.32} = 3.12$$

and

$$\omega_n \sqrt{1 - \zeta^2} = 3.12\sqrt{1 - (0.5)^2} = 2.702.$$

The second-order system:

$$F(s) = \frac{1}{1 + 0.32s + 0.10s^2}$$

satisfies the constraints (5.4), (5.5), (5.6) and (5.7).

The results above may be used to derive a controller $R(s)$ for a process $G(s)$ where the control objective is specified by the conditions 1–4. The controller is given by:

$$\frac{G(s)\,R(s)}{1 + G(s)\,R(s)} = F(s) \Rightarrow R(s) = \frac{F(s)}{G(s)\,(1 - F(s))}.$$

PROBLEM 5.7. Consider the following control objective:

(a) $\varepsilon(\infty) = 0$ for $y_r(t) = 1(t)$;

(b) $\varepsilon(\infty) < 0.2$ for $y_r(t) = t$;

(c) overshoot $OS \leq 15\%$;

(d) the frequency of free oscillations satisfy $2.5 < \omega_n \sqrt{1 - \zeta^2} < 3$.

1. Does a second-order stable system fulfill these conditions?

2. If the answer is no, include a zero in a second-order system.

SOLUTION 5.7. 1. On the basis of the solution of the previous problem, the conditions (b) and (c) are fulfilled if:

$$\frac{2\zeta}{\omega_n} < 0.2, \quad \zeta > 0.5,$$

which yields:

$$\omega_n > 10\zeta > 5.$$

For $\omega_n = 0.5$, from condition (d), we derive:

$$\omega_n \sqrt{1 - \zeta^2} = 5\sqrt{1 - (0.5)^2} = 4.3301.$$

Therefore, this condition is not fulfilled. The conditions (a), (b), (c) and (d) are not compatible.

2. Consider the following system:

$$F(s) = \frac{k(z + s)}{1 + 2\zeta s/\omega_n + s^2/\omega_n^2}, \quad z > 0,$$

where $-z$ represents the zero of this system.

(a) The Laplace transform of the step response of this system is given by:

$$Y(s) = \frac{1}{s} \frac{k(z + s)}{1 + 2\zeta s/\omega_n + s^2/\omega_n^2},$$

for which a partial fraction expansion[3] yields:

$$Y(s) = k\frac{z}{s} - k\frac{-1 + 2z\zeta/\omega_n + z/\omega_n^2}{1 + 2\zeta s/\omega_n + s^2/\omega_n^2}.$$

---

3. The partial fraction expansions of:

$$\frac{k(a + s)}{s(1 + bs + cs^2)}, \quad \frac{k(a + s)}{s^2(1 + bs + cs^2)}$$

are:

$$\frac{k(a + s)}{s(1 + bs + cs^2)} = k\frac{a}{s} - k\frac{-1 + ab + acs}{1 + bs + cs^2},$$

$$\frac{k(a + s)}{s^2(1 + bs + cs^2)} = -k\frac{-1 + ab}{s} + k\frac{a}{s^2} + k\frac{-b + ab^2 - ac + c(ab - 1)s}{1 + bs + cs^2}.$$

The inverse Laplace transform is given by:

$$y(t) = kz1(t) + v(t),$$

where $v(t)$ tends to zero as $t \to \infty$. The error is equal to:

$$\varepsilon(t) = y_r(t) - y(t) = (1 - kz)1(t) - v(t),$$

and $\varepsilon(\infty) = 0$ if:

$$kz = 1 \Rightarrow z = \frac{1}{k}.$$

(b) The Laplace transform of the response of this system to a ramp input is given by:

$$Y(s) = \frac{1}{s^2} \frac{k(z+s)}{1 + 2\zeta s/\omega_n + s^2/\omega_n^2}.$$

Its partial fraction expansion is given by:

$$\frac{1}{s^2} \frac{k(z+s)}{1 + 2\zeta s/\omega_n + s^2/\omega_n^2}$$

$$= -k\frac{-1 + 2z\zeta/\omega_n}{s} + k\frac{z}{s^2} + \frac{-2\zeta/\omega_n + z(2\zeta/\omega_n)^2 - z/\omega_n^2 - (1/\omega_n^2 - 2z\zeta/\omega_n^3)s}{1 + 2\zeta s/\omega_n + s^2/\omega_n^2},$$

which leads to the error:

$$\varepsilon(t) = y_r(t) - y(t) = t + k\left(-1 + z\frac{2\zeta}{\omega_n}\right)1(t) - t - v(t)$$

$$= \left(\frac{2\zeta}{\omega_n} - \frac{1}{z}\right)1(t) - v(t),$$

where $v(t) \to 0$ for $t \to \infty$ and $kz = 1$. Therefore, the steady-state error associated with a ramp input is equal to:

$$\varepsilon(\infty) = \frac{2\zeta}{\omega_n} - \frac{1}{z}. \tag{5.8}$$

From Equation (5.3), we deduce that $\varepsilon(\infty)$ has decreased.

(c) For a damping factor equal to $\zeta = 0.6$, the overshoot is less than 15%.

(d) The condition on the frequency of free oscillations,

$$2.5 < 0.8\omega_n < 3,$$

may be satisfied for $\omega_n = 3.2$. Equation (5.8) yields:

$$\varepsilon\left(\infty\right) = \frac{2\left(0.6\right)}{3.2} - \frac{1}{z} < 0.2,$$

which implies:

$$z < 10.$$

It is clear that in practice (because the magnitude increases and the system $(z + s)$ is not proper), the transfer function of the second-order system will be modified as follows:

$$F\left(s\right) = \frac{k}{\left(1 + 2\zeta s/\omega_n + s^2/\omega_n^2\right)} \frac{\left(z + s\right)}{\left(s - p_1\right)\left(s - p_2\right)}.$$

In this case, the Laplace transform of the error induced by a unit step is given by:

$$\Xi\left(s\right) = \left(1 - F\left(s\right)\right)\frac{1}{s}$$
$$= \frac{\left(1 + 2\zeta s/\omega_n + s^2/\omega_n^2\right)\left(s - p_1\right)\left(s - p_2\right) - k\left(z + s\right)}{\left(1 + 2\zeta s/\omega_n + s^2/\omega_n^2\right)\left(s - p_1\right)\left(s - p_2\right)}\frac{1}{s}$$

and the steady-state error is equal to:

$$\begin{aligned}
\varepsilon\left(\infty\right) &= \lim_{s \to 0} s\frac{\left(1 + 2\zeta s/\omega_n + s^2/\omega_n^2\right)\left(s - p_1\right)\left(s - p_2\right) - k\left(z + s\right)}{\left(1 + 2\zeta s/\omega_n + s^2/\omega_n^2\right)\left(s - p_1\right)\left(s - p_2\right)}\frac{1}{s} \\
&= \frac{p_1 p_2 - kz}{p_1 p_2}.
\end{aligned}$$

This error is equal to zero for:

$$kz = p_1 p_2. \tag{5.9}$$

The error due to a ramp input is given by:

$$\Xi\left(s\right) = \left(1 - F\left(s\right)\right)\frac{1}{s^2}$$
$$= \frac{\left(1 + 2\zeta s/\omega_n + s^2/\omega_n^2\right)\left(s - p_1\right)\left(s - p_2\right) - k\left(z + s\right)}{\left(1 + 2\zeta s/\omega_n + s^2/\omega_n^2\right)\left(s - p_1\right)\left(s - p_2\right)}\frac{1}{s^2}$$
$$= \frac{s\left(2\zeta/\omega_n + s/\omega_n^2\right)\left(s - p_1\right)\left(s - p_2\right) + s^2 - \left(p_1 + p_2 + k\right)s + p_1 p_2 - kz}{\left(1 + 2\zeta s/\omega_n + s^2/\omega_n^2\right)\left(s - p_1\right)\left(s - p_2\right)}\frac{1}{s^2}.$$

On the basis of Equation (5.9), we derive:

$$\Xi\left(s\right) = \frac{s\left(2\zeta/\omega_n + s/\omega_n^2\right)\left(s - p_1\right)\left(s - p_2\right) + s^2 - \left(p_1 + p_2 + k\right)s}{\left(1 + 2\zeta s/\omega_n + s^2/\omega_n^2\right)\left(s - p_1\right)\left(s - p_2\right)}\frac{1}{s^2}.$$

The steady-state error induced by a ramp input is equal to:

$$\varepsilon\left(\infty\right) = \lim_{s\to 0} s\,\frac{s\left(2\zeta/\omega_n + s/\omega_n^2\right)\left(s - p_1\right)\left(s - p_2\right) + s^2 - \left(p_1 + p_2 + k\right)s}{\left(1 + 2\zeta s/\omega_n + s^2/\omega_n^2\right)\left(s - p_1\right)\left(s - p_2\right)}\,\frac{1}{s^2},$$

$$\varepsilon\left(\infty\right) = \frac{2\zeta}{\omega_n} - \frac{1}{z} - \left(\frac{1}{p_1} + \frac{1}{p_2}\right).$$

Observe that this result can also be obtained from the partial fraction expansion of the Laplace transform of the output.

For $\zeta = 0.6$ and $\omega_n = 3.2$, we obtain:

$$\frac{2\zeta}{\omega_n} = 0.375, \quad \text{and } 2.5 < \omega_n\sqrt{1 - \zeta^2} = 2.56 < 3.$$

If we choose z to be equal to 2.5, and:

$$0.375 - 0.4 - \left(\frac{1}{p_1} + \frac{1}{p_2}\right) = -0.025 - \left(\frac{1}{p_1} + \frac{1}{p_2}\right) < 0.2,$$

$$-\left(\frac{1}{p_1} + \frac{1}{p_2}\right) < 0.2,$$

condition (b) is fulfilled. For example, p_1 and p_2 may be selected as follows:

$$p_1 = -10 \quad \text{and } p_2 = -20.$$

From Equation (5.9), we derive the static gain:

$$k = \frac{p_1 p_2}{z} = \frac{200}{2.5} = 80.$$

Finally, the system:

$$F\left(s\right) = \frac{80\left(s + 2.5\right)}{\left(1 + 0.375s + 0.09765s^2\right)\left(s + 10\right)\left(s + 20\right)}$$

fulfills the conditions related to the control objective.

PROBLEM 5.8. Consider a system with the following transfer function[4]:

$$G\left(s\right) = G_1\left(s\right)\exp\left(-\tau s\right), \quad G_1\left(s\right) = \frac{k}{1 + Ts}.$$

---

4. In process industries, first-order systems (with a dominant time constant plus a static gain) and a dead time are commonly used as models.

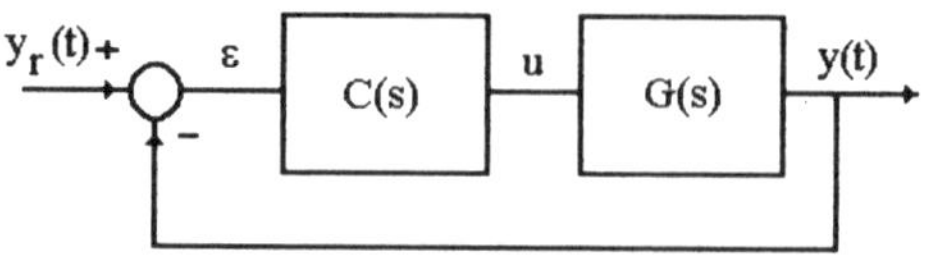

Figure 5.4. *Closed-loop control system including time delay*

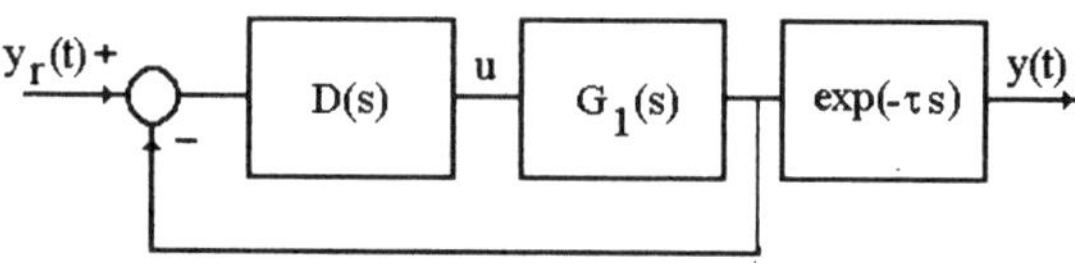

Figure 5.5. *Closed-loop control system connected to a time delay*

1. Draw its step response for $k = 2$, $T = 1\,s$ and $\tau = 1\,s$.

2. We now modify the structure of this system by the addition, in series, of a system with transfer function $C(s)$ as shown in Figure 5.4. Show that the resulting system is equivalent to the system depicted in Figure 5.5, where $C(s)$ is related to $D(s)$, $G_1(s)$ and $\exp(-\tau s)$ by a relation to be determined.

3. On the basis of the system depicted in Figure 5.5, determine an expression for $D(s)$ such that the global system is equivalent to a system with a time constant λT in series with a time delay $\exp(-\tau s)$.

4. Show that $C(s)$ can be realized by the diagram given in Figure 5.6, and determine $D_1(s)$ for $D_2(s) = 1/\lambda$.

SOLUTION 5.8. 1. The step response is given by:

$$Y(s) = \frac{k}{1 + Ts}\exp(-\tau s)\,Y_r(s), \quad Y_r(s) = \frac{1}{s},$$
$$Y(s) = \frac{k}{s(1 + Ts)}\exp(-\tau s). \tag{5.10}$$

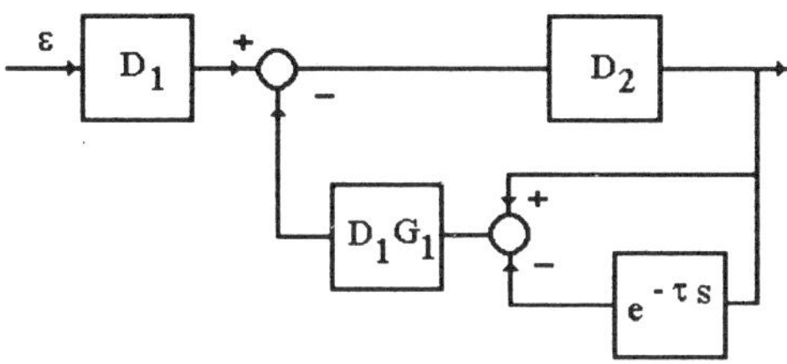

Figure 5.6. *Diagram of a compensator*

Consider the Laplace transform:

$$Y^* (s) = \frac{k}{1 + Ts},$$

which yields:

$$y^* (t) = L^{-1} \{Y^* (s)\} = k \left[1 - \exp\left(-\frac{t}{T}\right)\right].$$

Then, from Equation (5.10), we obtain:

$$y (t) = \mathcal{L}^{-1} \{Y (s)\} = y^* (t - \tau).$$

Here, $y (t)$ is derived from $y^* (t)$ by moving the latter along the time axis (a translation of offset τ). For $k = 2$ and $\tau = 1\ s$, the step response is plotted in Figure 5.7.

2. The equivalence between the systems depicted in Figures 5.4 and 5.5 leads to the equality of their transfer functions:

$$\frac{C (s) G (s)}{1 + C (s) G (s)} = \frac{D (s) G_1 (s)}{1 + D (s) G_1 (s)} \exp (-\tau s),$$

$$C (s) G (s) [1 + D (s) G_1 (s)] = D (s) G_1 (s) \exp (-\tau s) [1 + C (s) G (s)],$$

$$C (s) [1 + D (s) G_1 (s) (1 - \exp (-\tau s))] = D (s),$$

$$C (s) = \frac{D (s)}{1 + D (s) G_1 (s) [1 - \exp (-\tau s)]}. \tag{5.11}$$

3. The transfer function of a system with a time constant λT in series with a time delay $\exp (-\tau s)$ is:

$$\frac{k}{1 + \lambda Ts} \exp (-\tau s).$$

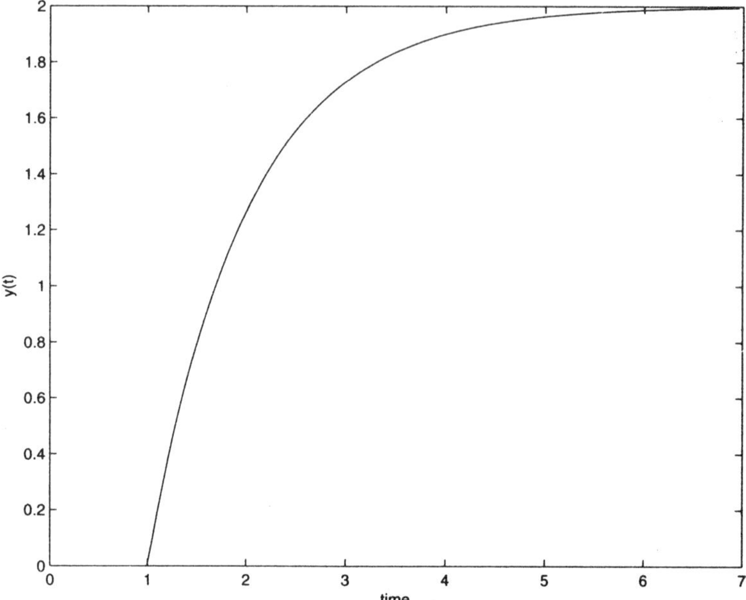

Figure 5.7. *Step response of a first-order system with transport delay*

The transfer function of the system depicted in Figure 5.5 is:

$$\frac{D\left(s\right)G_1\left(s\right)}{1+D\left(s\right)G_1\left(s\right)}\exp\left(-\tau s\right).$$

The equivalence between these systems leads to:

$$\frac{D\left(s\right)G_1\left(s\right)}{1+D\left(s\right)G_1\left(s\right)}\exp\left(-\tau s\right)=\frac{k}{1+\lambda Ts}\exp\left(-\tau s\right),$$

$$D\left(s\right)=\frac{1+Ts}{1-k+\lambda Ts}.$$

Observe that for $\lambda < 1$ the regulator described here reduces the process time response by a factor $1/\lambda$, and the time delay remains unchanged. To reduce the time delay it is necessary to design a new process with a lower residence time (see Appendix A).

4. The transfer function of the system represented in Figure 5.6 is:

$$\frac{D_1\left(s\right)D_2\left(s\right)}{1+D_1\left(s\right)D_2\left(s\right)G_1\left(s\right)\left(1-\exp\left(-\tau s\right)\right)}. \tag{5.12}$$

From Equations (5.11) and (5.12), we derive:

$$D\left(s\right) = D_1\left(s\right) D_2\left(s\right),$$

and then we obtain:

$$D_1\left(s\right) = \frac{D\left(s\right)}{D_2\left(s\right)} = \lambda D\left(s\right) = \frac{\lambda\left(1 + Ts\right)}{1 - k + \lambda Ts}.$$

The remainder of this chapter is devoted to PID tuning methods.

5.3. PID tuning

"PID" stands for "proportional plus integral plus derivative". A PID controller models, in some sense, the reaction of an operator when he/she observes a difference between the desired and the measured outputs. The operator's reaction may be of the following types.

1) *More or less strong with respect to the error. This reaction may be modeled by:*

$$k_c \varepsilon\left(t\right), \quad \text{proportional effect.}$$

2) *Based on the operator's experience. This experience is related to the period of time during which the operator has observed the process. It corresponds to an accumulation of information. Recall that in Chapter 1 we observed that the accumulation of water in a tank can be modeled by an integral. Therefore, this type of reaction of the operator may be modeled by:*

$$\frac{1}{\tau_i} \int y(t)\, dt.$$

As the desired output is a priori known, we may write this integral in the following form:

$$\frac{1}{\tau_i} \int \left(y_r(t) - y(t)\right) dt = \frac{1}{\tau_i} \int \varepsilon(t)\, dt, \quad \text{integral effect.}$$

3) *A function of the operator's ability to predict the future behavior of the process. In the deterministic case, the derivative gives an indication of the future evolution of a given function as its argument increases. As the derivative of the desired output is a priori known, this behavior may therefore be modeled by:*

$$\tau_d \frac{d}{dt} \varepsilon(t), \quad \text{derivative effect.}$$

The *proportional effect*, when increased, leads to faster control. It can be used for any stable system. If the system is underdamped, one may add a proportional controller to decrease the rise time (the overshoot will increase). Observe, however, that measurement noise causes loss of performance, and it may occur that the controller output exceeds the plant capacity, leading to control saturation. In the case of the *integral effect*, the control action is proportional to the accumulated error over time. This action induces zero steady-state error. The bandwidth[5] and the argument decrease. If an integral action is to be added to a well-tuned proportional controller, a reduction in the proportional gain is necessary in order to keep the system in a stable operating state. The *derivative effect* which is proportional to the derivative of the error introduces a predictive action. It must not be used in the presence of high-frequency noise. It is commonly used for temperature control because of the presence of low-frequency noise. It increases the bandwidth[6] and the argument (phase angle). The derivative effect enhance the transient response. The integral and derivative effects are rotation transformations which preserve scalar products. The goal of the integral effect is to rotate a given point by an angle of $-\pi/2$, while the derivative effect rotates a point by an angle of $\pi/2$.

$$M \xrightarrow{rotation\ \pi/2} M', \quad M \xrightarrow{rotation\ -\pi/2} M''.$$

Figure 5.8 shows graphically the integral and derivative effects. Let $s = x + jy$, $s' = x' + jy'$ and $s'' = x'' + jy''$ be the complex numbers associated with the points $M(x, y)$, $M'(x', y')$ and $M''(x'', y'')$, respectively. We derive:

$$s' = js, \quad s'' = -js,$$

$$\arg s' = \arg s + \arg j = \arg s + \frac{\pi}{2},$$

$$\arg s'' = \arg s + \arg(-j) = \arg s - \frac{\pi}{2}.$$

Observe that in order for the PID transfer function to be proper, a high-frequency pole has to be added in practice. This partly explains the existence of many PID parameterizations. Many industrial processes are under PID control. According to

---

5. The bandwidth of a system corresponds to a frequency range over which the magnitude ratio of the output to the input does not vary by more than $-3\,dB$. In other words, harmonic inputs with frequencies less than ω_b are practically not attenuated by the system, while sinusoidal inputs of high frequency ($\omega > \omega_b$) are attenuated by a coefficient greater than or equal to 0.7 ($20\log(0.7) \simeq 20\log(\sqrt{2}/2) \simeq -3\,dB$).

6. The magnitudes of the derivative and integral effects are given by $20\log\omega$ and $-20\log\omega$, respectively. They correspond to lines of slopes equal to $20\,dB/decade$ and $-20\,dB/decade$, respectively.

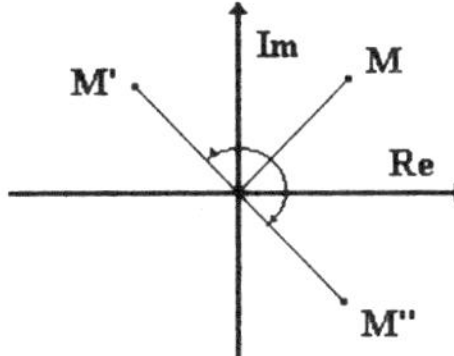

Figure 5.8. *Graphical representation of the effects of both integral and derivative control*

Koivo and Tanttu [KOI 91], there exist perhaps only 5% to 10% of industrial systems that cannot be controlled by a PID controller.

There exist a plethora of tuning rules, which are based on different model structures and different control objectives. There exists also a set of PID parameterizations. Goodwin and co-authors [GOO 91] rightly mention that caution must be exercised when applying PID tuning rules. For example, the same PID parameterization may be called "parallel", "ideal parallel", "noninteracting", etc. The PID settings that work for a process equipped with a controller designed by manufacturer A may not work the the same process if the PID controller is replaced by another designed by manufacturer B. The control objective must correspond to the criterion on which the PID tuning technique is based. For example, the two criteria:

$$\int_0^\infty \varepsilon(t)\,dt \quad \text{and} \quad \int_0^\infty |\varepsilon(t)|\,dt,$$

which are different ($\int_0^\infty \varepsilon(t)\,dt \leq \int_0^\infty |\varepsilon(t)|\,dt$), will lead to different PID settings. The second criterion takes more account of oscillations than does the first one. The first criterion takes into account mainly the change which follows the application of a step input. An important set of PID tuning techniques are presented in [ODW 03, KOI 91]. PID tuning can be considered as a tailoring of the Nyquist diagram (in the frequency domain) in order to achieve the desired specifications.

Unfortunately, there is no ready-made machinery that the reader can put a control problem into to produce the values of the PID design parameters by turning the handle, and there exists no tuning technique that is well adapted to all processes. We can compare the job of a PID designer to an operation done by an ophthalmic surgeon in order to correct the curvature of the eyes of a person with myopia with a laser. Alternatively, in a thermal power plant, it may happens that the turbine is damaged, and it

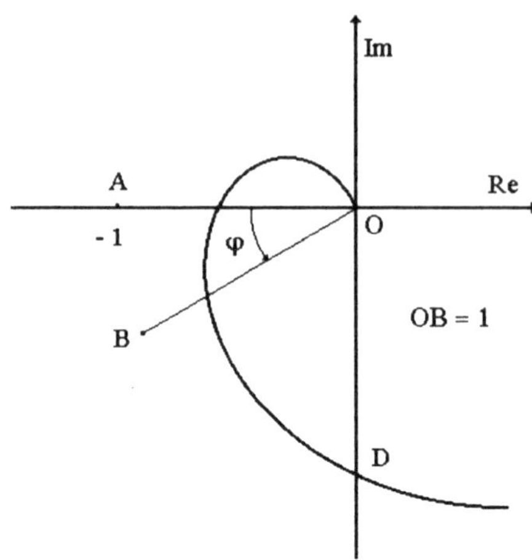

Figure 5.9. *PI regulator based on Nyquist diagram*

is necessary to heat the turbine at specific points to straighten it. These specific points, as well as the amount of energy to be used for heating the turbine, depend on the know-how of the technician who performs this job. This know-how cannot be found in books on heat transfer and materials. These jobs can be compared to the job of a PID designer (open-loop shaping by manipulating the gain, zeros and poles) who has to select the PID settings in order to modify the diagram (Bode, Nyquist or Nichols) of the uncompensated system such that the diagram (frequency-response curve) of the compensated system will correspond to the desired diagram (a curve which meets the control specifications). In summary, knowledge and engineering judgement must be applied to determine the best PID tuning for any given process. The engineer has to take into account all aspects of the controller design and of the desired control performance in order to decide whether one particular solution is better than another. The engineer has to consider what K. J. Åström calls the *"Gang of Six"*, which corresponds to the transfer functions relating the reference, load disturbances and the measurement noise to the controller output, the process output and the measured output. Finally, the designer has to think hard before introducing the derivative effect.

PROBLEM 5.9. Consider the system for which the Nyquist diagram is depicted in Figure 5.9. Determine the parameters k and τ_i of the PI regulator:

$$R\left(s\right) = k_c \left(1 + \frac{1}{\tau_i s}\right)$$

such that the point D which corresponds to the frequency ω_d moves to the point B.

SOLUTION 5.9. The open-loop transfer function of the compensated system is given by:

$$T(s) = k_c \left(1 + \frac{1}{\tau_i s}\right) G(s)$$

where $G(s)$ represents the transfer function of the uncompensated system (the process). For the point B, we have:

$$\arg\left[k_c\left(1 + \frac{1}{\tau_i s}\right)G(s)\right] = \arg\left(1 + \frac{1}{j\tau_i\omega_d}\right) + \arg G(j\omega_d) = -\pi + \varphi,$$

$$\arg\left(1 + \frac{1}{j\tau_i\omega_d}\right) - \frac{\pi}{2} = -\pi + \varphi,$$

$$\arg\left(1 - \frac{j}{\tau_i\omega_d}\right) = \varphi - \frac{\pi}{2}. \tag{5.13}$$

Equation (5.13) leads to:

$$\tan\left(\varphi - \frac{\pi}{2}\right) = -\frac{1}{\tau_i\omega_d} = -\frac{1}{\tan\varphi},$$

$$\tau_i = \frac{1}{\omega_d}\tan\varphi.$$

This tuning method was suggested by Åström and Hägglund [AST 84]. It ensures a phase margin equal to φ for the compensated system.

PROBLEM 5.10. Consider the Nyquist diagram of a system $G(s)$ which intersects the negative real axis at a frequency ω_c. Determine the PID settings necessary to obtain a phase margin equal to φ, and a unit gain for the open-loop transfer function at $\omega = \omega_c$.

SOLUTION 5.10. At the frequency ω_c, we have:

$$G(j\omega_c) = Re\left(G(j\omega_c)\right) + Im\left(G(j\omega_c)\right) = Re\left(G(j\omega_c)\right),$$

$$Re\left(G(j\omega_c)\right) < 0 \quad \text{and} \quad \arg G(j\omega_c) = -\pi.$$

Let us consider the following PID parameterization:

$$R(s) = k_c\left[1 + \frac{1}{\tau_i s} + \tau_d s\right], \quad R(j\omega) = k_c\left[1 + j\left(\tau_d\omega - \frac{1}{\tau_i\omega}\right)\right].$$

The phase φ introduced by the PID controller is given by:

$$\arg R(j\omega) = \arg k_c\left[1 + j\left(\tau_d\omega - \frac{1}{\tau_i\omega}\right)\right],$$

$$\varphi = \arctan\left(\tau_d\omega - \frac{1}{\tau_i\omega}\right). \tag{5.14}$$

This condition is satisfied by many values of the pair τ_d and τ_i Åström and Hägglund [AST 84] have suggested that τ_d and τ_i should be selected as follows:

$$\tau_i = \alpha \tau_d$$

Other conditions can be also considered (see for instance Problem 3.24). From Equation (5.14), we obtain:

$$\alpha \omega_c^2 \tau_d^2 - (\tan \varphi)\, \alpha \omega_c \tau_d - 1,$$

$$\tau_d \quad = \quad \frac{\tan \varphi \pm \sqrt{(\tan \varphi)^2 + 4/\alpha}}{2\omega_c}$$

The solution $\left(\tan \varphi - \sqrt{(\tan \varphi)^2 + 4/\alpha} \right) / 2\omega_c$ is negative. It has to be rejected. Finally, we obtain:

$$\tau_d = \frac{\tan \varphi + \sqrt{(\tan \varphi)^2 + 4/\alpha}}{2\omega_c}.$$

Let us now calculate the gain of the open-loop transfer function:

$$|G(j\omega_c)\, R(j\omega_c)| = k_c\, |G(j\omega_c)| \left| 1 + j \left(\tau_d \omega_c - \frac{1}{\tau_i \omega_c} \right) \right|$$

$$= k_c\, |G(j\omega_c)| \sqrt{1 + \left(\tau_d \omega_c - \frac{1}{\tau_i \omega_c} \right)^2}.$$

Taking Equation (5.14) into account, we derive:

$$|G(j\omega_c)\, R(j\omega_c)| = k_c\, |G(j\omega_c)| \sqrt{1 + (\tan \varphi)^2} = k_c\, |G(j\omega_c)| \frac{1}{\cos \varphi} = 1.$$

Therefore:

$$k_c = \frac{\cos \varphi}{|G(j\omega_c)|}.$$

PROBLEM 5.11. The dynamic behavior of a given process is represented by the transfer function:

$$G(s) = \frac{k}{1 + Ts}.$$

Consider a PI regulator:

$$R(s) = k_c \left[1 + \frac{1}{\tau_i s} \right].$$

Determine the parameters k and τ_i of this PI regulator in order that the closed-loop characteristic function will be equal to:

$$s^2 + 2\zeta\omega_n s + \omega_n^2 = 0, \tag{5.15}$$

where ζ and ω_n represent the damping factor and the natural frequency, respectively.

SOLUTION 5.11. The transfer function of the closed-loop system is given by:

$$F(s) = \frac{G(s)\,R(s)}{1 + G(s)\,R(s)}.$$

The characteristic function is:

$$s^2 + \left(\frac{1}{T} + \frac{kk_c}{T}\right)s + \frac{kk_c}{T\tau_i} = 0. \tag{5.16}$$

From Equations (5.15) and (5.16), we obtain:

$$\omega_n^2 = \frac{kk_c}{T\tau_i}, \quad 2\zeta\omega_n = \frac{1}{T} + \frac{kk_c}{T}.$$

The PI parameters are then given by:

$$k_c = \frac{2\zeta\omega_n T - 1}{k}, \quad \tau_i = \frac{2\zeta\omega_n T - 1}{\omega_n^2 T}.$$

In order that the gain of the PI regulator is positive, we must have:

$$\omega_n > \frac{1}{2\zeta T}.$$

This method for tuning a PI regulator belongs to the class of pole placement techniques.

PROBLEM 5.12. Given the control system depicted in Figure 5.10, with transfer functions given by:

$$G(s) = \frac{2}{s^3 + 7.5s^2 + 3.0s - 3.5}, \quad R(s) = \alpha_0 + \alpha_1 s + \alpha_2 s^2 + \alpha_3 s^3,$$

determine the parameters α_i, $i = \overline{0,3}$, of the controller $R(s)$ such that the closed-loop poles are the solutions of the following equation:

$$s^3 + 10s^2 + 31s + 30 = 0.$$

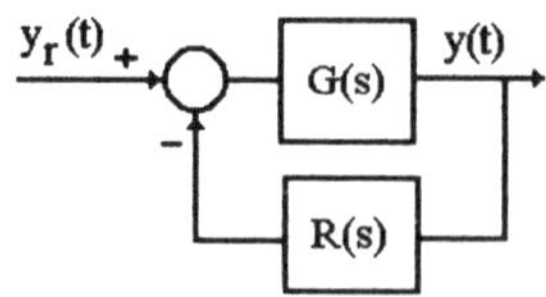

Figure 5.10. *Closed-loop system: state feedback controller*

SOLUTION 5.12. The characteristic equation of the closed-loop system is given by:

$$1 + G(s) R(s) = 0,$$

$$s^3 + 7.5s^2 + 3s - 3.5 + 2\left(\alpha_0 + \alpha_1 s + \alpha_2 s^2 + \alpha_3 s^3\right) = 0,$$

$$(1 + 2\alpha_3) s^3 + (7.5 + 2\alpha_2) s^2 + (3 + 2\alpha_1) s + 2\alpha_0 - 3.5 = 0.$$

Equating the coefficients of the powers of s in the desired characteristic equation and the characteristic equation of the compensated system, we obtain:

$$\alpha_3 = 0, \quad \alpha_2 = 1.25, \quad \alpha_1 = 14 \quad \text{and } \alpha_0 = 16.75$$

The poles of $G(s)$ are:

$$p_1 = -7.0, \quad p_2 = .5 \quad \text{and } p_3 = -1.$$

This system is unstable. However, the closed-loop poles are:

$$p_1 = -2, \quad p_2 = -3 \quad \text{and } p_3 = -5.$$

The control strategy presented in this problem is commonly used in the framework of the state-space representation of systems. It corresponds to a state feedback. We have:

$$R(s) Y(s) = \alpha_0 Y(s) + \alpha_1 sY(s) + \alpha_2 s^2 Y(s),$$

where $Y(s)$, $sY(s)$ and $s^2 Y(s)$ represent the output, the first derivative of the output and the second derivative of the output, respectively[7]. These quantities correspond to a state representation of the system. Let us introduce the following variables:

$$y(t) = x_1(t), \quad x_2(t) = \frac{d}{dt}x_1(t), \quad x_3(t) = \frac{d}{dt}x_2(t).$$

---

7. In some cases, the first and second derivatives of the output can be provided by tachometers and accelerometers, respectively.

From the transfer function $G(s)$, we obtain:

$$\frac{d^3}{dt^3}y(t) + 7.5\frac{d^2}{dt^2}y(t) + 3\frac{d}{dt}y(t) - 3.5y(t) = 2u(t),$$

which yields:

$$\frac{d}{dt}x_3(t) = -7.5x_3(t) - 3x_2(t) + 3.5x_1(t) + 2u(t).$$

The state-space representation of this system is given by:

$$\dot{\mathbf{x}} = \mathbf{A}\mathbf{x} + \mathbf{B}u(t), \quad u(t) = \mathbf{C}^T\mathbf{x}, \tag{5.17}$$

where:

$$\mathbf{x} = \begin{bmatrix} x_1(t) \\ x_2(t) \\ x_3(t) \end{bmatrix}, \quad \mathbf{A} = \begin{bmatrix} 0 & 1 & 0 \\ 0 & 0 & 1 \\ 3.5 & -3 & -7.5 \end{bmatrix},$$

$$\mathbf{B} = \begin{bmatrix} 0 \\ 0 \\ 2 \end{bmatrix}, \quad \mathbf{C}^T = \begin{bmatrix} 1 & 0 & 0 \end{bmatrix}.$$

Equation (5.17) corresponds to one possible state representation of this system.[8]

Taking the Laplace transform of Equation (5.17), we derive the transfer function associated with this system:

$$s\mathbf{x} = \mathbf{A}\mathbf{x} + \mathbf{B}U(s) \implies (s\mathbf{I} - \mathbf{A})\mathbf{x} = \mathbf{B}U(s),$$

$$\mathbf{x} = (s\mathbf{I} - \mathbf{A})^{-1}\mathbf{B}U(s) \implies Y(s) = \mathbf{C}^T(s\mathbf{I} - \mathbf{A})^{-1}\mathbf{B}U(s),$$

$$F(s) = \frac{Y(s)}{U(s)} = \mathbf{C}^T(s\mathbf{I} - \mathbf{A})^{-1}\mathbf{B}.$$

---

8. The following variable change:

$$\mathbf{x}' = \mathbf{M}\mathbf{x}, \quad \det\mathbf{M} \neq 0, \quad \Rightarrow \mathbf{x} = \mathbf{M}^{-1}\mathbf{x}'$$

yields the following state-space representation:

$$\dot{\mathbf{x}}' = \mathbf{M}\mathbf{A}\mathbf{M}^{-1}\mathbf{x}' + \mathbf{M}\mathbf{B}u(t), \quad y(t) = \mathbf{C}^T\mathbf{M}^{-1}\mathbf{x}'.$$

PROBLEM 5.13. Consider again Problem 5.11, with a second-order system:

$$G\left(s\right) = \frac{k}{\left(1 + T_1 s\right)\left(1 + T_2 s\right)}$$

and a PID regulator:

$$R\left(s\right) = k_c \left[1 + \frac{1}{\tau_i s} + \tau_d s\right],$$

where the desired characteristic equation is:

$$\left(s + a\omega\right)\left(s^2 + 2\zeta\omega_n s + \omega_n^2\right) = 0$$

SOLUTION 5.13. For:

$$T(s) = \frac{k k_c}{\left(1 + T_1 s\right)\left(1 + T_2 s\right)} \left[1 + \frac{1}{\tau_i s} + \tau_d s\right],$$

the closed-loop characteristic equation is given by:

$$1 + T(s)$$

$$= T_1 T_2 \tau_i s^3 + \left[\tau_i \left(T_1 + T_2\right) + k k_c\, \tau_d \tau_i\right] s^2 + \tau_i \left(k k_c + 1\right) s + k k_c = 0.$$

By comparison with the desired characteristic equation:

$$s^3 + \left(2\zeta\omega_n + a\omega\right) s^2 + \omega_n \left(a\omega 2\zeta + \omega_n\right) s + a\omega\omega_n^2 = 0,$$

we obtain:

$$T_1 T_2 \tau_i = 1, \quad \tau_i \left(T_1 + T_2\right) + k k_c \tau_d \tau_i = 2\zeta\omega_n + a\omega,$$

$$\tau_i \left(k k_c + 1\right) = \omega_n \left(2 a\omega\zeta + \omega_n\right), \quad k k_c = a\omega\omega_n^2,$$

and solve to obtain:

$$k_c = \frac{a\omega\omega_n^2}{k}, \quad \tau_i = \frac{\omega_n \left(2 a\omega\zeta + \omega_n\right)}{a\omega\omega_n^2 + 1},$$

$$\tau_d = \frac{\left(2\zeta\omega_n + a\omega\right)\left(a\omega\omega_n^2 + 1\right)}{a\omega\omega_n^3 \left(2 a\omega\zeta + \omega_n\right)} - \frac{\left(T_1 + T_2\right)}{a\omega\omega_n^2}.$$

PROBLEM 5.14. Consider a unity-feedback system with an open-loop transfer function equal to:

$$G\left(s\right) = \frac{375}{s^3 + 13s^2 + 61s + 225}, \quad H\left(s\right) = 1.$$

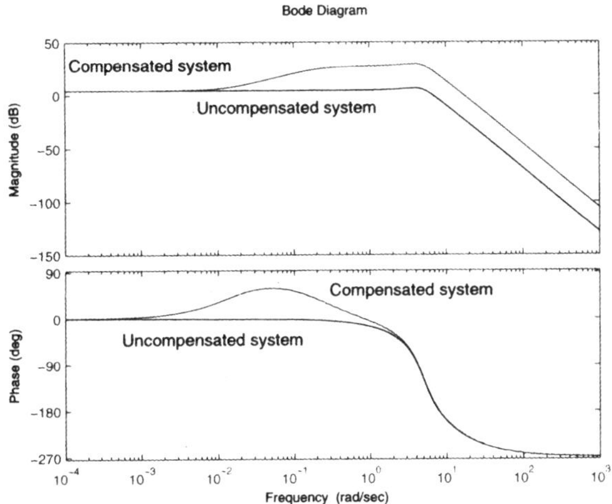

Figure 5.11. *Bode diagrams of an uncontrolled and a controlled system*

1. Draw the Bode diagram and determine the phase margin.

2. In order to increase the phase margin, we introduce the following regulator (a lead network or lead compensator):

$$R(s) = \frac{1 + aTs}{1 + Ts}.$$

Determine the values of the design parameters a and T in order that the phase margin of the compensated system will be equal to $\varphi_d = 45°$.

3. Draw the Bode diagrams of both the uncompensated and the compensated system.

SOLUTION 5.14. 1. The Bode diagrams are plotted in Figure 5.11. From this figure, we deduce the critical frequency $\omega_c = 6.7\,rad/s$, the phase margin $\varphi_m = 17°$ and the gain margin $k_m = 3.61\,dB$. The resonant frequency ω_r is equal to $3.9\,ad/s$.

2. Recall that the phase-lead compensator introduces an extra phase (it is a phase-advance network). The desired phase margin is $\varphi_d = 45°$. We obtain $\varphi_m = 1.2\varphi_d$ to $1.3\varphi_d$. Recall (see Problem 3.16) that:

$$\varphi_m = \arcsin\frac{a-1}{a+1} = 54 \text{ to } 58.5 \simeq 60,$$

which leads to:

$$\frac{a-1}{a+1} = \sin\varphi_m \simeq \sin 60° = \frac{\sqrt{3}}{2} = 0.866,$$

$$a = 13.928.$$

The maximal value of the phase is obtained for $\omega_m = \dfrac{1}{T\sqrt{a}}$. This frequency has to be adapted to the resonant frequency ω_r, i.e., we select ω_m according to:

$$\omega_m = 1.3 \text{ to } 1.5 \left(\omega_r\right)_{us}, \quad \left(\omega_r\right)_{us} = 3.9\, rad/s,$$

$$\omega_m = 5.07 \text{ to } 5.85\, rad/s,$$

where $\left(w_r\right)_{us}$ the resonant frequency of the uncompensated system. We obtain:

$$T = \frac{1}{\omega_m\sqrt{a}} = \begin{cases} 0.0528 \text{ for } \omega_m = 5.07 rad/s, \\ 0.0458 \text{ for } \omega_m = 5.85 rad/s. \end{cases}$$

3. For $T = 0.052$, the Bode diagram of the compensated system is depicted in Figure 5.11. As expected, the bandwidth has increased.

4. The difference between the numbers of poles and zeros of the open-loop transfer function remains unchanged when a lead compensator is introduced. The intersection of the asymptotes of the root locus with the real axis becomes:

$$\delta = \frac{\sum\limits_{i=1}^{n} polessys - \sum\limits_{i=1}^{m} zerossys - (1/T - 1/aT)}{n - m},$$

where *polessys* and *zerossys* represent the poles and the zeros of the uncompensated system. As $a > 1$, the intersection of the asymptotes is moved far from the origin, and the entire root locus is shifted to the left.

PROBLEM 5.15. 1. For the system:

$$G(s) = \frac{s + 0.003}{s^3 + 2.01s^2 + 23.02s + 0.23}, \quad H(s) = 1,$$

determine the parameters of the lag compensator:

$$R(s) = b\frac{1 + Ts}{1 + bTs}$$

required in order to reduce the steady state error (position error) to one-tenth and to obtain a decrease in phase equal to $\varphi = 45°$ around the resonant frequency.

2. What is the effect of the lag compensator on the root locus?

SOLUTION 5.15. 1. The Laplace transform of the error is given by:

$$\Xi\left(s\right) = \frac{1}{1+G\left(s\right)}Y_r\left(s\right) = \frac{s^3 + 2.01s^2 + 23.02s + 0.23}{s^3 + 2.01s^2 + 24.02s + 0.233}\frac{1}{s}, \quad Y_r\left(s\right) = \frac{1}{s},$$

and the steady-state error is equal to:

$$\varepsilon\left(\infty\right) = \lim_{s\to 0} s\Xi\left(s\right) = \frac{0.23}{0.233} \simeq 1.$$

The resonant frequency ω_r is equal to 4.63. For the compensated system, the Laplace transform of the error is given by:

$$\Xi\left(s\right) = \frac{1}{1+R\left(s\right)G\left(s\right)}Y_r\left(s\right), \quad R\left(s\right) = b\frac{1+Ts}{1+bTs},$$

and

$$\varepsilon\left(\infty\right) = \frac{0.23}{0.23 + 0.003b}.$$

From the condition on the steady-state error, we obtain:

$$\varepsilon\left(\infty\right) = \frac{0.23}{0.23 + 0.003b} = \frac{1}{10}\left(\frac{0.23}{0.233}\right) \simeq 0.1 \implies b = 690.$$

The desired phase decrease around the resonant frequency is given by:

$$\arg\left(b\frac{1+jT\omega}{1+jbT\omega}\right)_{\omega=\omega_r} = \arctan(T\omega_r) - \arctan\left(bT\omega_r\right).$$

For a large value of $b\,(b = 690)$, we derive:

$$\arctan\left(bT\omega_r\right) \simeq \frac{\pi}{2},$$

which implies:

$$\arg\left(b\frac{1+jT\omega}{1+jbT\omega}\right)_{\omega=\omega_r} = \arctan(T\omega_r) - \frac{\pi}{2} = \varphi.$$

Let us interpret this relation graphically.

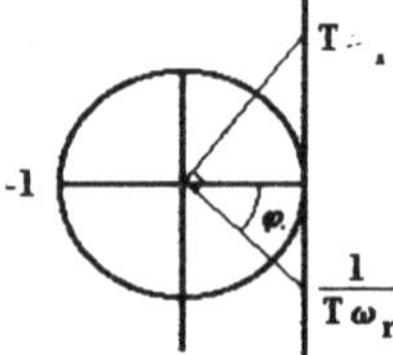

Figure 5.12. *Graphical representation of the phase of a lag compensator*

From Figure 5.12, we derive[9]:

$$\tan \varphi = \frac{1}{T\omega_r}.$$

for $\varphi = 45°$, we get:

$$\frac{1}{T\omega_r} = 1 \Rightarrow T = \frac{1}{\omega_r} = \frac{1}{4.64} = 0.215.$$

The Bode diagrams of the uncompensated and the compensated system are depicted in Figure 5.13. The bandwidth has been reduced.

2. As in the previous problem, the difference between the numbers of poles and zeros in the open-loop transfer function remains unchanged. The intersection of the asymptotes is given by:

$$\delta = \frac{\sum\limits_{i=1}^{n} polessys - \sum\limits_{i=1}^{m} zerossys + (1/T - 1/bT)}{n - m}.$$

As $b > 1$, the term $(1/T - 1/bT) > 0$, and therefore the intersection of the asymptotes is moved close to the origin and the root locus is shifted to the right.

In summary, the lag compensator permits us to increase the gain for low frequencies, and therefore to reduce the steady-state error. The drawback of this system is the decrease of the argument (it can destabilize some systems). Note that a lead–lag compensator exhibits the advantages of both lead and lag compensators. For the synthesis of such a compensator, one solution would be to first design the lead compensator,

---

9. $\tan (\theta - \pi/2) = -1/\tan \theta.$

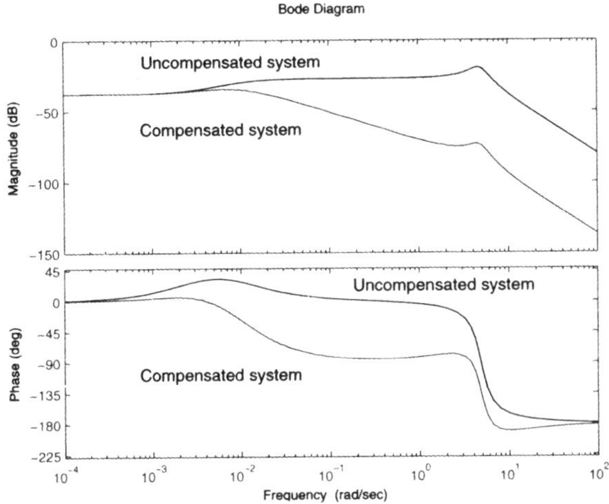

Figure 5.13. *Bode diagrams of an uncompensated and a compensated system*

consider the system plus the lead compensator as a new system, and then tune the lag compensator for this new system. As the lead compensator increases the bandwidth and the lag compensator decreases it, this lead–lag compensator has practically a negligible effect on the bandwidth globally.

PROBLEM 5.16. Determine the PID parameters for the system:

$$G\left(s\right) = \frac{5\left(1 + s\right)}{\left(9s + 1\right)\left(2s + 1\right)^{3}\left(0.5s + 1\right)}$$

on the basis of the Ziegler–Nichols method.

SOLUTION 5.16. The ultimate gain and period are given by:

$$k_{\mathrm{lim}} = 1.4877, \quad \omega_{\mathrm{lim}} = 0.40016, \quad T_{\mathrm{lim}} = \frac{\omega}{2\pi} = 6.3687 \times 10^{-2}.$$

Note that a simple relay controller (see Figure 5.14) can be used to determine the ultimate gain (see Appendix A) and the ultimate period [KRY 47, TSY 74].

Using the Ziegler–Nichols method, we obtain:

$$P \Longrightarrow R\left(s\right) = k_{c} \longrightarrow k_{c} = 0.5k_{\mathrm{lim}} = 0.669,$$

$$PI \Longrightarrow R\left(s\right) = k_{c}\left(\frac{\tau_{i}s + 1}{\tau_{i}s}\right) \longrightarrow k_{c} = 0.45k_{\mathrm{lim}} = 3.346,$$

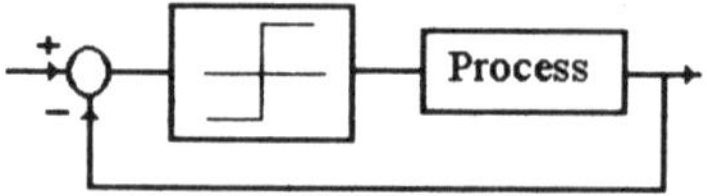

Figure 5.14. *Relay controller for determining the ultimate gain and ultimate period*

$$\tau_i = 0.83 T_{\lim} = 0.0528,$$

$$PID \implies R(s) = k_c \left(\frac{\tau_i s + 1}{\tau_i s} + \tau_d s \right) \longrightarrow k_c = 0.6 k_{\lim} = 0.8926,$$

$$\tau_i = 0.5 T_{\lim} = 0.03184 \quad \text{and} \quad \tau_d = 0.12 T_{\lim} = 7.642 \times 10^{-3}.$$

Two successive step responses of this compensated system are plotted in Figure 5.15.

We observe an overshoot of 46%, which can be reduced by reducing the Ziegler–Nichols PID settings.

The Ziegler–Nichols method is based on the minimization of the following criterion:

$$J = \int_0^\infty |\varepsilon(t)| \, dt.$$

This method[10] gives good disturbance rejection, high overshoot (see Figure 5.15) and a high control signal. The latter may lead to undesirable behavior such as actuator saturation. Note that the task of finding relations between the PID parameters and the parameters of the process model in order to minimise a performance index yields an optimization problem. Many authors [AND 04, TAV 03, VIS 01] have used genetic

---

10. On the basis of stability and robustness considerations, it has been shown in [SIL 03], that for first-order models with time delays, the range of τ/T values that ensures controller robustness for the Ziegler–Nichols tuning technique is given by:

$$0 < \frac{\tau}{T} < 1.07,$$

where τ and T represent the time delay and the time constant, respectively.

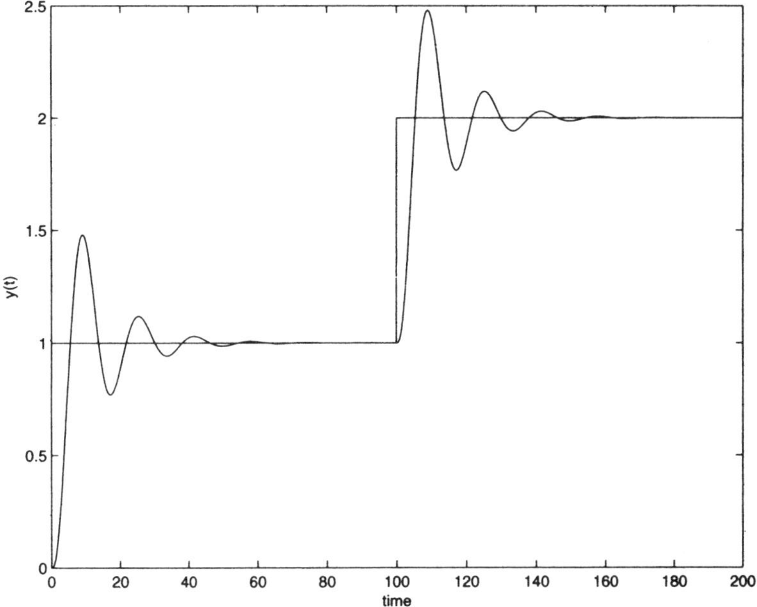

Figure 5.15. *Successive step responses of the compensated system in Problem 5.16*

algorithms to solve this optimization problem. The main performance indices used are:

$$J^1 \;=\; \int_0^\infty \varepsilon\,(\cdot)^2\,dt, \quad J^2 = \int_0^\infty t\,|\varepsilon\,(\cdot)|\,dt,$$

$$J^3 \;=\; \int_0^\infty \left[\varepsilon^2\,(\cdot) + T^2 \left(\frac{d\varepsilon\,(t)}{dt} \right)^2 \right]\,dt, \quad J^4 = \int_0^\infty \left[\varepsilon^2\,(\cdot) + \lambda u^2\,(t) \right].$$

The criterion J^1 (integrated quadratic error (IQE), or integral of the squared error (ISE)) is mainly used to deal with sudden changes in the set point. PID settings based on this criterion lead to a resonant ratio[11] greater than 1.3. The criterion J^2 (integral

---

11. The resonant ratio is equal to:
$$\frac{\max_{\omega} |F\,(j\omega)|}{F\,(0)}.$$

of time multiplied by the absolute value of error (ITAE)) is used to take account of long-transient-time behavior. Here, the term t can be considered as a weighting factor. The criterion J^3 contains the square of the derivative of the error. Therefore, it reduces the variation of the error. The criterion J^4 leads to a trade-off between energy (control effort) and error reduction. Using optimization techniques [NAJ 04], anyone can derive their own PID tuning method.

Many chemical processes can be modeled by an integrator or a time constant in series with a dead time [CHI 90]. This kind of model can be obtained by approximating models using many approaches. We shall present two approaches to obtaining an approximation for a given system in the form of a time constant in series with a dead time.

First approach:

1) *We neglect the small time constants and include them and opposite for the negative zeros in the dead time.*

2) *We keep the largest time constant and the remaining time constants using what is called the half method.*

Second approach (*the Broida method*). Broida has proposed a method based on two values of the response, at t_1 and t_2 such that:

$$y\left(t_1\right) = 0.28y\left(\infty\right) \quad \text{and } y\left(t_2\right) = 0.4y\left(\infty\right).$$

The delay and the time constant in the approximate model are given by:

$$\tau = 2.8t_1 - 1.8t_2, \quad T = 5.5\left(t_2 - t_1\right).$$

The following example illustrates these approximation methods. Let us consider the system:

$$G\left(s\right) = k\frac{\left(0.01s + 1\right)\left(0.1s - 1\right)}{\left(2s + 1\right)\left(0.5s + 1\right)\left(0.03s + 1\right)^4\left(4s + 1\right)},$$

to be approximated by the system:

$$G_a\left(s\right) = k\frac{\exp\left(-\tau s\right)}{\left(Ts + 1\right)}.$$

For a second-order system, this is equal to:

$$\frac{1}{2\zeta\sqrt{\left(1 - \zeta^2\right)}}.$$

The maximum gain $\max\limits_{\omega}\left|F\left(j\omega\right)\right|$ of a unity-feedback system is given by the value of the magnitude contours (M-contours) which are tangential to the Nyquist diagram of the open-loop transfer function.

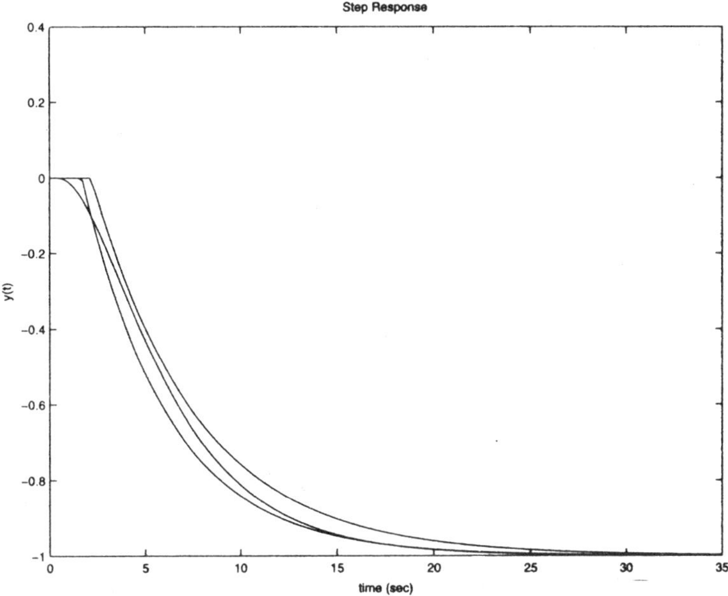

Figure 5.16. *Step responses of a system and its approximation*

The first approach leads to:

$$T = 4 + \frac{2}{2} = 4.5, \quad \tau = \frac{2}{2} + 0.5 + 4\,(0.03) - 0.01 + 0.1 = 1.71.$$

From the step response, we obtain using the second approach:

$$t_1 \simeq 4, \quad t_2 \simeq 5,$$

which implies:

$$T = 5.5, \ \tau = 2.2.$$

The step responses of the original system and its approximations are depicted in Figure 5.16.

These approximate models are acceptable. Note that the means $((4.5 + 5.5)\,/2,\ (1.71 + 2.2)\,/2\ )$ of the values of T and τ obtained by these two methods leads to a better approximation.

The next problems are dedicated to PID tuning for such dead-time systems.

PROBLEM 5.17. The transfer functions of an integrating dead-time system and a PID controller are given by:

$$G(s) = k\frac{\exp(-\tau s)}{s}, \quad R(s) = k_c\left(1 + \frac{1}{\tau_i s} + \tau_d s\right).$$

1. Calculate the closed-loop transfer function and use the Padé approximation to approximate the term $\exp(-\tau s)$ appearing in the denominator.

2. Derive the parameters of this controller such that the process output $y(t)$ follows a desired output $y_d(t - \tau)$, i.e., the closed-loop transfer function is equal to a time delay. Calculate the PID settings for the following system:

$$G(s) = 0.45\frac{\exp(-3s)}{s}.$$

SOLUTION 5.17. We shall present a solution to this control problem based on the method described in [CHI 03].

1. As in [CHI 03], let us denotes τs by q. The closed-loop transfer function is given by:

$$
\begin{aligned}
F(s) &= \frac{kk_c(1 + 1/\tau_i s + \tau_d s)\exp(-\tau s)}{s + kk_c(1 + 1/\tau_i s + \tau_d s)\exp(-\tau s)} \\
&= \frac{(K_1 q + K_2 + K_3 q^2)\exp(-q)}{q^2 + (K_1 q + K_2 + K_3 q^2)\exp(-q)},
\end{aligned}
\tag{5.18}
$$

where:

$$K_1 = kk_c\tau, \quad K_2 = \frac{K_1}{\tau_i/\tau} \quad \text{and } K_3 = K_1\frac{\tau_d}{\tau}. \tag{5.19}$$

Let us use the Padé approximation[12] for the term $\exp(-\tau s)$ which appears in the denominator of the expression (5.18). We obtain:

$$\frac{Y(s)}{Y_r(s)} = \frac{(K_1 q + K_2 + K_3 q^2)(1 + 0.5q)\exp(-\tau s)}{q^2(1 + 0.5q) + (K_1 q + K_2 + K_3 q^2)(1 - 0.5q)}.$$

Observe that the remaining term $\exp(-\tau s)$ operates as a shift in the output.

---

12. [SMI 75] were the first to approximate the dead-time term in the denominator by a first-order Taylor series expansion.

2. In order that the output $y(t)$ should follow the desired response $y_d(t-\tau)$, this transfer function must be equal to $\exp(-\tau s)$. It follows that:

$$\left(K_1 q + K_2 + K_3 q^2\right)\left(1 + 0.5q\right)$$

$$= q^2\left(1 + 0.5q\right) + \left(K_1 q + K_2 + K_3 q^2\right)\left(1 - 0.5q\right),$$

$$K_2 + \left(K_1 + 0.5K_2\right)q + \left(K_3 + 0.5K_1\right)q^2 + 0.5K_3 q^3$$

$$= K_2 + \left(K_1 - 0.5K_2\right)q + \left(1 + K_3 - 0.5K_1\right)q^2 + \left(0.5 - 0.5K_3\right)q^3.$$

By equating the coefficients of the powers of q, we obtain:

$$K_2 = 0, \quad K_1 = 1 \quad \text{and} \quad K_3 = 0.5.$$

From Equation (5.19), we derive:

$$kk_c\tau = 1, \quad \frac{\tau_i}{\tau} = \infty \quad \text{and} \quad \frac{\tau_d}{\tau} = 0.5,$$

which corresponds to a PD controller.

The value of the ratio $Y(t)/Y_d(t)$ can exceed one (overshoot). To deal with this problem, Chidambaram and Padma Sree [CHI 03] have suggested that one should equate the coefficients of the power of q in the numerator to α times the coefficients of the same power of the denominator ($\alpha > 1$). This procedure leads to:

$$\left(K_1 + 0.5K_2\right)q + \left(K_3 + 0.5K_1\right)q^2 + 0.5K_3 q^3,$$

$$\alpha\left(K_1 - 0.5K_2\right)q + \alpha\left(1 + K_3 - 0.5K_1\right)q^2 + \alpha\left(0.5 - 0.5K_3\right)q^3,$$

which yields:

$$K_1\left(1 - \alpha\right) + 0.5K_2\left(1 + \alpha\right) = 0, \tag{5.20}$$

$$K_3\left(1 - \alpha\right) + 0.5K_1\left(1 + \alpha\right) = \alpha, \tag{5.21}$$

$$K_3\left(1 + \alpha\right) = \alpha. \tag{5.22}$$

The solution of these equations leads to:

$$K_1 = \frac{4\alpha^2}{\left(\alpha + 1\right)^2}, \quad \frac{K_1}{K_2} = 0.5\left(\frac{\alpha + 1}{\alpha - 1}\right) \tag{5.23}$$

$$\text{and } \frac{K_3}{K_1} = 0.25\left(\frac{\alpha + 1}{\alpha}\right).$$

This in turn implies (the design parameters of the PID controller):

$$k_c = \frac{K_1}{k_p\tau} = \frac{4\alpha^2}{k\tau\left(\alpha + 1\right)^2}, \quad \tau_i = \tau\frac{K_1}{K_2} = 0.5\tau\left(\frac{\alpha + 1}{\alpha - 1}\right),$$

$$\tau_d = \tau\frac{K_3}{K_1} = 0.25\tau\left(\frac{\alpha + 1}{\alpha}\right). \tag{5.24}$$

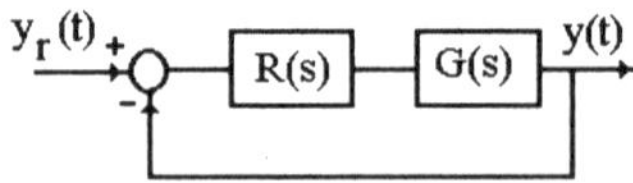

Figure 5.17. *Feedback system*

The closed-loop transfer function resulting from the implementation of a PI controller is given by:

$$F(s) = \frac{\left(k_c k \tau q + k_c k^2 \tau / \tau_i\right) \exp\left(-\tau s\right)}{q^2 + \left(k_c k \tau q + k_c k^2 \tau / \tau_i\right) \exp\left(-\tau s\right)}$$

$$= \frac{\left(k_1 q + k_2\right)\left(1 + 0.5q\right) \exp\left(-\tau s\right)}{q^2 + \left(k_1 q + k_2\right)\left(1 - 0.5q\right)}$$

$$= \frac{\left(0.5 k_1 q^2 + \left(k_1 + 0.5 k_2\right) q + k_2\right) \exp\left(-\tau s\right)}{\left(1 - 0.5 k_1\right) q^2 + \left(k_1 - 0.5 k_2\right) q + k_2}.$$

Bearing in mind the analysis performed earlier for a PID controller, we derive the design parameters of the PI controller as follows:

$$0.5 k_1 = \alpha \left(1 - 0.5 k_1\right) \implies k_1 = k_c k \tau = \frac{2\alpha}{\alpha + 1},$$

$$k_1 + 0.5 k_2 = \alpha \left(k_1 - 0.5 k_2\right) \implies \frac{k_1}{k_2} = \frac{\tau_i}{\tau} = 0.5 \frac{\alpha + 1}{\alpha - 1}. \tag{5.25}$$

For the system considered, we have:

$$k = 0.45, \quad \tau = 3;$$

$\alpha = 1.25$ and Equation (5.24) yield the PID design parameters:

$$k_c = 0.91449, \quad \tau_i = 13.5, \quad \tau_d = 1.35.$$

PROBLEM 5.18. 1. Consider the following system:

$$G(s) = \frac{k}{\left(1 + T_1 s\right)\left(1 + T_2 s\right)} \exp\left(-\tau s\right).$$

On the basis of the feedback system depicted in Figure 5.17, derive the parameters of the PID controller:

$$R(s) = k_c \left(\frac{\tau_i s + 1}{\tau_i s}\right)\left(\tau_d s + 1\right)$$

in order that:

$$\frac{Y(s)}{Y_r(s)} = \frac{1}{1 + T_c s} \exp(-\tau s).$$ (5.26)

2. On the basis of this PID tuning method, derive the PID settings for the following systems:

$$G_1(s) = \frac{7 \exp(-3s)}{(5s+1)(s+1)}, \quad \left.\frac{Y(s)}{Y_d(s)}\right|_{desired} = \frac{\exp(-3s)}{4s+1},$$

$$G_2(s) = \frac{5(1+s)}{(9s+1)(2s+1)^3(0.5s+1)}, \quad T_c = 5.$$

SOLUTION 5.18. 1. The closed-loop transfer function is given by:

$$\frac{Y(s)}{Y_r(s)} = \frac{G(s)R(s)}{1 + G(s)R(s)},$$

which leads to:

$$R(s) = \frac{Y(s)/Y_r(s)}{G(s)(1 - Y(s)/Y_r(s))}.$$

On the basis of Equation (5.26), we derive:

$$R(s) = \frac{(1 + T_1 s)(1 + T_2 s)}{k((1 + T_c s) - \exp(-\tau s))}.$$

As in many tuning approaches, let us approximate the time delay by a first-order Taylor series, i.e., $\exp(-\tau s) = 1 - \tau s$:

$$R(s) = \frac{(1 + T_1 s)(1 + T_2 s)}{k(\tau + T_c)s}.$$

Identifying the parameters of this controller with the PID parameterization adopted here [SKO 03],

$$R(s) = k_c \left(\frac{\tau_i s + 1}{\tau_i s}\right)(\tau_d s + 1) = \frac{k_c}{\tau_i s}(\tau_i s + 1)(\tau_d s + 1),$$

we obtain the following settings:

$$\tau_i = T_1, \quad \tau_d = T_2, \quad k_c = \frac{T_1}{k(T_c + \tau)}.$$

This PID tuning method was developed by Skogestad [SKO 03]. It requires some trial and error in order to specify a closed-loop time constant T_c that will lead to an acceptable closed-loop behavior.

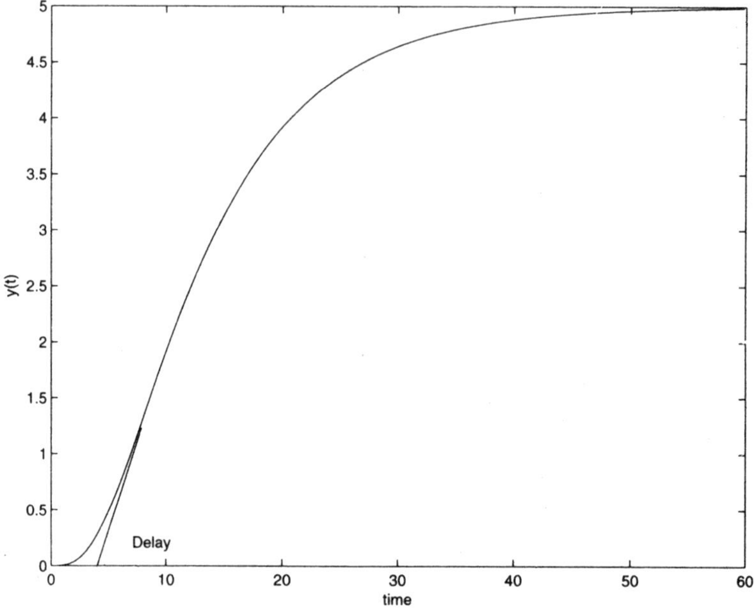

Figure 5.18. *Step response of the original second system in Problem 5.18*

2. For the first system, the PID settings are:

$$k \;=\; 7, \quad T_1 = 5, T_2 = 1, \quad \tau = 3,$$

$$k_c \;=\; \frac{1}{7}\frac{5}{4+3} = 0.102, \tau_i = T_1 = 5, \quad \tau_d = T_2 = 1.$$

The step response of the second system is depicted in Figure 5.18. From this response, we can derive an estimate of the time delay in the approximate system. Using the "half method", we derive:

$$T \;=\; 9 + 3\frac{1}{2}, \quad \tau = 4.2,$$

$$G\left(s\right) \;=\; \frac{5}{11.5s + 1}\, \exp(-4.2s).$$

The step responses of both the original system and its approximation are plotted in Figure 5.19. Another method would be to draw the tangent to the step response at the

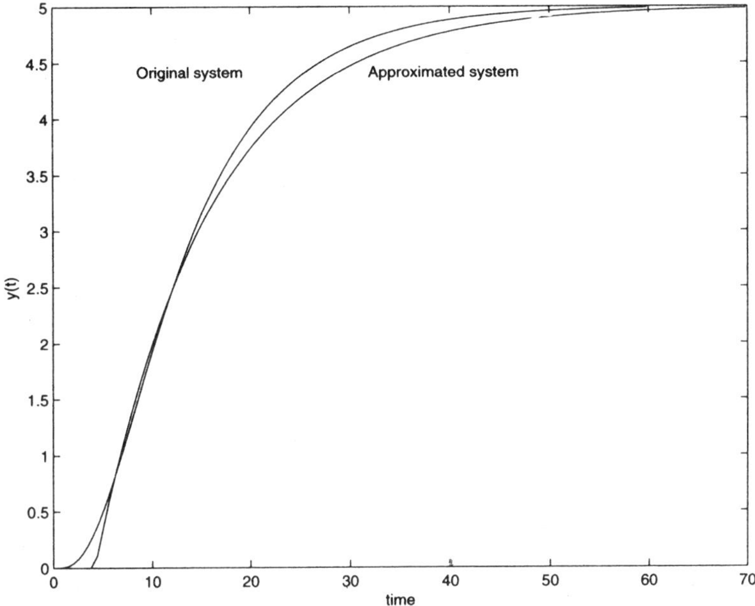

Figure 5.19. *Step responses of the original system and its approximation*

inflection point and to determine the time constant as if the system was a first-order system. In general, the estimated value is relatively high and has to be reduced. The values of the time delay and the time constant obtained must be adapted on the basis of observation of the response of the process to some step input.

In order to obtain a second-order system with a dead time as an approximate model, we proceed as follows. We keep the dominant time constant; in this case this time constant is equal to 9. As the transfer function of the system contains a term:

$$\frac{1}{(2s+1)^3},$$

the second time constant is close to 2×3. We start with these values:

$$T_1 = 9 \quad \text{and } T_2 = 6.$$

We obtain the steps response and adjust the dead time. The final value of the dead time is 1.8. With these values, we observe that the approximate model has slightly slower dynamics. Therefore, we decrease the value of the second time constant. Finally, we obtain:

$$T_1 = 9, \quad T_2 = 4 \quad \text{and } \tau = 1.8.$$

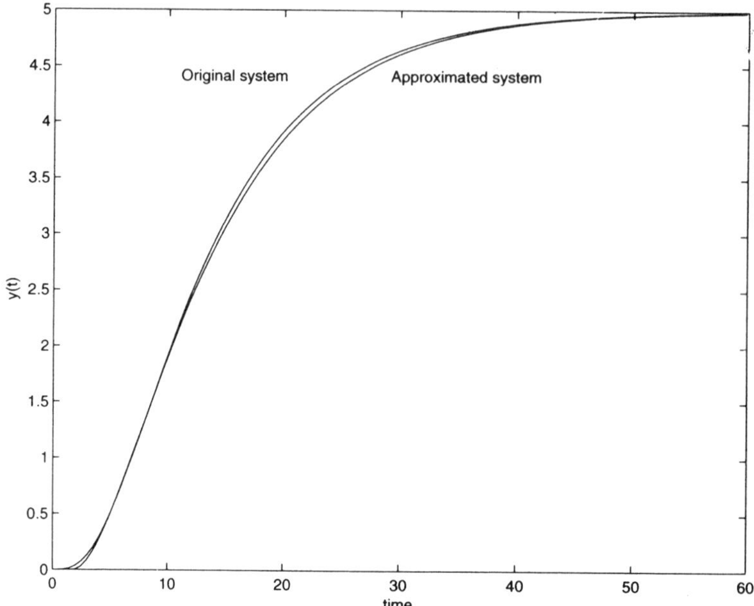

Figure 5.20. *Step responses of the system and its approximation by a second-order dead time model*

The step responses for the system and its approximation by a second-order system plus a delay are plotted in Figure 5.20.

The PID tuning technique presented above leads to the following settings:

$$\tau_i = T_1 = 9, \quad \tau_d = T_2 = 4, \quad k_c = \frac{T_1}{k\left(T_c + \tau\right)} = 0.29.$$

The reader may consider the following systems:

$$G_1\left(s\right) = \frac{\left(0.2s + 1\right)\left(s - 1\right)}{\left(7s + 1\right)\left(0.7s + 1\right)^2\left(0.01s + 1\right)\left(2.5s + 1\right)},$$

$$G_2\left(s\right) = \frac{\left(0.1s + 1\right)^2\left(1.5s + 1\right)}{\left(3s + 1\right)\left(0.45s + 1\right)\left(0.01s + 1\right)\left(s + 1\right)}.$$

Their first- and second-order approximations by a system with dead time are:

$$G_1(s) \simeq -\frac{\exp(-2.9s)}{9s+1}, \quad G_1(s) \simeq -\frac{\exp(-2.2s)}{(7s+1)(2.2s+1)},$$

$$G_2(s) \simeq \frac{\exp(-0.1s)}{2.65s+1}, \quad G_2(s) \simeq \frac{\exp(-0.1s)}{(2.8s+1)(0.02s+1)}.$$

PROBLEM 5.19. Consider the system:

$$G(s) = \frac{2}{(1+6s)^2}$$

and a PI controller:

$$R(s) = k_c\left(1 + \frac{1}{\tau_i s}\right) = k_c\frac{1/\tau_i + s}{s} = k_c\frac{s+z}{s}.$$

On the basis of the step responses and the root locus approach, determine the values of the PI settings such that the compensated system will be critically damped.

SOLUTION 5.19. The root locus of the compensated system consists of a vertical line which intersects the real axis at the point of abscissa $-1/6$. For $z < 1/6$, $z = 1/6$ and $z > 1/6$, we obtain the root loci depicted in Figure 5.21.

For $z < 1/6$ and high values of the gain, the closed-loop poles are close to the imaginary axis. Therefore, the compensated system will not have a good stability margin. Let us now look at the step responses, which are plotted in Figure 5.22. Both step responses are underdamped. The value $z = 1/6$ leads to a greater damping factor. This value will be selected. Now, we have to tune the gain k_c.

For this setting, the closed-loop transfer function is given by:

$$F(s) = \frac{k_c}{18s^2 + 3s + k_c} = \frac{k_c/18}{s^2 + s/6 + k_c/18}.$$

The damping factor and the natural frequency are given by:

$$2\zeta\omega_n = \frac{1}{6}, \omega_n^2 = \frac{k_c}{18} \Longrightarrow \omega_n = \sqrt{\frac{k_c}{18}}, \quad \zeta = \frac{1}{12\sqrt{k_c/18}}.$$

If we select $\zeta = 1$, we obtain:

$$\zeta = \frac{1}{12\sqrt{k_c/18}} = 1 \Longrightarrow k_c = \frac{1}{8}, \quad \omega_n = \frac{1}{12}.$$

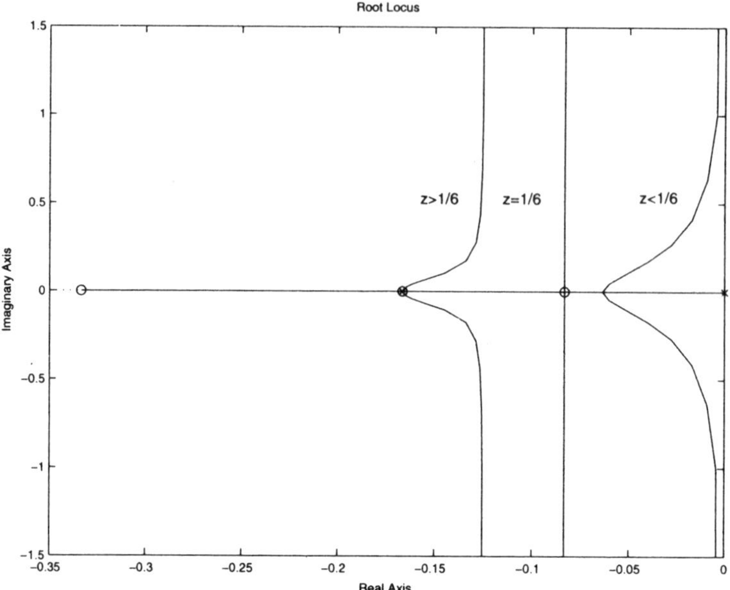

Figure 5.21. *Root locus for different values of z*

PROBLEM 5.20. Consider the following PID parameterization:

$$R(s) = k_c \left(1 + \frac{1}{\tau_i s} + \tau_d s \right).$$

1. Derive the PID settings for the non-minimum-phase (non-invertible) system:

$$G(s) \;=\; G_-(s)\, G_+(s) = \frac{k(1 - cs)}{(T_1 s + 1)(T_2 s + 1)}, \quad c = 0.2,$$

$$T_1 \;=\; 1, \quad T_2 = 3, \quad k = 5, G_-(s) = \frac{5}{(T_1 s + 1)(T_2 s + 1)}$$

in order that the closed-loop transfer function is equal to:

$$\frac{Y(s)}{Y_r(s)} = G_+(s)\, \frac{1}{1 + \alpha s}, \quad \alpha > 0.$$

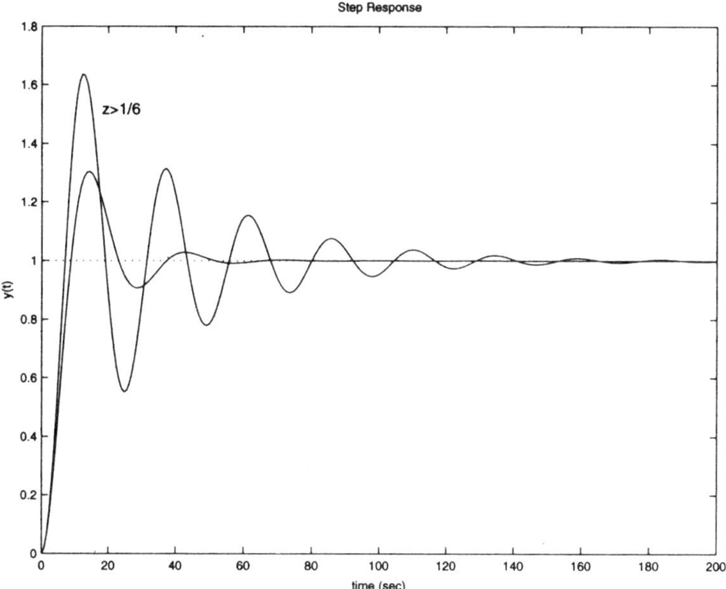

Figure 5.22. *Step responses for two values of z*

2. Answer the same question for:

$$\frac{Y\left(s\right)}{Y_r\left(s\right)} = G_+\left(s\right)\frac{1}{1 + as + bs^2}$$

and with a PID having a filtered derivative action,[13] i.e.,

$$R\left(s\right) = k_c\left[1 + \frac{1}{\tau_i s} + \frac{\tau_d s}{1 + \delta s}\right].$$

---

13. Industrial PID controllers have the following transfer function:

$$R\left(s\right) = k_c\left[1 + \frac{1}{\tau_i s} + \frac{\tau_d s}{1 + \beta \tau_d s}\right].$$

The parameter β is, however, usually predetermined by the manufacturer and cannot be changed.

SOLUTION 5.20. 1. Let us denote by $R(s)$ the transfer function of the PID controller. The closed-loop transfer function is given by:

$$F(s) = \frac{Y(s)}{Y_r(s)} = \frac{R(s)\,G(s)}{1 + R(s)\,G(s)}.$$

The transfer function of the controller is related to the desired closed-loop transfer function by:

$$R(s) = \frac{Y(s)/Y_r(s)}{G(s)\,(1 - Y(s)/Y_r(s))}.$$

As the system is non-invertible, we shall consider only the "invertible part" $G_-(s)$. We obtain:

$$R(s) = \frac{1/(1 + \alpha s)}{G_-(s)\,(1 - G_+(s)/(1 + \alpha s))}$$

$$= \frac{1/(1 + \alpha s)}{k\,(1 - (1 - cs)/(1 + \alpha s))\,/\,[(T_1 s + 1)(T_2 s + 1)]}$$

$$= \frac{(T_1 s + 1)(T_2 s + 1)}{k\,(\alpha + c)\,s} = \frac{T_1 + T_2}{k\,(\alpha + c)}\left[1 + \frac{1}{(T_1 + T_2)\,s} + \frac{T_1 T_2}{T_1 + T_2}s\right]$$

$$= k_c\left(1 + \frac{1}{\tau_i s} + \tau_d s\right),$$

which leads to:

$$k_c = \frac{T_1 + T_2}{k\,(\alpha + c)}, \quad \tau_i = T_1 + T_2, \quad \tau_d = \frac{T_1 T_2}{T_1 + T_2}.$$

For $\alpha = 0.3$, we obtain:

$$k_c = 1.6, \quad \tau_i = 4, \quad \tau_d = 0.75.$$

Two successive step responses of this compensated system are plotted in Figure 5.23.

The simulation shown in Figure 5.23 was carried out without filtering the derivative action. Of course, the non-minimum-phase behavior remains. The Nyquist diagrams associated with the sensitivity functions of the uncompensated and compensated systems are depicted in Figure 5.24; we observe that the maximum sensitivity has increased.

2. By selecting a "reference model" of the form $G_+(s)\,/\,(1 + as + bs^2)$, and a PID controller with a filtered derivative action, we obtain:

$$R(s) = \frac{(T_1 s + 1)(T_2 s + 1)}{k\,(a + c + bs)\,s} = k_c\left[1 + \frac{1}{\tau_i s} + \frac{\tau_d s}{1 + \delta s}\right]$$

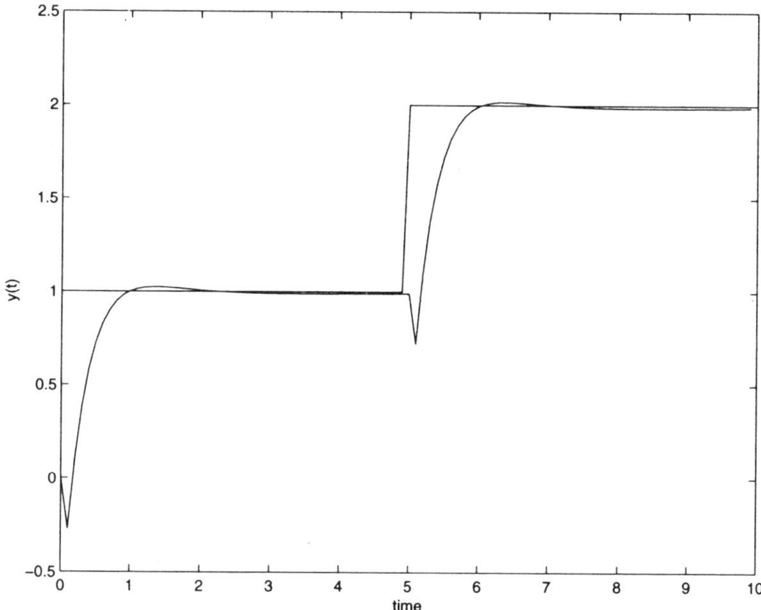

Figure 5.23. *PID control of anon-minimum-phase system*

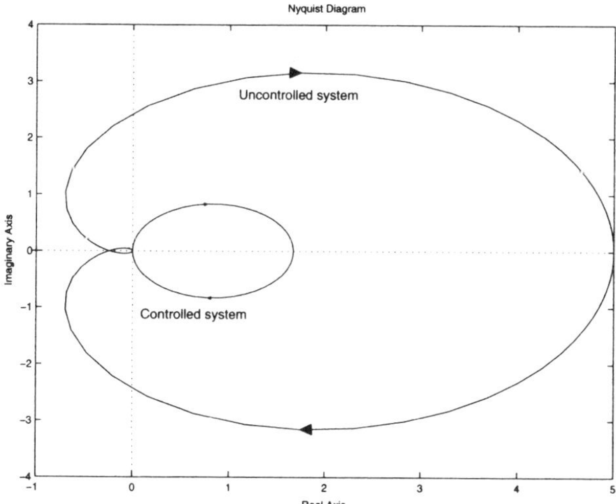

Figure 5.24. *Nyquist diagrams of an uncontrolled and a controlled system*

$$= \frac{k_c \left[1/\tau_i + (1 + \delta/\tau_i)\, s + (\tau_d + \delta)\, s^2 \right]}{s\,(1 + \delta s)}.$$

By identification, we derive the PID settings:

$$\delta = \frac{b}{a + c}, \quad k_c = \frac{T_1 + T_2 - \delta}{k\,(a + c)},$$

$$\tau_i = T_1 + T_2 - \delta, \quad \tau_d = \frac{T_1 T_2}{T_1 + T_2 - \delta} - \delta,$$

which correspond to a PID controller with a filtered derivative action. For a and b such that $a = 2\alpha$ and $b = \alpha^2$ (there is a double pole $1/\left(1 + \alpha s\right)^2$), we obtain:

$$\delta = \frac{\alpha^2}{2\alpha + c}, \quad k_c = \frac{T_1 + T_2 - \delta}{k\,(2\alpha + c)},$$

$$\tau_i = T_1 + T_2 - \delta, \quad \tau_d = \frac{T_1 T_2}{T_1 + T_2 - \delta} - \delta.$$

Before we end this chapter, we shall present a brief introduction to the integrator wind-up [BAK 00] and a class of non-linear systems. When actuator saturation occurs, the control input does not correspond to the control provided by the controller. In other words, there is a mismatch between the controller output and the system input; a variation of the controller output has no effect on the controlled system. Under these conditions, the loop gain becomes equal to zero, and the integrator continues to integrate a non-zero error and will therefore produce a high controller output (this effect is known as "wind-up"). Saturation of integrators was observed long ago in electronic amplifiers and there has existed for a long period a set of techniques for desaturating amplifiers (use of a zener diode, etc.). Also, another problem appears when a change in controller occurs, as when we switch from one control mode to another (such as from automatic control to manual control) or when we substitute one controller with another, as in the gain-scheduling control approach. It is necessary that the outputs of the two controllers coincide. This corresponds to what is called "bumpless transfer". Because of its practical significance, there has been a considerable interest in the literature in these topics; see, for instance, [BAK 00].

PROBLEM 5.21. Derive the equations governing the behavior of the system depicted in Figure 5.25.

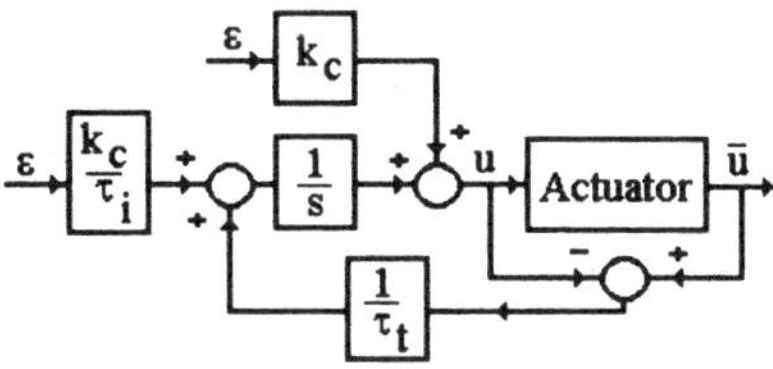

Figure 5.25. *Anti-wind-up system for PI controller*

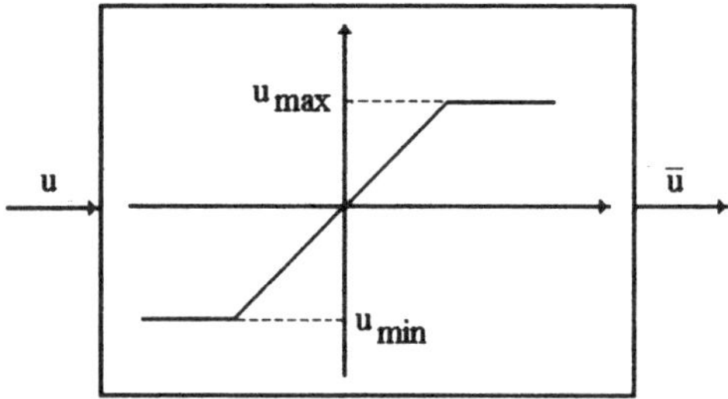

Figure 5.26. *Actuator saturation*

SOLUTION 5.21. When the actuator saturates,[14] a change in the controller output has no effect on the process input (see Figure 5.26):

$$U\left(s\right) = \frac{1}{s}\left[\frac{k_c}{\tau_i}\Xi\left(s\right) + \frac{1}{\tau_t}\left(\overline{U}\left(s\right) - U\left(s\right)\right)\right] + k_c\Xi\left(s\right),$$

$$U\left(s\right) = k_c\frac{\tau_t s}{1 + \tau_t s}\frac{\tau_i s + 1}{\tau_i s}\Xi\left(s\right) - \frac{1}{1 + \tau_t s}\overline{U}\left(s\right).$$

This equation shows that the integral effect has been replaced by a first-order system with a time constant equal to τ_t.

---

14. Large inputs to the plant may occur that exceed the capacity of the plant.

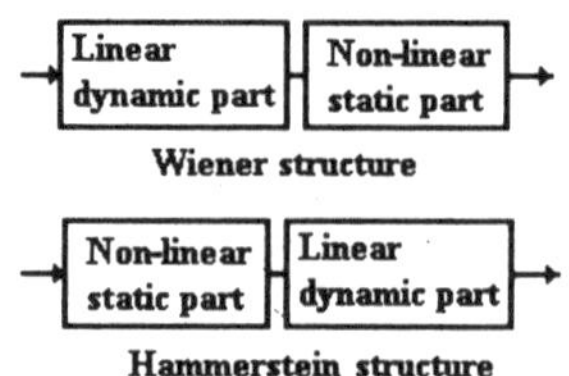

Figure 5.27. *Wiener and Hammerstein structures*

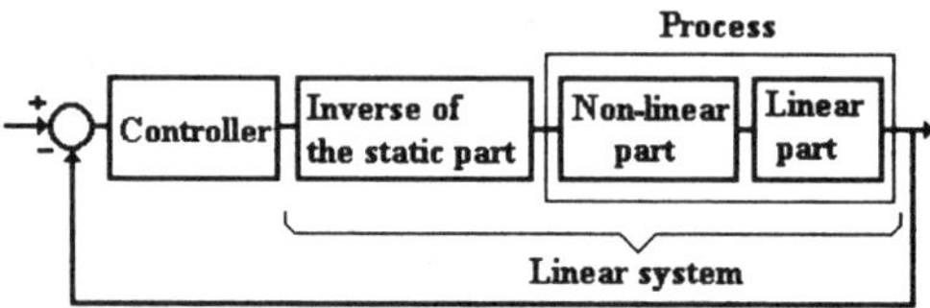

Figure 5.28. *ammerstein structure in series with the inverse of the non-linear
static part*

The reader may find an extensive treatment and many references on the subject of
the control of systems with constraints in [BAK 00].

The Volterra series and the Wiener and Hammerstein models (5.27) are quite general non-linear representations. The Wiener and Hammerstein systems are composed
of various combinations of dynamic linear and static non-linear elements. They have
proved to be useful descriptions of non-linear dynamic systems [IKO 02].

In a Wiener structure, the linear dynamic part is followed by the non-linear, mem-
oryless static part, and the Hammerstein structure, the non-linear part, precedes the
linear dynamic part. Sometimes it is possible to obtain information on the non-linear
part from a physical study of the process. In general, a functional approximation
technique is used to approximate the non-linear system. This technique maps the non-
linearity with a set of basis functions (orthogonal polynomials, etc.). Control based
on these structures is reduced to the design of controllers for linear systems. If the
inverse of the non-linear static part is available, a PID controller or any linear control
strategy can be easily implemented, as shown in Figures 5.28 and 5.29.

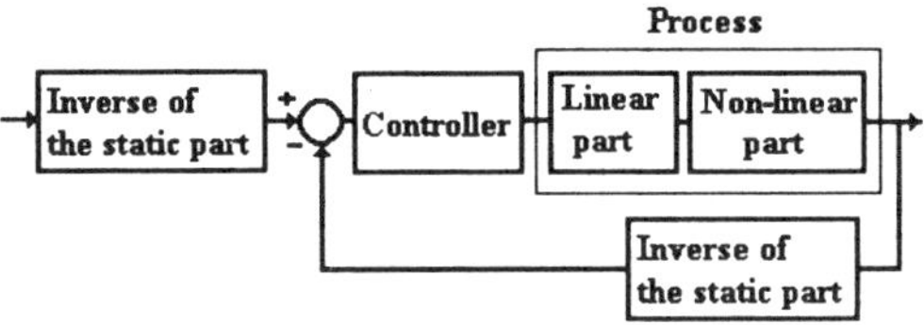

Figure 5.29. *Wiener structure and the inverse of the non-linear part*

With the inverse of the non-linear static part, we obtain a linear system which corresponds to the linear dynamic part of the structure considered. Many control applications based on such structures are presented in [IKO 02].

PROBLEM 5.22. Calculate the closed-loop transfer functions for the non-linear systems depicted in Figures 5.28 and 5.29.

SOLUTION 5.22. For the Hammerstein structure, we obtain:

$$(Y_r(s) - Y(s)) R(s) Nl^{-1} Nl G(s) = Y(s),$$
$$F(s) = \frac{Y(s)}{Y_r(s)} = \frac{R(s) G(s)}{1 + R(s) G(s)},$$

where $R(s)$ and $G(s)$ represent the transfer functions of the controller and of the linear part of the Hammerstein structure. Nl and Nl^{-1} denote the direct and the inverse model, respectively, of the static non-linear part of this structure.

For the Wiener structure, we derive:

$$(Y_r(s) - Y(s)) Nl^{-1} R(s) G(s) Nl = Y(s),$$

$$Nl^{-1} R(s) G(s) Nl Y_r(s) = \left(1 + Nl^{-1} R(s) G(s) Nl\right) Y(s),$$
$$Nl^{-1} Nl = 1,$$

$$R(s) G(s) Nl Y_r(s) = (Nl + R(s) G(s) Nl) Y(s),$$
$$R(s) G(s) Nl Y_r(s) = (1 + R(s) G(s)) Nl Y(s),$$
$$F(s) = \frac{Y(s)}{Y_r(s)} = \frac{R(s) G(s)}{1 + R(s) G(s)}.$$

The closed-loop transfer function obtained corresponds to the classical transfer function of a linearly controlled system.

Appendix A

On Theoretical Aspects

In this Appendix we quote a number of results dealing with the Dirac impulse, residence times, step inputs, stability and the root locus.

A.1. The Dirac impulse

Here we describe the Dirac impulse function (also known as the Dirac delta function).

Let us consider a symmetric function which decreases quickly, has its maximal value at $x = 0$ and is such that:

$$\int_{-\infty}^{+\infty} f(x)\, dx = 1.$$

In other words, the area under the graph is equal to 1. The functions[1]:

$$h(x) = \frac{1}{\sqrt{\pi}} \exp\left(-x^2\right), \quad g(x) = \frac{1}{\pi}\frac{1}{1 + x^2} \tag{A.1}$$

____

1. Recall the following:

$$erf(x) = \frac{2}{\sqrt{\pi}} \int_0^x \exp\left(-z^2\right) dz, \quad erf(\infty) = 1.$$

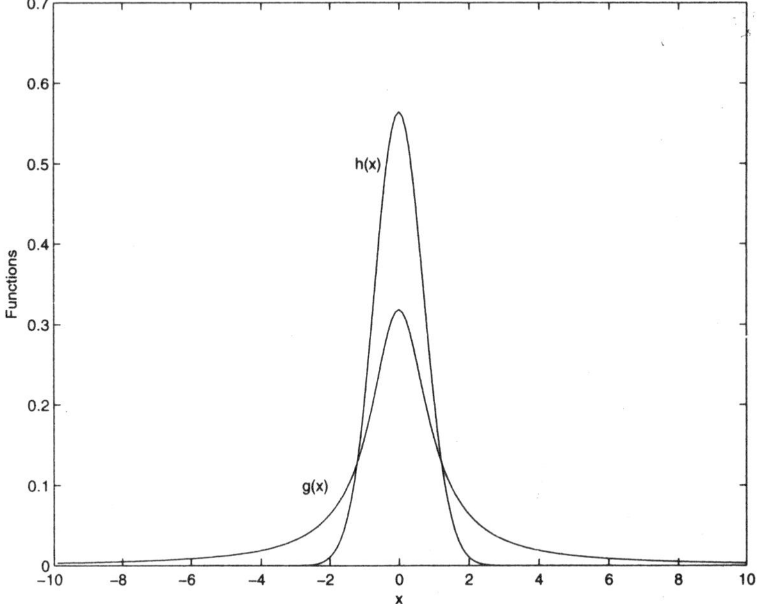

Figure A.1. *Surrounding functions of the Dirac impulse*

satisfy these conditions. They are depicted in Figure A.1. Let us now consider the following transformations:

$$f \to f_a(x) = af(ax).$$

For the function $h(x)$, the effect of these transformations is shown in Figure A.2.

Consider the following variable change:

$$u = -z$$

$$\Rightarrow \int_0^x \exp\left(-x^2\right) dx = \int_0^{-x} -\exp\left(-u^2\right) du = \int_{-x}^0 \exp\left(-u^2\right) du$$

$$\Rightarrow \int_{-\infty}^\infty \exp\left(-x^2\right) dx = 2 \int_0^\infty \exp\left(-x^2\right) dx = 2\frac{\sqrt{\pi}\,erf\left(\infty\right)}{2} = \sqrt{\pi},$$

which leads to:

$$\int_{-\infty}^{+\infty} h(x) = \int_{-\infty}^{+\infty} \frac{1}{\sqrt{\pi}} \exp\left(-x^2\right) = 1.$$

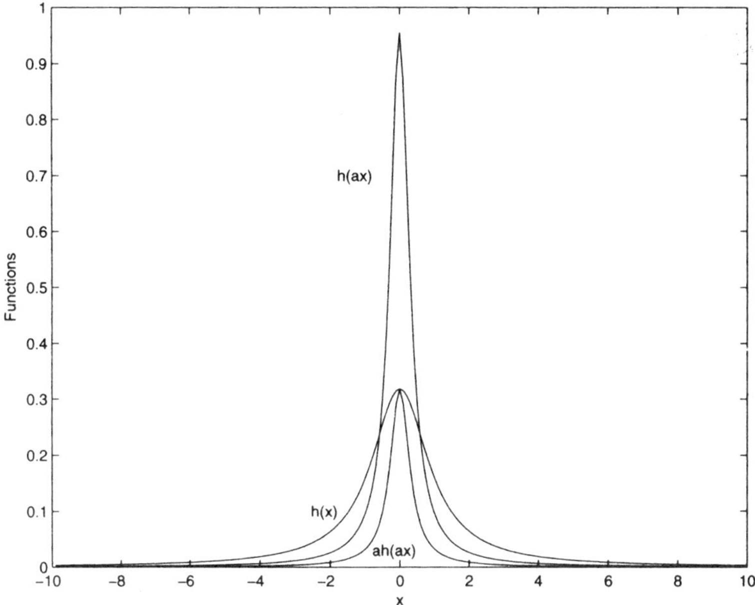

Figure A.2. *Effect of the transformations* $mh\,(x)$ *and* $h\,(mx)$ *on* $h\,(x)$

In this case the integral (the area between the graphical representation of this function and the x axis) remains unchanged:

$$\int_{-\infty}^{+\infty} f_a\,(x)\,dx = \int_{-\infty}^{+\infty} af\,(ax)\,dx = \int_{-\infty}^{+\infty} f\,(ax)\,d\,(ax) = \int_{-\infty}^{+\infty} f\,(x)\,dx.$$

A question now arises. What happens when $a \to \infty$? For the function $ag\,(ax)$, we obtain:

$$ag\,(ax) = \frac{a}{\pi}\frac{1}{1 + a^2x^2} \simeq \frac{1}{\pi ax^2},$$

which decreases indefinitely as a increases. If we now consider the transformed function $h\,(x)$, we obtain:

$$ah\,(ax) = \frac{a}{\sqrt{\pi}}\exp\left(-a^2x^2\right).$$

This function also decreases as a increases, but it decreases more quickly than $ag\,(ax)$. Therefore, by increasing the parameter a indefinitely, we obtain a function which is equal to zero for all values of x such that $x < 0$ and $x > 0$, and equal to

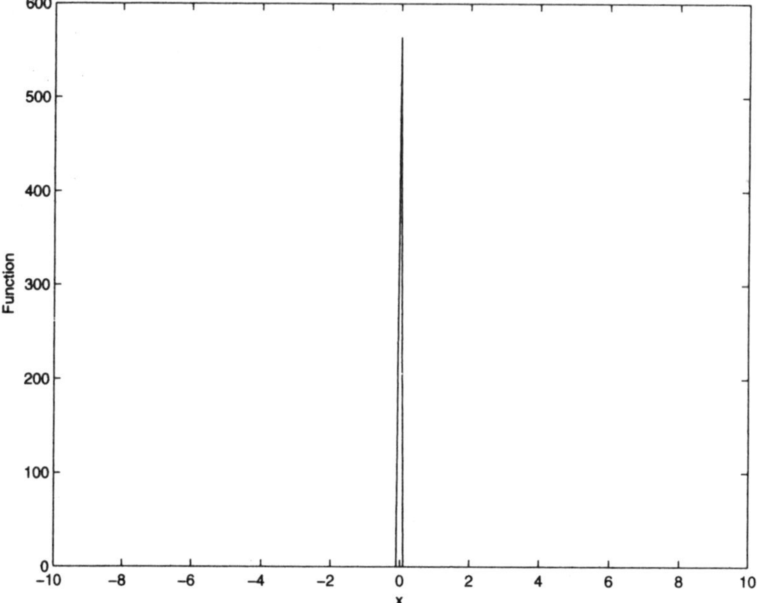

Figure A.3. *The function $ah\,(ax)$ for $a = 10^3$*

infinity for $x = 0$; its integral from $-\infty$ to $+\infty$ is equal to one. The function obtained is the Dirac impulse function. Figure A.3 shows the function $ah\,(ax)$ for $a = 10^3$.

The Dirac impulse exhibits many interesting properties. We shall present some of them here.

Let us consider the integral:

$$I = \int\limits_{-\infty}^{+\infty} \delta\,(x)\,f\,(x)\,dx.$$

Recall that $\delta\,(x)$ is equal to zero for all $x \neq 0$. Therefore, for small values of ε, we obtain:

$$I = \int\limits_{-\varepsilon}^{+\varepsilon} \delta\,(x)\,f\,(x)\,dx$$

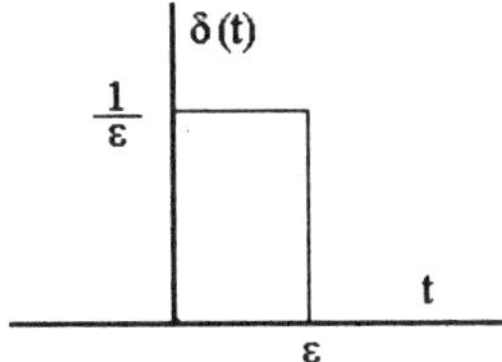

Figure A.4. *Approximation of the Dirac impulse*

Since the integration interval is very small (its length is equal to 2ε), $f(x) \simeq f(0)$. Therefore, we obtain:

$$I_0 = \int_{-\infty}^{+\infty} \delta(x) f(x)\, dx = f(0).$$

From the preceding considerations, we derive:

$$I_\alpha = \int_{-\infty}^{+\infty} \delta(x - \alpha) f(x)\, dx = f(\alpha).$$

The Dirac impulse may also be considered as the function depicted in Figure A.4, when $\varepsilon \to 0$. This approximation yields:

$$\delta_\varepsilon(t) = \frac{1}{\varepsilon} 1(t) - \frac{1}{\varepsilon} 1(t - \varepsilon), \quad \delta(t) = \lim_{\varepsilon \to 0} \delta_\varepsilon(t),$$

where:

$$1(t) = \left\{ \begin{array}{l} 1 \text{ for } t \geq 0, \\ 0 \text{ for } t < 0. \end{array} \right.$$

We may calculate the Laplace transform of the Dirac impulse:

$$\mathcal{L}\left[\delta(t)\right] = \lim_{\varepsilon \to 0} \mathcal{L}\left[\frac{1(t) - 1(t - \varepsilon)}{\varepsilon}\right] = \lim_{\varepsilon \to 0} \left(\frac{1 - \exp(-s\varepsilon)}{s\varepsilon}\right).$$

Taking into account the Taylor expansion of $\exp(-s\varepsilon)$, we derive:

$$\mathcal{L}\left[\delta(t)\right] = \lim_{\varepsilon \to 0} \left(\frac{1 - (1 - s\varepsilon + \cdots)}{s\varepsilon}\right) = 1.$$

A.1.1. *Residence time*

Process engineers are often confronted with the calculation of residence times in continuous-flow systems (reactors, columns, etc.). The residence time is the time needed for a fluid to travel from one end of the system (process) to the other. The residence time is a convenient time base for normalization (usually, the state variables are made dimensionless and scaled to take a value of unity at their target value). The residence time is also directly related to the efficiency and productivity of a chemical process. Some examples are as follows:

– For a heat exchanger, the efficiency of energy transfer from one fluid to another depends on the residence time. Several heat exchanger configurations exist which allow the residence time to be increased by multiplying the number of passes of the fluids in the device.

– The relative degree of conversion (a measure of the extent of a chemical reaction) of a reactant is directly related to the residence time. It increases with the residence time.

– In adistillation column, the selectivity and the energy efficiency are functions of the residence time.

Tracer tests (isotopic, etc.) are commonly used in chemical engineering for determining the residence time. An amount of tracer is fed into the process as quickly as possible (an impulse input). The output is then measured and interpreted as the impulse response of the process.

For a linear system, the residence time can be directly calculated from the impulse response or from the parameters of the transfer function. In fact, a linear system can be defined by its impulse response $g(t)$. Its output equation is given by:

$$y(t) = \int_{\tau=0}^{\infty} g(t - \tau)u(\tau)\,d\tau = g * u, \quad Y(s) = G(s)\,U(s), \qquad \text{(A.2)}$$

where $y(t)$ and $u(t)$ represent the output and the input, respectively. The residence time is given by:

$$\tau_{res} = \frac{\int_{\tau=0}^{\infty} t\,g(t)\,dt}{\int_{\tau=0}^{\infty} g(t)\,dt}. \qquad \text{(A.3)}$$

In the case of a continuous flow system, the residence time can be interpreted as the expected time that it takes for a molecule to pass through the flow system.

The residence time can also be related to the input and output signals without using a description of the process by a phenomenological model. On the basis of the transfer function, the residence time can be calculated as follows for a continuous system:

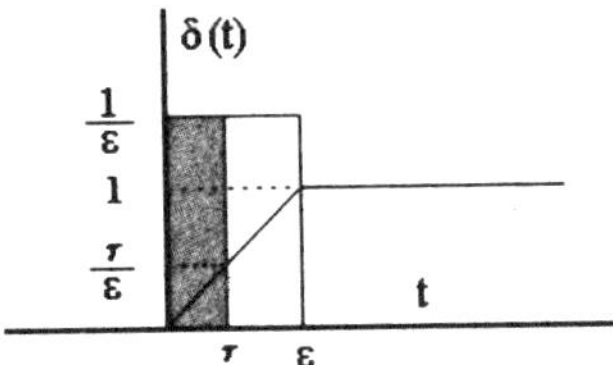

Figure A.5. *Approximation of the unit step*

$$\tau_{res} = -\frac{G'(0)}{G(0)},$$

where $G(s)$ is the Laplace transform of the impulse response $g(t)$, and $G'(\cdot)$ is the derivative of $G(\cdot)$ with respect to s. From Equation (A.2), we derive:

$$G(s) = \mathcal{L}\left[g(t)\right] = \int_0^\infty g(t)\exp\left(-st\right)dt,$$

$$\frac{G(s)}{ds} = -\int_0^\infty tg(t)\exp\left(-st\right)dt,$$

$$G(0) = \int_0^\infty g(t)\,dt, \quad G'(0) = \left.\frac{G(s)}{ds}\right|_{s=0} = -\int_0^\infty tg(t)\,dt,$$

and

$$\frac{d^n G(s)}{ds^n} = (-1)^n \int_0^\infty t^n g(t)\exp\left(-st\right)dt,$$

$$\left.\frac{d^n G(s)}{ds^n}\right|_{s=0} = (-1)^n \int_0^\infty t^n g(t)\,dt.$$

A.2. The unit step

In many industrial processes, variation of the operating point is done by making a change of the relevant variables in a form close to that of a step. From a mathematical point of view, the unit step is intimately related to the Dirac impulse.

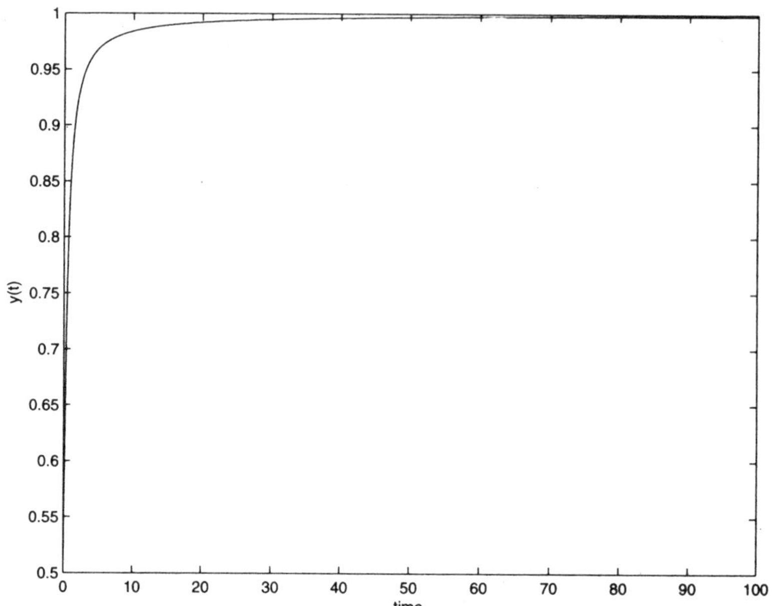

Figure A.6. *Approximation of a step response by* $(1/\pi)\arctan 2x + 1/2$

From Figure A.5, we obtain the unit step:

$$1(t) = \int_{-\infty}^{t} \delta(t) f \, dt. \tag{A.4}$$

It is clear from Equation (A.4) that:

$$1(t) = 0 \text{ for } t < 0, \, 1(t) = 1 \quad \text{for } t > 0.$$

On the basis of the approximation $g(x)$ (A.1), we obtain:

$$v(t) = \int_{-\infty}^{t} \frac{1}{\pi} \frac{a}{1 + a^2 t^2} \, dt = \left[\frac{1}{\pi} \arctan(at) \right]_{-\infty}^{t} = \frac{1}{\pi} \arctan(at) + \frac{1}{2}.$$

Figure A.6 shows $v(t)$ for $a = 2$.

The derivative $dv(t)/dt$ of $v(t)$ is equal to $\delta(t)$. The Laplace transform of $1(t)$ is equal to $1/s$.

A.3. The Routh–Hurwitz criterion

In what follows, we shall present the Routh–Hurwitz criterion and a proof of it.

Theorem 1 *All the roots of $f(s)$:*

$$f(s) = a_0 s^n + a_1 s^{n-1} + \cdots + a_n, \text{ where } a_0 > 0, a_i \in \mathcal{R}, \quad i = \overline{1, n}, \qquad (A.5)$$

lie in the left half-plane if and only if all the determinants:

$$\Delta_1 = |a_1|, \Delta_2 = \begin{vmatrix} a_1 & a_0 \\ a_3 & a_2 \end{vmatrix}, \quad \Delta_3 = \begin{vmatrix} a_1 & a_0 & 0 \\ a_3 & a_2 & a_1 \\ a_5 & a_4 & a_3 \end{vmatrix}, \cdots,$$

$$\Delta_n = \begin{vmatrix} a_1 & a_0 & 0 & \cdots & 0 \\ a_3 & a_2 & a_1 & \cdots & 0 \\ \cdots & \cdots & \cdots & \cdots & \cdots \\ a_{2n-1} & a_{2n-2} & a_{2n-3} & \cdots & a_n \end{vmatrix} \qquad (A.6)$$

where $a_i = 0$ for $i > n$, are positive.

Proof. The proof is performed by induction. For $n = 1$, the conditions of the Routh–Hurwitz criterion are fulfilled. From the condition $a_0 > 0$, we deduce that the zero of the polynomial:

$$f(s) = a_0 s + a_1 = 0$$

is negative $(-a_1/a_0)$ if $a_1 > 0$. In this case conditions (A.6) reduce to $a_1 > 0$.

Let us assume that the conditions (A.6) hold for polynomials of degree $\leq n - 1$, and we shall show that they hold also for a polynomial of degree n.

The decomposition of the polynomial $f(s)$ into odd and even terms leads to:

$$f(s) = f_1(s) + f_2(s), \qquad (A.7)$$

where:

$$f_1(s) = a_0 s^n + a_2 s^{n-2} + \cdots,$$

$$f_2(s) = a_1 s^{n-1} + a_3 s^{n-3} + \cdots. \qquad (A.8)$$

Let us consider the following polynomial $g(s)$ of degree $\leq n - 1$:

$$g(s) = a_1 f_1(s) + (a_1 - a_0 s) f_2(s)$$
$$= (a_1)^2 s^{n-1} + (a_1 a_2 - a_0 a_3) s^{n-2} + a_1 a_3 s^{n-3} + (a_1 a_4 - a_0 a_5) s^{n-4} + \cdots.$$
$$(A.9)$$

The determinants (A.6) associated with this polynomial will be denoted by Λ_i, $i = 1, 2, \cdots, n-1$.

We shall now determine an expression relating Λ_i to Λ_{i+1}.

Let us consider the following determinant:

$$a_0 a_1 \Lambda_i = a_0 a_1 \begin{vmatrix} a_1 a_2 - a_0 a_3 & a_1^2 & 0 & 0 & \cdots & 0 \\ a_1 a_4 - a_0 a_5 & a_1 a_3 & a_1 a_2 - a_0 a_3 & a_1^2 & \cdots & 0 \\ a_1 a_6 - a_0 a_7 & a_1 a_5 & a_1 a_4 - a_0 a_5 & a_1 a_3 & \cdots & 0 \\ \cdots & \cdots & \cdots & \cdots & \cdots & \cdots \end{vmatrix}.$$

Observe that the determinant:

$$\begin{vmatrix} a_0 a_1 & 0 & \cdots & \cdots & \cdots & \cdots & 0 \\ a_0 a_3 & a_1 a_2 - a_0 a_3 & a_1^2 & 0 & \cdots & \cdots & 0 \\ a_0 a_5 & a_1 a_4 - a_0 a_5 & a_1 a_3 & a_1 a_2 - a_0 a_3 & a_1^2 & \cdots & 0 \\ a_0 a_7 & a_1 a_6 - a_0 a_7 & a_1 a_5 & a_1 a_4 - a_0 a_5 & a_1 a_5 & \cdots & 0 \\ \cdots & \cdots & \cdots & \cdots & \cdots & \cdots & \cdots \end{vmatrix}.$$

is equal to $a_0 a_1 \Lambda_i$. If we expand this determinant with respect to the first row, we obtain $a_0 a_1 \Lambda_i$. Now let us add the elements of the first column to the second column, add the elements of the third column multiplied by a_0 / a_1 to the fourth column, and so on. We obtain:

$$a_0 a_1 \Lambda_i = \begin{vmatrix} a_0 a_1 & 0 + a_0 a_1 & \cdots & \cdots & \cdots & \cdots & 0 \\ a_0 a_3 & \underbrace{a_1 a_2 - a_0 a_3 + a_0 a_3}_{=0} & a_1^2 & \underbrace{0 + a_1^2 \dfrac{a_0}{a_1}}_{a_1 a_0} & \cdots & \cdots & 0 \\ a_0 a_5 & \underbrace{a_1 a_4 - a_0 a_5 + a_0 a_5}_{=0} & a_1 a_3 & \underbrace{a_1 a_2 - a_0 a_3 + a_1 a_3 \dfrac{a_0}{a_1}}_{=0} & \cdots & \cdots & 0 \\ a_0 a_7 & \underbrace{a_1 a_6 - a_0 a_7 + a_0 a_7}_{=0} & a_1 a_5 & \underbrace{a_1 a_4 - a_0 a_5 + a_1 a_5 \dfrac{a_0}{a_1}}_{=0} & \cdots & \cdots & 0 \\ \cdots & \cdots & \cdots & \cdots & \cdots & \cdots & \cdots \end{vmatrix}$$

We have two common factors, namely a_0 (in the first column) and a_1 (in the other columns). Consequently, we obtain:

$$a_0 a_1 \Lambda_i = \begin{vmatrix} a_0 a_1 & a_0 a_1 & \cdots & \cdots & \cdots & \cdots & 0 \\ a_0 a_3 & a_1 a_2 & a_1^2 & a_1 a_0 & \cdots & \cdots & 0 \\ a_0 a_5 & a_1 a_4 & a_1 a_3 & a_1 a_2 & \cdots & \cdots & 0 \\ a_0 a_7 & a_1 a_6 & a_1 a_5 & a_1 a_4 & \cdots & \cdots & 0 \\ \cdots & \cdots & \cdots & \cdots & \cdots & \cdots & \cdots \end{vmatrix} = a_0 a_1^i \Delta_{i+1}. \qquad \text{(A.10)}$$

Now we shall show that the roots of the polynomial $f(s)$ are located in the left half-plane if and only if the roots of $g(s)$ satisfy the same condition.

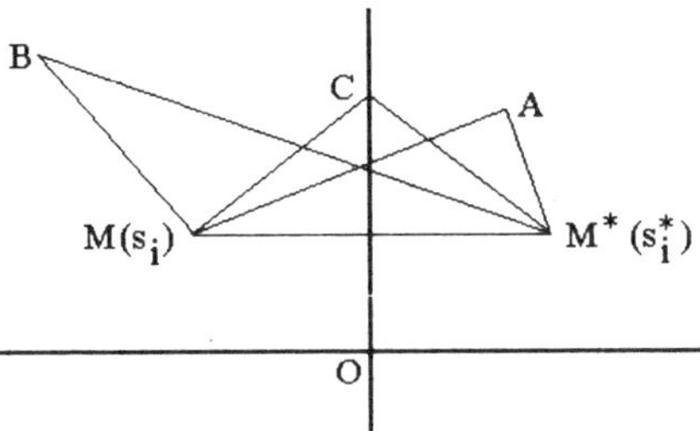

Figure A.7. *Relation between* $|s - s_i|$ *and* $|s + s_i^*|$

Let us now assume the following.

H0. The roots $s_i = x_i + jy_i$ of the polynomial $f(s)$ are located in the left half-plane ($x_i < 0$), where:

$$f(s) = a_0 \prod_{i=1}^{n} (s - s_i) = a_0 s^n + a_1 s^{n-1} + \cdots + a_n. \tag{A.11}$$

We introduce the polynomial:

$$f_*(s) = a_0 \prod_{i=1}^{n} (s + s_i^*) = a_0 s^n - a_1 s^{n-1} + \cdots + (-1)^n a_n,$$
$$s_i^* = x_i - jy_i, \tag{A.12}$$

the roots of which are located in the right half-plane. We have:

$$-s_i^* = -(x_i - jy_i) = -x_i + jy_i, \quad Re(-s_i^*) = -x_i > 0.$$

It follows that:

$$f(s) + f_*(s) = 2f_1(s) \quad \text{and } f(s) - f_*(s) = 2f_2(s). \tag{A.13}$$

Observe that the points M and M^* representing s_i and s_i^* are symmetric relative to the imaginary axis. In other words, the imaginary axis represents the perpendicular bisector of the line segment MM^*. From Figure A.7, we derive:

$$AM > AM^*, CM = CM^*, \quad BM < BM^*.$$

It follows that:

$$|s - s_i| \gtreqless |s + s_i^*| \quad \text{for} Re s \gtreqless 0.$$

Therefore, from Equations (A.11) and (A.12), we obtain:

$$|f(s)| \gtreqless |f_*(s)| \quad \text{for } Res \gtreqless 0. \tag{A.14}$$

Now we shall prove that the roots of $f_2(s)$ are located on *the imaginary axis* $(s = \pm j\omega)$. Consider the roots of the polynomial $f_2(s)$, which are also the roots of (see Equation A.13):

$$f(s) - f_*(s) = 0. \tag{A.15}$$

Hence, from Equation (A.15), we derive:

$$f(s) = f_*(s),$$

which leads to:

$$|f(s)| = |f_*(s)|.$$

In view of Equation (A.14), this result holds only for $Res = 0$, which means that *the roots of $f(s) - f_*(s)$ are imaginary*. Therefore, we deduce from (A.13) that *the roots of $f_2(s)$ are also imaginary*.

We shall now show that these roots (roots of $f_2(s)$) are *simple*. We proceed by contradiction. We assume the following.

H1. $f(j\omega) = f_*(j\omega)$ and $f'_*(j\omega) = f'_*(j\omega)$ $(j\omega$ is a root at least of order 2), where $f'_*(.)$ denotes the derivative.

Using this assumption and the fact that $f(.)$ and $f_*(.)$ have no zeros on the imaginary axis, we obtain:

$$\frac{f'(j\omega)}{f(j\omega)} = \frac{f'_*(j\omega)}{f_*(j\omega)} \quad \text{or } (\ln f(j\omega))' = (\ln f_*(j\omega))'. \tag{A.16}$$

Using Equations (A.11) and (A.12), we derive:

$$\sum_{i=1}^{n} \frac{1}{j\omega - s_i} = \sum_{i=1}^{n} \frac{1}{j\omega + s_i^*}, \tag{A.17}$$

but[2]:

$$Re \frac{1}{j\omega - s_i} > 0 \quad \text{and } Re \frac{1}{j\omega + s_i^*} < 0 \text{ for all } i,$$

which contradicts Equation (A.17). Consequently, the assumption **H1** is not true. Therefore, *the roots of $f_2(s)$ are simple*.

2.

$$Re \frac{1}{s} = Re \left(\frac{s^*}{|s|^2} \right) = \frac{Re(s^*)}{|s|^2}.$$

The roots of the polynomial $f_2(s)$ are such that:

$$\frac{a_1}{a_0} = -\sum_{i=1}^{n} s_i > 0. \tag{A.18}$$

The truth of Equation (A.18) is obvious. Therefore, $a_1 \neq 0$. From Equation (A.8), the degree of $f_2(x)$ is equal to $n-1$. In the above, we have shown that the roots of the polynomial $f_2(s)$ are imaginary and simple, and its degree is equal to $n-1$. Observe that $\deg f_1(x) = n$. Using the previous results, we obtain the following expansion:

$$\frac{f_1(s)}{f_2(s)} = \frac{a_0}{a_1}s + \sum_{i=1}^{n-1}\frac{\delta_i}{s - j\gamma_i} + C, \quad C = \begin{cases} 0 \text{ for } n \text{ odd,} \\ \dfrac{\beta_i}{s} \text{ for } n \text{ even.} \end{cases} \tag{A.19}$$

Next, note that:

$$\frac{f_1(s)}{f_2(s)} = \frac{f(s) + f_*(s)}{f_2(s) - f_*(s)} = \frac{1 + f_*(s)/f(s)}{1 - f_*(s)/f(s)} = \frac{1 + x + jy}{1 - x - jy}, \tag{A.20}$$

where $x = Ref_*(s)/f(s)$ and $y = Imf_*(s)/f(s)$. Consider the sign of the real part of the homographic function[3]:

$$\frac{1 + f_*(s)/f(s)}{1 - f_*(s)/f(s)} = \frac{(1 + x + jy)(1 - x + jy)}{(1 - x - jy)(1 - x + jy)} = \frac{1 - x^2 - y^2 + 2jy}{(1 - x)^2 + y^2},$$

$$Re\left(\frac{1 + f_*(s)/f(s)}{1 - f_*(s)/f(s)}\right) = \frac{1 - x^2 - y^2}{(1 - x)^2 + y^2} = \frac{1 - |f_*(s)/f(s)|}{(1 - x)^2 + y^2}.$$

---

3. The geometric transformations $\mathcal{T}$ which associate a point $M(s)$ of coordinates $x = Res$ and $y = Ims$ with a point $M'(x', y')$, where $x' = g(x, y)$ and $y' = h(x, y)$, can be characterized by functions $f(s)$ of the complex variable s, i.e.,

$$x' \quad - \quad y(x, y) = Ref(s),$$

$$y' \quad = \quad h(x, y) = Imf(s).$$

A homographic function (a fractional transformation or Möbius transformation) is defined by:

$$f(s) = \frac{\alpha s + \beta}{\delta s + \gamma},$$

where α, β, δ and γ are complex constants such that:

$$\alpha\gamma - \beta\delta \neq 0.$$

For $\alpha\gamma - \beta\delta = 0$, we obtain $\alpha/\delta = \beta/\gamma$, and the expression for $f(s)$ reduces to:

$$f(s) = \frac{\alpha s + \beta}{\delta s + \gamma} = \frac{\alpha s/\delta\gamma + \beta/\delta\gamma}{\delta s/\delta\gamma + \gamma/\delta\gamma} = \frac{\alpha}{\delta}\frac{s/\gamma + 1/\delta}{s/\gamma + 1/\delta} = \alpha/\delta = const.$$

A homographic fonction transforms a line or a circle into a line or a circle.

we obtain:

$$sgn Re \left(\frac{1 + f_* (s) / f (s)}{1 - f_* (s) / f (s)} \right) = sgn \left(1 - |f_* (s) / f (s)| \right). \qquad \text{(A.21)}$$

From Equations (A.20) and (A.14), we have:

$$Re \frac{f_1 (s)}{f_2 (s)} \gtreqless 0 \quad \text{for } Re s \gtreqless 0. \qquad \text{(A.22)}$$

Observe that:

$$Re \frac{1}{s - j\gamma_i} \gtreqless 0 \quad \text{for } Re s \gtreqless 0,$$

and, in the neighborhood of $s = j\gamma_i$, the sign of $Re f_1 (s) / f_2 (s)$ is determined by the sign of $Re 1 / (s - j\gamma_i)$. The latter observation is derived readily from the expansion (A.19). It follows that in the expansion (A.19) all the δ_i are positive, and C is a pure imaginary constant.

From Equations (A.9) and (A.19), we derive:

$$\begin{aligned}
g (s) &= a_1 f_1 (s) + (a_1 - a_0 s) f_2 (s) = f_2 (s) \left[a_1 \frac{f_1 (s)}{f_2 (s)} + (a_1 - a_0 s) \right] \\
&= f_2 (s) \left[a_1 \left(\frac{a_0}{a_1} s + \sum_{i=1}^{n-1} \delta_i / (s - j\gamma_i) + C \right) + (a_1 - a_0 s) \right] \qquad \text{(A.23)} \\
&= a_1 f_2 (s) \left[1 + \sum_{i=1}^{n-1} \frac{\delta_i}{s - j\gamma_i} + C \right].
\end{aligned}$$

The real part of the term $\left[1 + \sum_{i=1}^{n-1} \delta_i / (s - j\gamma_i) + C \right]$ is positive for $Re s \geq 0$, and the roots of $f_2 (s)$ are $s = j\gamma_i$, for which $g (s) \neq 0$. If we had $g (j\gamma_i) = 0$, we would deduce $f_1 (j\gamma_i) = 0$, which would mean that $f (j\gamma_i) = 0$ (see equation (A.7)), which is in contradiction with the assumption **H0**.

Therefore, the roots of $g (s)$ are located to the left of the imaginary axis ($Re s < 0$).

Let us now assume the following.

H2. *All the roots of the polynomial $g (s)$ are located to the left of the imaginary axis, and $a_1 > 0$.*

From Equation (A.9), we deduce that for $g (s)$, the polynomials $g_1 (s)$ and $g_2 (s)$ analogous to $f_1 (s)$ and $f_2 (s)$ for $f (s)$ can be written in the form:

$$g_1 (s) = a_1 f_2 (s), \quad g_2 (s) = a_1 f_1 (s) - a_0 s f_2 (s),$$

which leads to:

$$\frac{f_1(s)}{f_2(s)} = \frac{g_2(s)}{g_1(s)} + \frac{a_0}{a_1} s. \tag{A.24}$$

In the same fashion as for $f(s)$, we obtain for the function $g(s)$:

$$Re \frac{g_1(s)}{g_2(s)} \gtreqless 0 \quad \text{for } Res \gtreqless 0.$$

Observe that $a_0/a_1 > 0$ and that the sign of $Re f_2(s)/f_1(s)$ coincides with the sign of $Re f_1(s)/f_2(s)$. From equation (A.24), we deduce that the inequalities (A.22) are fulfilled. Thus, using equations (A.20) and (A.21) again, we have:

$$|f(s)| > |f_*(s)| \quad \text{for } Res > 0.$$

Therefore, it follows that $f(s)$ has no roots located to the right of the imaginary axis.

Observe also that $f(s)$ cannot have pure imaginary roots. On the imaginary axis we have:

$$|f(s)| = |f_*(s)|,$$

and, consequently, each pure imaginary root of $f(s)$ would also be a root of the polynomials $f_1(s)$ and $f_2(s)$, and consequently a root of $g(s)$ which contradicts the assumption made before. That is why all the roots of $f(s)$ are located to the left of the imaginary axis.

In summary, we have proved the following:

A. $\Lambda_i = a_1^{i-1} \Delta_{i+1}$;

B. *roots of* $\in LHP \iff$ *roots of* $g(s) \in LHP$, *where* LHP *denotes the left half-plane.*

We shall denote the determinants associated with the polynomial $f(s)$ by $\Delta(f)_i$, $i = \overline{1, n}$. On the basis of these results, we can easily iterate from $n - 1$ to n as follows.

Let us assume that this theorem is valid for any polynomial $p(s)$ such that $\deg p(s) \leq n - 1$ and $a_0 > 0 \implies$ the roots of $p(s) \in LHP \iff \Delta(p)_i > 0$, $i = \overline{1, n-1}$. For a given polynomial $f(s)$ of degree n such that $\Delta(f)_i > 0$, $i = \overline{1, n} \implies a_1 > 0$ $(-a_1/a_0 = \sum \text{roots})$, we obtain from **A** the result that $\Delta(g)_i > 0 \overset{\mathbf{B}}{\implies}$ roots of $f(s) \in LHP$.

Now, the roots of a given polynomial $f(s) \in LHP \overset{\mathbf{B}}{\implies}$ the roots of the polynomial $g(s) \in LHP \implies \Delta(g)_i > 0$. As $a_1 > 0$, this implies that $\Delta(f)_{i+1} > 0$. Here, the notation $\overset{\mathbf{D}}{\implies}$ means "the statement **D** implies". ∎

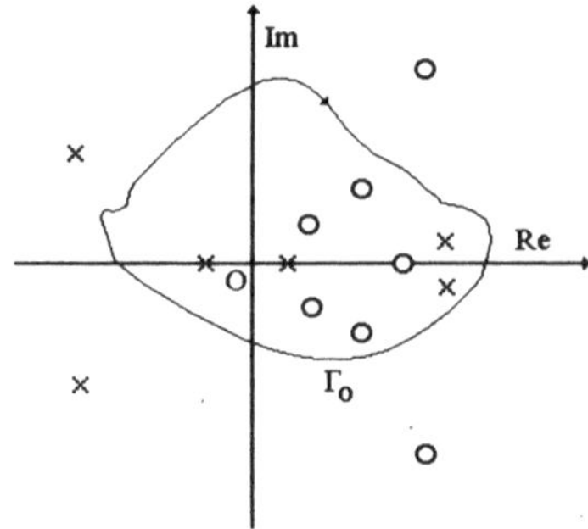

Figure A.8. *Contour enclosing a set of zeros and poles of the mapping function $f(s)$*

This proof is relatively long, but it constitutes a good exercise in the use of polynomials, determinants and complex functions. It it is very difficult to find this proof in the literature, and practically all existing books present a relatively complex form (based on the number of changes in sign of some coefficients) of the **Routh–Hurwitz** criterion.

The next subsection presents the Nyquist criterion.[4]

A.4. The Nyquist criterion

Let us consider a complex function $f(s)$. Any point M of coordinates x and y will be denoted here by $M(s)$, $s = x + jy$. This function is a transformation which associates any point $M(s)$ with a point $M'(f(s))$ of coordinates $x' = Re f(s)$ and $y' = Im f(s)$. In other words, M' represents the image of the original M under the mapping $f(s)$. Let us consider a curve Γ_o "original", where Z zeros and P poles of $f(s)$ are located inside it (see Figure A.8). Its mapping under $f(s)$ will be denoted by Γ_i "image".

The following theorem, which is known as the argument principle, is the main tool for establishing the Nyquist criterion. It is based on Cauchy's residue theorem.

---

4. The idea behind this criterion was originally due to the Russian I. Vychnégradski in 1877, and it was developed later by the American engineer H. Nyquist. Its rigorous mathematical proof was given by the Soviet mathematician N. Meymann in 1949.

Theorem 2 *If a function $f(s)$ is analytic[5] within a domain defined by the interior of the curve Γ_o except for a finite number of points at most, then the number N of clockwise encirclements of the origin by Γ_i, which represents the mapping of Γ_o under $f(s)$, is equal to $Z - P$, where Z and P represent the numbers of zeros and poles, respectively, of $f(s)$ enclosed by the closed contour Γ_o.*

For the example presented above (see Figure A.8), where $Z = 6$ and $P = 4$, we derive $N = 2$ using the argument principle.

Proof. We shall present the main lines of the proof of this theorem. In order to relate the argument principle to a stability analysis, let us consider the open-loop transfer function $T(s)$ as the mapping $f(s)$. The open-loop transfer function can be written in the form:

$$T(s) = \frac{\prod_{i=1}^{m}(s - z_i)}{\prod_{i=1}^{n}(s - p_i)},$$

where z_i and p_i are the zeros and the poles of $T(s)$. The argument of $T(s)$ is given by:

$$\arg T(s) = \sum_{i=1}^{m} \arg(s - z_i) - \sum_{i=1}^{n} \arg(s - p_i). \tag{A.25}$$

Let us associate each complex number s with a point M, and the zeros z_i and poles p_i with the points Z_i and P_i, respectively. Equation (A.25) can be written as follows:

$$\arg T(s) = \arg \sum_{i=1}^{m} \overrightarrow{Z_i M} - \arg \sum_{i=1}^{n} \overrightarrow{P_i M}.$$

---

5. A function $f(s)$, $s = x + jy$, is said to be analytic if it has a derivative. The real part $u(x, y)$ and the imaginary part $v(x, y)$ of an analytic function are related by what are called the Cauchy–Riemann conditions,

$$\frac{\partial u(x, y)}{\partial x} = \frac{\partial v(x, y)}{\partial y}, \quad \frac{\partial v(x, y)}{\partial x} = -\frac{\partial u(x, y)}{\partial y}.$$

For example, the function $f(s) = s^* = x - jy$ is not analytic. In this case:

$$\frac{\partial u(x, y)}{\partial x} = 1, \quad \frac{\partial v(x, y)}{\partial x} = 0, \quad \frac{\partial u(x, y)}{\partial y} = 0, \quad \frac{\partial v(x, y)}{\partial y} = -1.$$

The Cauchy–Riemann conditions are not fulfilled. The reader can also easily verify that the function:

$$f(s) = s^2 = (x + jy)^2 = x^2 - y^2 + 2jxy$$

is not analytic.

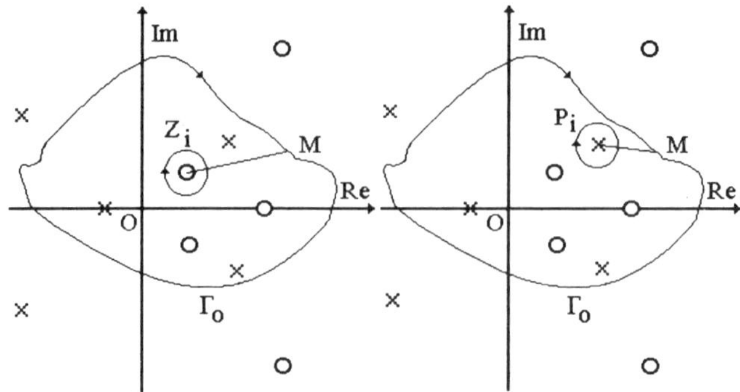

Figure A.9. *Contributions of the poles and zeros located inside the curve Γ_0*

We shall classify the zeros and poles of the function $T\left(s\right)$ into two classes:

Class 1: the zeros and the poles located inside the curve Γ_o.

From Figure A.9, we deduce the contributions of these zeros and poles to the variation of the argument of $T\left(s\right)$ when the point M moves around the curve Γ_o to be:

$$\arg \overrightarrow{Z_i M} = -2\pi, \quad \arg \overrightarrow{P_i M} = -2\pi. \tag{A.26}$$

Class 2: the zeros and poles located outside the curve Γ_o.

From Figure A.10, we deduce the contributions of these zeros and poles to the variation of the argument of $T\left(s\right)$ when the point M moves around the curve Γ_0 to be:

$$\arg \overrightarrow{Z_i M} = 0, \quad \arg \overrightarrow{P_i M} = 0. \tag{A.27}$$

In summary, the variation of the argument of the function $T\left(s\right)$ when a clockwise trip is made around the contour Γ_0 is:

$$\Delta \arg T\left(s\right) = 2\pi Z - 2\pi P = 2\pi\left(Z - P\right), \tag{A.28}$$

and $N = Z - P$, which corresponds to the desired results. ∎

In stability analysis, we are interested in finding the poles of the closed-loop transfer function of the system, i.e., the zeros of the characteristic equation:

$$1 + T\left(s\right) = 0,$$

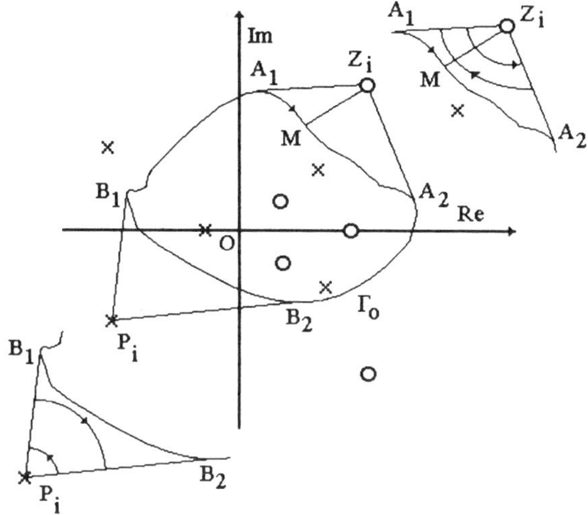

Figure A.10. *Contributions of the poles and zeros located outside the curve Γ_0*

that lie in the right half-plane. In order to apply the argument principle to stability analysis, let us consider a contour Γ_0 which coincides with the instability domain (the Nyquist contour), namely the right side of the imaginary axis ($Re p_i > 0$, where the p_i represent the poles of the system considered). To achieve this objective, we have first to consider the domain defined by a semi-circle of radius equal to infinity centred at the origin (see Figure A.11). This domain coincides with the instability domain.

It may happen that the mapping function has poles or zeros on the imaginary axis. In this case the mapping function will not be defined for these poles or zeros. We must loop (wind) around the points representing these singularities (singular points) when we are plotting the contour Γ_0 (see Figure A.12). The radius ε approaches zero.

Second, we observe that the transformation $f(s) = 1 + T(s)$ is a composition of two transformations: a translation specified by a vector $\overrightarrow{V}$ of components $(1, 0)$, and the transformation defined by the mapping $T(s)$. This translation transforms the origin $O(0, 0)$ into the point $A(-1, 0)$, which is called the "critical point". Recall that a translation is an isometry. In other words, a translation preserves distances (rigid motion); a translation moves every point by a fixed distance in the same direction. Taking these comments into account, the argument principle can be used directly for the transformation $T(s)$ by changing the statement "the number N of clockwise encirclements of the origin by Γ_i, which represents the mapping of Γ_0 under $f(s)$, is equal to $Z - P$" to "the number N of clockwise encirclements of the *critical point*

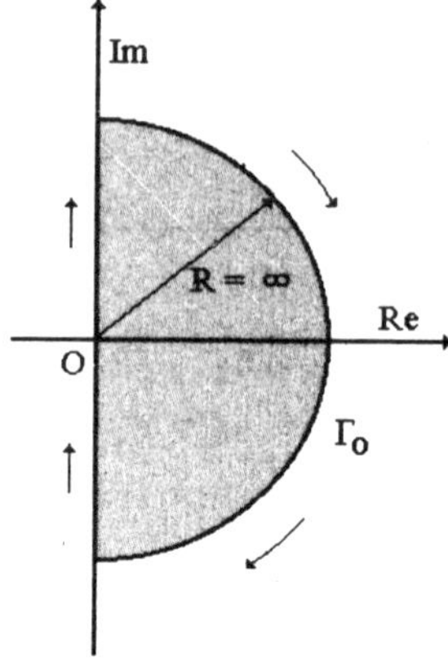

Figure A.11. *Instability domain*

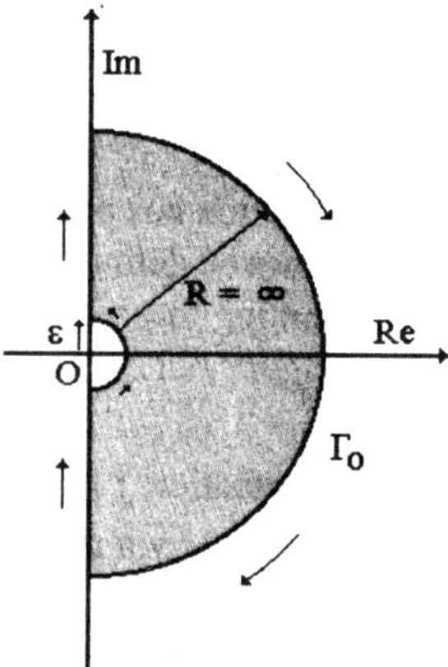

Figure A.12. *Contour, and a singularity at the origin*

$A(-1;0)$ by Γ_i, which represents the mapping of Γ_0 under $f(s)$, is equal to $Z - P$". This corresponds to the statement of the Nyquist criterion.

In what follows, we shall present proofs of the main rules related to the geometric (graphical) approach for drawing the loot locus presented in Chapter 4.

A.5. The root locus

Recall that we can associate the complex number $s = x + jy$ with a point M of coordinates (x, y). If O represents the origin $O(0,0)$ of the axes, then $OM = \sqrt{x^2 + y^2} = |s|$, and the position θ ($\arctan \theta = y/x$) of the line OM with respect to the real axis is given by $\theta = \arg s$. An analogous association can be made between vectors and complex numbers. Let us consider three points A, B and C, of coordinates (x_a, y_a), (x_b, y_b) and (x_c, y_c), respectively. We can make the following associations:

$$\overrightarrow{AB} \to s = s_b - s_a, \quad \overrightarrow{CA} \to s = s_a - s_c, \quad \overrightarrow{CB} \to s = s_b - s_c,$$

$$\overrightarrow{CA}\overrightarrow{CB} \to s = (s_a - s_c)(s_b - s_c), \quad \frac{\overrightarrow{CA}}{\overrightarrow{CB}} \to s = \frac{s_a - s_c}{s_b - s_c}, \tag{A.29}$$

where:

$$s_a = x_a + jy_a, \quad s_b = x_b + jy_b \quad \text{and} \quad s_c = x_c + jy_c.$$

From Equation (A.29), we deduce:

$$\arg \overrightarrow{AB} = \arg s, \quad \left|\overrightarrow{AB}\right| = |s|,$$

$$\arg \frac{\overrightarrow{CA}}{\overrightarrow{CB}} = \arg \frac{s_a - s_c}{s_b - s_c} = \arg(s_a - s_c) - \arg(s_b - s_c),$$

$$\left|\frac{\overrightarrow{CA}}{\overrightarrow{DB}}\right| - \left|\frac{s_a - s_c}{s_b - s_d}\right| - \frac{|s_a - s_c|}{|s_b - s_d|},$$

$$\arg \overrightarrow{CA}\overrightarrow{DB} = \arg \overrightarrow{CA} + \arg \overrightarrow{DB} = \arg(s_a - s_c) + \arg(s_b - s_d),$$

$$\left|\overrightarrow{CA}\overrightarrow{DB}\right| = \left|\overrightarrow{CA}\right|\left|\overrightarrow{DB}\right| = |s_a - s_c||s_b - s_d|.$$

Taking these correspondences into account, we can write the characteristic equation $1 + kT(s) = 0$ as follows:

$$\frac{\prod_{i=1}^{m} \overrightarrow{Z_i M}}{\prod_{i=1}^{n} \overrightarrow{P_i M}} = -\frac{1}{k}, \tag{A.30}$$

where the points M, Z_i and P_i are associated with the complex numbers s, z_i and p_i, respectively.

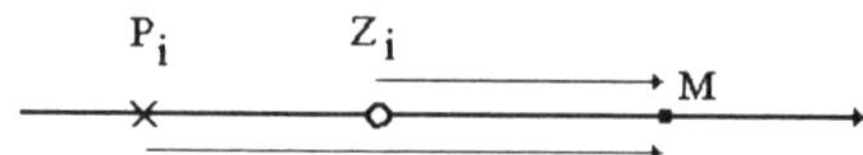

Figure A.13. *Poles and zeros located to the left of the point M*

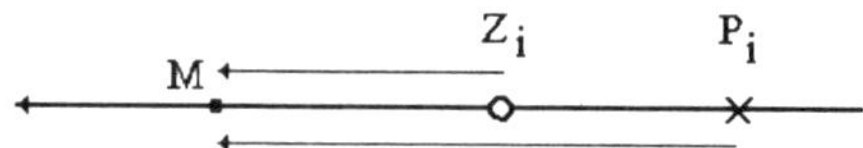

Figure A.14. *Poles and zeros located to the right of the point M*

From Equation (A.30), we derive:

$$\begin{cases} \sum_{i=1}^{m} \arg \overrightarrow{Z_i M} - \sum_{i=1}^{n} \arg \overrightarrow{P_i M} = \pm (2\lambda + 1)\,\pi, \\ \dfrac{\prod\limits_{i=1}^{m} |Z_i M|}{\prod\limits_{i=1}^{n} |P_i M|} = \dfrac{1}{k}, \quad k \in [0, +\infty[, \end{cases} \tag{A.31}$$

where λ is an integer. The latter expressions are the basis for the development of the geometric approach for drawing the root locus.

The proofs of rules 1 and 2 are given in section 4.2. Let us consider rules $3 - 9$.

Rule 3 (real-axis portions) Let us first consider the angular (argument) contributions of the real poles and zeros of the open-loop transfer function located to the left of a point M on the real axis. From Figure A.13, we obtain:

$$\arg \overrightarrow{Z_i M} = 0 \quad \text{and} \quad \arg \overrightarrow{P_i M} = 0.$$

For the real poles and zeros located to the right of the point M on the real axis (see Figure A.14), we obtain:

$$\arg \overrightarrow{Z_i M} = -\pi \quad \text{and} \quad \arg \overrightarrow{P_i M} = -\pi. \tag{A.32}$$

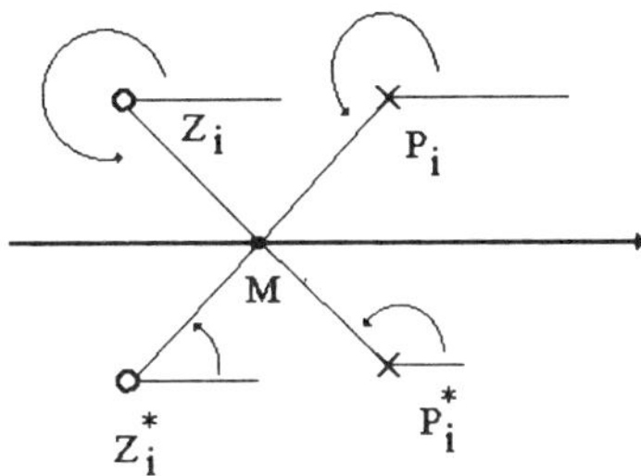

Figure A.15. *Complex poles and zeros*

Finally, let us consider the angular contributions of the complex poles and zeros of the open-loop transfer function. From Figure A.15, we derive:

$$\arg \overrightarrow{Z_i M} + \arg \overrightarrow{Z_i^* M} = 2\pi,$$

$$\arg \overrightarrow{P_i M} + \arg \overrightarrow{P_i^* M} = 2\pi.$$

Observe that the angular contribution of the complex poles and zeros and those located to the left of the point M on the real axis is equal to 0 (or 2π). As a consequence, these poles and zeros play no role in the argument condition (A.31). Rule 3 follows immediately from Equation (A.32).

Rule 4 (number of asymptotic directions $n - m$) It is well-known that the number of asymptotic directions of the fraction:

$$\frac{P(x)}{Q(x)} \simeq \frac{1}{x^{n-m}} \quad \text{when } x \to \pm\infty$$

is equal to $n - m$, where n and m are the degrees of the polynomials $P(x)$ and $Q(x)$, respectively. Thus, the system has $(n - m)$ zeros at $s = \infty$.

Rule 5 (position of asymptotic directions) Let us consider a point M belonging to the root locus and situated at infinity in the direction of a given asymptotic direction. We can immediately derive the result that the vectors $\overrightarrow{Z_i M}$ and $\overrightarrow{P_i M}$ are parallel[6] and

---

6. Two parallel lines can be characterized by the fact that their intersection is located at infinity.

form an angle β with the real axis:

$$\arg \overrightarrow{Z_i M} = \arg \overrightarrow{P_i M} = \beta,$$

$$\sum_{i=1}^{m} \arg \overrightarrow{Z_i M} - \sum_{i=1}^{n} \arg \overrightarrow{P_i M} = m\beta - n\beta = \pm (2\lambda + 1)\, \pi, \quad \lambda = 1, 2, \cdots,$$

$$\beta = \frac{(2\lambda + 1)\, \pi}{n - m}.$$

Rule 6 (intersections of the asymptotic directions with the real axis) Let us rewrite the characteristic function in a more convenient form as follows:

$$1 + k \frac{\displaystyle\prod_{i=1}^{m} (s - z_i)}{\displaystyle\prod_{i=1}^{n} (s - p_i)} = 1 + k \frac{s^m + b_1 s^{m-1} + b_2 s^{m-2} + \cdots + b_m}{s^n + a_1 s^{n-1} + a_2 s^{n-2} + \cdots + a_n} \tag{A.33}$$

$$= 1 + k \frac{P(s)}{Q(s)} = 0.$$

From the well-known expression:

$$\prod_{i=1}^{l} (s - z_i) = s^l - \left(\sum_{i=1}^{l} z_i \right) s^{l-1} + \cdots + (-1)^l \prod_{i=1}^{l} z_i,$$

we deduce that:

$$1 + k \frac{s^m - \displaystyle\sum_{i=1}^{m} z_i s^{m-1} + b_2 s^{m-2} + \cdots + b_m}{s^n - \displaystyle\sum_{i=1}^{n} p_i s^{n-1} + a_2 s^{n-2} + \cdots + a_n} = 0.$$

Recall that we are concerned with the asymptotic directions. For large values of s, we shall take into account only the first two terms of the result of division of the polynomial $Q(s)$ by the polynomial $P(s)$. We obtain:

$$\frac{Q(s)}{P(s)} = s^{n-m} - \left(\sum_{i=1}^{n} p_i - \sum_{i=1}^{m} z_i \right) s^{m-1} + \cdots \tag{A.34}$$

Taking into account only the first two terms of Equations (A.34) and (A.33) leads to:

$$1 + \frac{k}{s^{n-m} - \left(\displaystyle\sum_{i=1}^{n} p_i - \displaystyle\sum_{i=1}^{m} z_i \right) s^{n-m-1}} = 0. \tag{A.35}$$

For s large, the poles and zeros are located on the real axis at the points of abscissa δ. Let us now consider the approximation:

$$\frac{k}{(s - \delta)^{n-m}},$$

where the parameter δ represents the centroid[7]. Taking into account only the first two terms of the result of this division, we derive from the characteristic equation:

$$1 + \frac{k}{s^{n-m} - (n-m)\,\delta s^{n-m-1}} = 0. \tag{A.36}$$

Comparing Equations (A.35) and (A.36), we obtain:

$$\delta\,(n-m) = \sum_{i=1}^{n} p_i - \sum_{i=1}^{m} z_i \Longrightarrow \delta = \frac{\sum\limits_{i=1}^{n} p_i - \sum\limits_{i=1}^{m} z_i}{n-m}.$$

In what follows, we shall present a rigorous proof of this result, and we shall show that the intersection of the asymptotes with the real axis is unique. This proof is due to Professor P. Thomas, of UPS Toulouse, France. The root locus corresponds to the set:

$$C = \left\{ s \in \mathbb{C} : 1 + k\frac{P\,(s)}{Q\,(s)} = 0, \quad k \in \mathcal{R}_+ \right\}$$

where:

$$P\,(s) = \prod_{i=1}^{m} (s - z_i), \quad Q\,(s) = \prod_{i=1}^{n} (s - p_i), \quad n > m,$$

and $kP\,(s)\,/Q\,(s)$ represents the open-loop transfer function. We shall prove the following proposition.

Proposition 1 *C contains $n-m$ infinite branches, having as asymptotes the half-lines whose parametric representation is given by*

$$s = wt + \frac{1}{n-m}\left(\sum_{i-1}^{n} p_i - \sum_{i-1}^{m} z_i\right), \quad t \in \mathcal{R}_+, \quad w^{n-m} = -1.$$

Proof. To simplify the computations, let us introduce the following notation:

$$k = t^{n-m}, \quad \text{with } t \in \mathcal{R}_+.$$

As Q and P are polynomials of degree n and m, respectively, there exist two constants c_1 and c_2, and $R > 0$, such that for:

$$|s| \geq R,$$

---

7. The centroid of the poles and zeros can be determined by associating a mass equal to 1 with each pole and a mass equal to -1 with each zero.

$$Q(s) \geq c_1 |s|^n, \quad P(s) \geq c_1 |s|^m.$$

As:

$$Q(s) + t^{n-m} P(s) = 0, \tag{A.37}$$

we get:

$$|s|^n \leq c_1^{-1} |Q(s)| = c_1^{-1} |t|^{n-m} |P(s)|$$
$$\leq \frac{c_2}{c_1} |t|^{n-m} |s|^m$$

Finally, for $|s| \geq R$, we obtain:

$$|s| \leq \left(\frac{c_2}{c_1} \right)^{1/(n-m)} |t|.$$

We have to consider $t \to \infty$ in order to obtain the infinite branches of C.

We shall work in a neighborhood of $0 \in \mathbb{C}$ by introducing:

$$\sigma = \frac{1}{s}, \quad \tau = \frac{1}{t}.$$

Let:

$$Q(s) = s^n + \sum_{j=0}^{n-1} \tilde{q}_j s^j, \quad P(s) = s^m + \sum_{j=0}^{m-1} \tilde{p}_j s^j,$$

$$\tilde{Q}(X) \quad : \quad = X^n Q\left(\frac{1}{X}\right) = 1 + \sum_{j=0}^{n-1} \tilde{q}_{n-j} X^j,$$

$$\tilde{P}(X) \quad : \quad = X^m P\left(\frac{1}{X}\right) = 1 + \sum_{j=0}^{n-1} \tilde{p}_{m-j} X^j,$$

where ":=" means "by definition". Equation (A.37) becomes:

$$Q\left(\frac{1}{\sigma}\right) + \tau^{m-n} P\left(\frac{1}{\sigma}\right) = 0, \quad \text{for } \sigma \neq 0,$$

which implies:

$$\sigma^n Q\left(\frac{1}{\sigma}\right) + \sigma^{n-m} \tau^{m-n} \sigma^m P\left(\frac{1}{\sigma}\right) = 0,$$
$$\iff \tilde{Q}(\sigma) = -\sigma^{n-m} \tilde{P}(\sigma) \left(\frac{1}{\tau}\right)^{n-m} \tag{A.38}$$
$$\iff \tau^{n-m} = -\sigma^{n-m} \frac{\tilde{P}(\sigma)}{\tilde{Q}(\sigma)} \quad \text{(for } |\tau| \text{ and } |\sigma| \text{ very close to 0).}$$

As:

$$\frac{\widetilde{P}(\sigma)}{\widetilde{Q}(\sigma)} = 1,$$

there exists a unique analytic function $F_0(\sigma)$ defined in a neighborhood of 0 such that:

$$F_0(\sigma)^{n-m} = \frac{\widetilde{P}(\sigma)}{\widetilde{Q}(\sigma)}, \tag{A.39}$$

and Equation (A.38) becomes:

$$\tau^{n-m} = (ej\sigma F_0(\sigma))^{n-m},$$

the solutions of which are given by:

$$e^{2j\pi\mu/(n-m)}\tau = e^{j\pi/(n-m)}\sigma F_0(\sigma), \quad 0 \le \mu < n - m. \tag{A.40}$$

Now, we want to solve:

$$F(\sigma) = \lambda$$

in a neighborhood of 0, where:

$$F(\sigma) := \sigma F_0(\sigma).$$

Taking into account the fact that:

$$F(0) = 0, \quad F'(0) = F_0(0) = 1,$$

and using the inverse-function theorem, we obtain a unique analytic function G such that:

$$G = F^{-1}, \quad \text{with } G(0) = 0.$$

Therefore,

$$F(G(\lambda)) = \lambda \implies F'(G(\lambda))G'(\lambda) = 1 \tag{A.41}$$

and:

$$F''(G(\lambda))G'(\lambda)^2 + F'(G(\lambda))G''(\lambda) = 0,$$

which leads to $G'(0) = 1$ and:

$$G''(0) = -F''(0).$$

Finally, if we express λ as:

$$F(\sigma) = \sigma + \alpha_2\sigma^2 + O(\sigma^3),$$

we derive[8]:

$$\sigma = G\left(\lambda\right) = \lambda - \alpha_2 \lambda^2 + O\left(\lambda^3\right).$$

The derivation of Equation (A.41) permits us to obtain all the desired coefficients for the expansion of G. Let us now calculate the coefficient α_2. We obtain:

$$F\left(\sigma\right) = \sigma\left(1 + \alpha_2 \sigma + O\left(\sigma^2\right)\right) = \sigma F_0\left(\sigma\right),$$

and:

$$F_0\left(\sigma\right) = \left(\frac{1 + \widetilde{p}_{n-1}\sigma + O\left(\sigma^2\right)}{1 + \widetilde{q}_{n-1}\sigma + O\left(\sigma^2\right)}\right)^{1/(n-m)} = 1 + \frac{\left(\widetilde{p}_{n-1} - \widetilde{q}_{n-1}\right)\sigma}{n-m} + O\left(\sigma^2\right).$$

Recall that:

$$\widetilde{p}_{n-1} = -\sum_{j=1}^{m} z_j, \quad \widetilde{q}_{n-1} = -\sum_{j=1}^{n} p_j,$$

which yields:

$$\alpha_2 = \frac{\displaystyle\sum_{j=1}^{n} p_j - \sum_{j=1}^{m} z_j}{n-m}. \tag{A.42}$$

Let us now express s as a function of t. Notice that:

$$s = \frac{1}{\sigma} = \frac{1}{G\left(\lambda\right)} = \frac{1}{\lambda}\frac{1}{1 - \alpha_2\lambda + O\left(\lambda^2\right)} = \frac{1}{\lambda}\left(1 + \alpha_2\lambda + O\left(\lambda^2\right)\right).$$

By selecting a branch:

$$\lambda = e^{-j(2k-1)\pi/(n-m)}\tau := w\tau$$

for $0 \leq k \leq n - m - 1$, we derive:

$$s = wt\left(1 + \frac{\alpha_2}{wt} + O\left(\frac{1}{t^2}\right)\right) = wt + \alpha_2 + O\left(\frac{1}{t}\right), \tag{A.43}$$

where w is any $(n-m)th$ root of -1. We have now described the infinite branches of the root locus C. Their asymptotes have the following parametric representation:

$$s = wt + \alpha_2.$$

They intersect in the unique point α_2 (given by Equation (A.42)). ∎

---

8. The notation $x = O\left(\lambda^3\right)$ means:

$$x < cst \times \lambda^3.$$

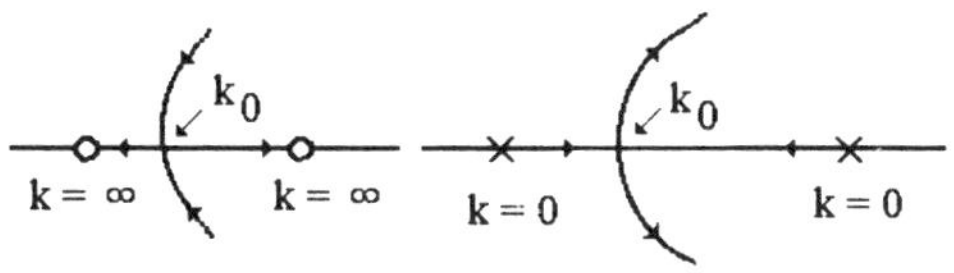

Figure A.16. *Intersection of the root locus with the real axis*

Rule 7 (intersections of the root locus with the real axis) The abscissae of the points M on the real axis belonging to the root locus (see Figure A.16) correspond to a double root of the characteristic equation:

$$1 + k\frac{P(x)}{Q(x)} = 0,$$

or:

$$\frac{P(x)}{Q(x)} = -\frac{1}{k}, \quad \frac{Q(x)}{P(x)} = -k.$$

Let us consider the real variable x and the functions:

$$f_1(x) = \frac{Q(x)}{P(x)}, \quad f_2(x) = -k.$$

The double root corresponds to the extremum (maximum or minimum) of $f_1(x)$, which is given by:

$$\frac{d}{dx}\left(\frac{Q(x)}{P(x)}\right) = 0,$$

$$P(x)\frac{dQ(x)}{dx} - Q(x)\frac{dP(x)}{dx} = 0,$$

$$\frac{dQ(x)/dx}{Q(x)} - \frac{dP(x)/dx}{P(x)} = 0,$$

$$\frac{d\left[\log Q(x)\right]}{dx} = \frac{d\left[\log P(x)\right]}{dx}.$$

Taking into account the fact that:

$$P(x) = \prod_{i=1}^{m}(x - z_i), \quad Q(x) = \prod_{i=1}^{m}(x - p_i),$$

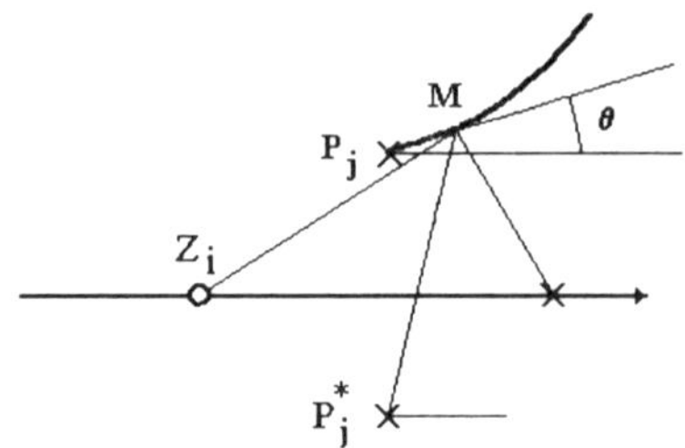

Figure A.17. *Root locus and tangent at a complex pole*

we derive:

$$\frac{d\left[\log Q\left(x\right)\right]}{dx} = \sum_{i=1}^{n} \frac{1}{x - p_i}, \qquad \frac{d\left[\log R\left(x\right)\right]}{dx} = \sum_{i=1}^{m} \frac{1}{x - z_i},$$
$$\sum_{i=1}^{n} \frac{1}{x - p_i} = \sum_{i=1}^{m} \frac{1}{x - z_i}. \tag{A.44}$$

The value of k_0 is equal to $P\left(x_0\right)/Q\left(x_0\right)$ where x_0 denotes the solution of Equation (A.44).

Rule 9 (positions of the tangents to the root locus at complex poles p_i (angles of departure) and zeros z_i (angles of arrival) with respect to the real axis) Recall the argument (phase) condition:

$$\sum_{i=1}^{m} \arg\left(s - z_i\right) - \sum_{i=1}^{n} \arg\left(s - p_i\right) = \pm\left(2\lambda + 1\right)\pi, \tag{A.45}$$

and consider a point M located on the tangent, very close to the complex pole p_j. From Equation (A.45), we derive (see Figure A.17):

$$\sum_{i=1}^{m} \arg\left(p_j - z_i\right) - \sum_{i=1,i\neq j}^{n} \arg\left(p_j - p_i\right) - \theta = -\pi = -180,$$

which leads to:

$$\theta = \sum_{i=1}^{m} \arg\left(p_j - z_i\right) - \sum_{i=1,i\neq j}^{n} \arg\left(p_j - p_i\right) + 180.$$

Similarly, for a complex zero z_j, we obtain:

$$\sum_{i=1,i\neq j}^{m} \arg\left(z_j - z_i\right) + \theta - \sum_{i=1,i\neq j}^{n} \arg\left(z_j - p_i\right) = +\pi = +180,$$

which gives:

$$\theta = 180 - \left(\sum_{i=1,i\neq j}^{m} \arg\left(z_j - z_i\right) - \sum_{i=1,i\neq j}^{n} \arg\left(z_j - p_i\right)\right).$$

A.6. Computation of integrals of the form J_2

Let us consider integrals of the form:

$$J_2 = \frac{1}{2\pi j} \int_{-j\infty}^{+j\infty} \frac{B\left(s\right) B\left(-s\right)}{A\left(s\right) A\left(-s\right)}\, ds,$$

where:

$$A\left(s\right) = a_0 s^n + a_1 s^{n-1} + \cdots + a_{n-1} s + a_n,$$
$$B\left(s\right) = b_1 s^{n-1} + \cdots + b_{n-1} s + b_n.$$

In what follows, we shall present an algorithm for computing this kind of integral. We assume that the roots of the polynomial $A\left(s\right)$ are located to the left of the imaginary axis. Let us introduce the following polynomials:

$$A\left(s\right) = \overline{A}\left(s\right) + \widetilde{A}\left(s\right), \overline{A}\left(s\right) = a_0 s^n + a_2 s^{n-2} + \cdots,$$
$$\widetilde{A}\left(s\right) = a_1 s^{n-1} + a_3 s^{n-3} + \cdots,$$
$$A_k\left(s\right) = a_0^k s^k + a_1^k s^{k-1} + \cdots + a_k^k,$$
$$B_k\left(s\right) = b_1^k s^{k-1} + b_2^k s^{k-2} + \cdots + b_k^k.$$

The polynomials $A_k\left(s\right)$ and $B_k\left(s\right)$ are defined by the following recursions:

$$A_{k-1}\left(s\right) = A_k\left(s\right) - \alpha_k s \widetilde{A}_k\left(s\right), \quad B_{k-1}\left(s\right) = B_k\left(s\right) - \beta_k \widetilde{A}_k\left(s\right),$$

where:

$$\alpha_k = \frac{a_0^k}{a_1^k}, \quad \beta_k = \frac{b_1^k}{a_1^k}, \quad A_n(s) = A(s) \quad \text{and } B_n(s) = B(s).$$

Consider now the following integral:

$$J^k = \frac{1}{2\pi j} \int_{-j\infty}^{+j\infty} \frac{B_k(s) B_k(-s)}{A_k(s) A_k(-s)} ds.$$

This integral is given by the following recursion:

$$J^k = J^{k-1} + \frac{\beta_k^2}{2\alpha_k}, \quad J^0 = 0, k = \overline{1,n},$$

and

$$J_2 = J^n.$$

A.7. On non-linear systems

Let us consider a non-linear system subject to a sinusoidal input $x(t) = x_1 \sin \omega t$. The non-linear output $y(t)$ may be expanded into a Fourier series:

$$y(t) = a_0 + \sum_{n=1}^{\infty} (a_n \sin n\omega t + b_n \cos n\omega t).$$

If we neglect the harmonics greater than one, the output of a relay is then given by the fundamental harmonic (quasi-linearization):

$$y(t) = a_0 + a_1 \sin \omega t + b_1 \cos \omega t,$$

where:

$$a_0 = \frac{1}{2\pi} \int_0^{2\pi} \varphi\left(x(t), \frac{d}{dt}x(t)\right) d(\omega t),$$

$$a_1 = \frac{1}{\pi} \int_0^{2\pi} \varphi(\cdot) \sin(\omega t) d(\omega t),$$

$$b_1 = \frac{1}{\pi} \int_0^{2\pi} \varphi(\cdot) \cos(\omega t) d(\omega t),$$

and $\varphi\left(\cdot\right)$ represents the characteristic of the non-linearity. For a symmetric non-linear system $a_0 = 0$. The non-linear system can be described as follows:

$$y\left(t\right) = a_1 \sin \omega t + b_1 \cos \omega t = q\left(x_1, \omega\right) x\left(t\right) + \frac{q'\left(x_1, \omega\right)}{\omega} s x\left(t\right),$$

$$q\left(x_1, \omega\right) = \frac{a_1}{x_1}, \quad q'\left(x_1, \omega\right) = \frac{b_1}{x_1}, \quad s = \frac{d}{dt}.$$

For $s = j\omega$, and on the basis of this approximation, we can define a generalized gain called the describing function,

$$N\left(x_1, \omega\right) = q\left(x_1, \omega\right) + jq'\left(x_1, \omega\right) = B\left(x_1\right) \exp\left(j\phi\left(x_1\right)\right),$$

where:

$$B\left(x_1\right) = \sqrt{q^2 + q'^2}, \quad \phi\left(x_1\right) = \arctan \frac{q'}{q} = \arctan \frac{b_1}{a_1}.$$

This representation of a non linear system is an attempt to extend the concept of a transfer function to non-linear systems. This approximation is reasonable, since many dynamic systems act as low-pass filters. It can be shown that the describing function minimizes the squared error in the linearizing approximation. In what follows, we shall compute the describing function (or linear descriptor) for two types of relay: an ideal relay and a relay with a dead zone and hysteresis.[9] A relay with a dead zone and hysteresis is depicted in Figure A.18. This relay can be used to decrease the influence of noise; inputs of amplitude less than $\left(\delta + \Delta\right)$ have no influence on the relay output. The input–output characteristic of this non-linear system is multivalued.

The responses of these relays to a sinusoidal input are plotted in Figures A.19 and A.20.

The non-linearities of these relays are widely used intentionally in process control. For an ideal relay, the output is equal to $M \times sgn\left(x\left(t\right)\right)$, and its characteristic is symmetric; therefore, $a_0 = 0$ and:

$$a_1 = \frac{1}{\pi} \int_0^{2\pi} M \times sgn\left(x\left(t\right)\right) \sin\left(\omega t\right) d\left(\omega t\right), \quad b_1 = 0.$$

We assume that $x_1 > M$. From the waveform depicted in Figure A.19, we obtain:

$$a_1 = \frac{1}{\pi}\left[\int_0^{\pi} M \sin\left(\omega t\right) d\left(\omega t\right) - \int_{\pi}^{2\pi} M \sin\left(\omega t\right) d\left(\omega t\right)\right] = \frac{4M}{\pi},$$

---

9. In electronics, Schmitt triggers are used as relays with hysteresis.

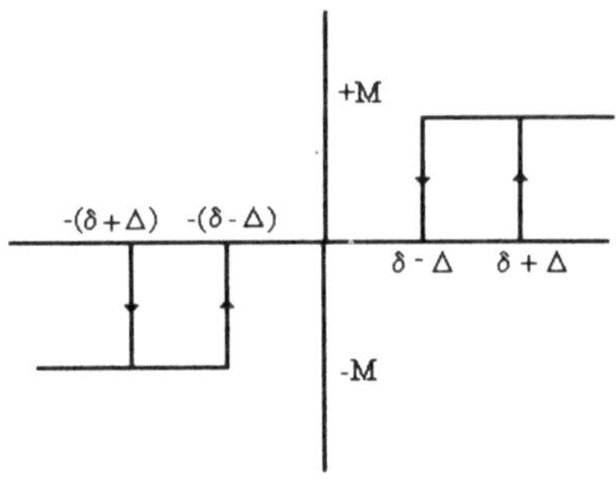

Figure A.18. *Relay with dead zone and hysteresis*

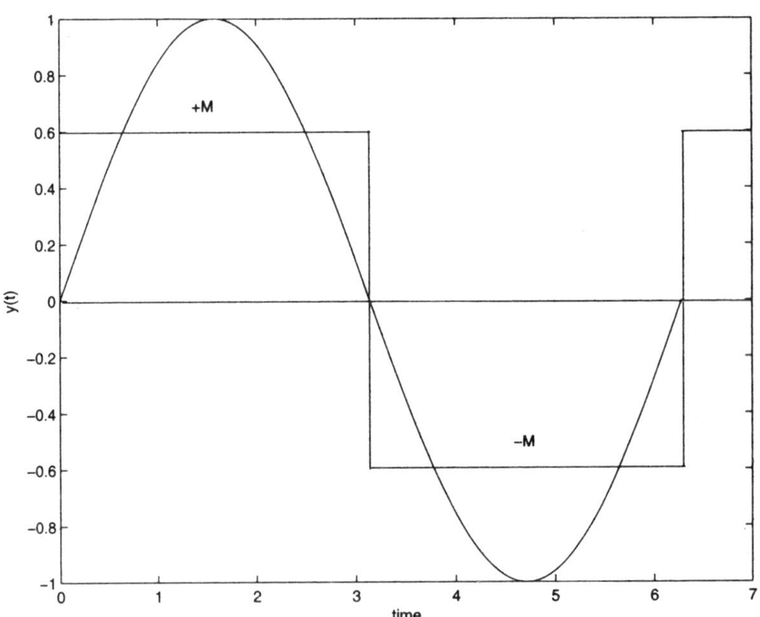

Figure A.19. *Response of an ideal relay*

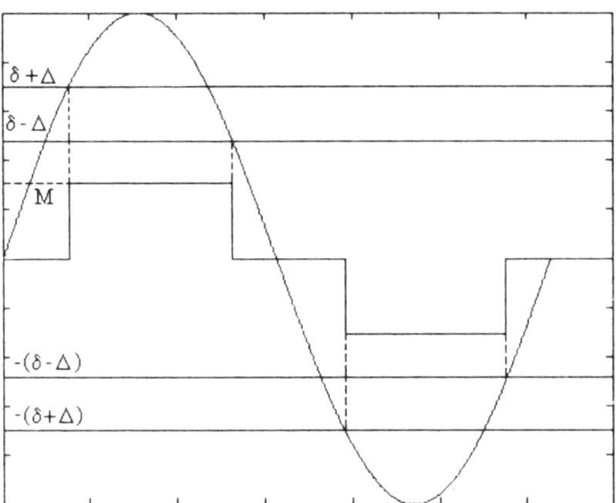

Figure A.20. *Response of a relay with a dead zone and hysteresis*

and

$$N\left(x_1, \omega\right) = \frac{4M}{\pi x_1}. \tag{A.46}$$

Note that inputs of amplitude x_1 such that $|x_1| < M$ (the relay amplitude) lead to a relay output equal to zero. As before, from the waveform plotted in Figure A.20, and under the assumption that $x_1 > \Delta + \delta$, we obtain:

$$a_1 = \frac{1}{\pi}\left[\int_{\alpha}^{\beta} M \sin\left(\omega t\right) d\left(\omega t\right) - \int_{\alpha+\pi}^{\beta+\pi} M \sin\left(\omega t\right) d\left(\omega t\right)\right],$$

$$b_1 = \frac{1}{\pi}\left[\int_{\alpha}^{\beta} M \cos\left(\omega t\right) d\left(\omega t\right) - \int_{\alpha+\pi}^{\beta+\pi} M \cos\left(\omega t\right) d\left(\omega t\right)\right]$$

where:

$$\alpha = \arcsin\left(\frac{\Delta + \delta}{x_1}\right), \quad \beta = \pi - \arcsin\left(\frac{\delta - \Delta}{x_1}\right) \quad (\cos\beta < 0).$$

Finally, we obtain:

$$a_1 = \frac{2M}{\pi}\left[\cos\alpha - \cos\beta\right], \quad b_1 = \frac{2M}{\pi}\left[\sin\beta - \sin\alpha\right].$$

Note that $\cos \beta < 0$. It follows that:

$$a_1 = \frac{2M}{\pi}\left[\sqrt{1 - \left(\frac{\Delta + \delta}{x_1}\right)^2} + \sqrt{1 - \left(\frac{\delta - \Delta}{x_1}\right)^2}\right],$$

$$b_1 = -\frac{4M}{\pi x_1}\Delta,$$

$$N(x_1, \omega) = \frac{2M}{\pi x_1}\left[\sqrt{1 - \left(\frac{\Delta + \delta}{x_1}\right)^2} + \sqrt{1 - \left(\frac{\delta - \Delta}{x_1}\right)^2}\right] - j\frac{4M}{\pi x_1^2}\Delta. \quad (A.47)$$

The frequency of oscillation (limit cycle) is given by:

$$1 + N(x_1, \omega)G(j\omega) = 0, \quad (A.48)$$

the solution of which is determined by the harmonic balance:

$$Re\left(1 + N(x_1, \omega)G(j\omega)\right) = 0 \quad \text{and} \quad Im\left(1 + N(x_1, \omega)G(j\omega)\right) = 0.$$

This equation may be solved graphically by plotting the Nyquist diagram of the linear system and of $-1/N(x_1, \omega)$. The solutions of equation (A.48) correspond to the intersections of these two plots. For an ideal relay, these solutions correspond to the intersection of the Nyquist plot with the real negative axis $]-\infty, 0]$. Notice that Equation (A.48) gives only the condition for the possible existence of a limit cycle (or oscillation).

In order to illustrate this method, let us consider the following linear system:

$$G(s) = \frac{k}{s(s+1)(s+3)}.$$

In the case of an ideal relay, from Equation (A.46) and (A.48), we obtain:

$$1 + N(x_1, \omega)G(j\omega) = 1 + \frac{4M}{\pi x_1}\frac{k}{j\omega(j\omega + 1)(j\omega + 3)} = 0,$$

$$4kM - 4\omega^2\pi x_1 + j\omega\pi x_1\left(3 - \omega^2\right) = 0,$$

$$\omega^2 = 3, \quad \omega = \sqrt{3}, \quad x_1 = \frac{kM}{3\pi}.$$

We can check the value of $\arg G\left(j\sqrt{3}\right)$:

$$\arg G\left(j\sqrt{3}\right) = \arg\left(\frac{k}{-4\omega^2 + j\omega\left(3-\omega^2\right)}\right)_{\omega=\sqrt{3}} = -\pi\ (2\pi).$$

REMARK A.1. Observe that if:

$$G\left(s\right) = \frac{1}{s}G^*\left(s\right)$$

and ω_c represents the frequency of oscillation, then:

$$\arg\frac{1}{j\omega_c}G^*\left(j\omega_c\right) = \arg\frac{1}{j\omega_c} + \arg G^*\left(j\omega_c\right) = -\pi$$

which leads to:

$$\arg G^*\left(j\omega_c\right) = -\frac{\pi}{2}.$$

In other words, if we want to obtain the frequency which leads to an argument equal to $-\pi/2$ (the intersection of the Nyquist diagram with the negative imaginary axis) we have to introduce an integrator in series with the system and an ideal relay. In order to obtain other points of the Nyquist diagram, additional dynamics $G_a\left(s\right)$ can be introduced in series with the process under consideration. The point:

$$\arg G_a\left(j\omega_c\right)G\left(j\omega_c\right) = -\pi\ (2\pi),$$
$$\arg G\left(j\omega_c\right) = -\pi - \arg G_a\left(j\omega_c\right),$$

will be obtained.

REMARK A.2. For a feedback system $F\left(s\right) = G\left(s\right)/\left(1+G\left(s\right)\right)$ in series with an ideal relay, we obtain the following for ω_c, the frequency of oscillation:

$$\arg\frac{G\left(j\omega_c\right)}{1+G\left(j\omega_c\right)} = -\pi\ (2\pi),$$
$$\frac{G\left(j\omega_c\right)}{1+G\left(j\omega_c\right)} = X\left(\omega_c\right) + jY\left(\omega_c\right) = X\left(\omega_c\right) < 0.$$

Let $G\left(j\omega_c\right) = U\left(\omega_c\right) + jV\left(\omega_c\right)$. It follows that:

$$\frac{G\left(j\omega_c\right)}{1+G\left(j\omega_c\right)} = \frac{U\left(\omega_c\right) + jV\left(\omega_c\right)}{1+U\left(\omega_c\right) + jV\left(\omega_c\right)}$$
$$= \frac{U\left(\omega_c\right)\left(1+U\left(\omega_c\right)\right) + V^2\left(\omega_c\right) + jV\left(\omega_c\right)}{\left(1+U\left(\omega_c\right)\right)^2 + V^2\left(\omega_c\right)},$$

which leads to:

$$V\left(\omega_c\right) = 0,\, G\left(j\omega_c\right) = U\left(\omega_c\right) \text{ and } F\left(j\omega_c\right) = \frac{U\left(\omega_c\right)}{1 + U\left(\omega_c\right)}. \tag{A.49}$$

As $F\left(j\omega_c\right) < 0$, we derive:

$$U\left(\omega_c\right)\left(1 + U\left(\omega_c\right)\right) < 0 \Rightarrow U\left(\omega_c\right) < 0. \tag{A.50}$$

Therefore, from Equations (A.50) and (A.49), we derive:

$$\arg G\left(j\omega_c\right) = -\pi\left(2\pi\right).$$

In other words, the frequency ω_c of oscillation corresponds to the ultimate frequency.

In the case of a relay with a dead zone, the describing function is obtained from Equation (A.47) by setting $\Delta = 0$:

$$N\left(x_1, \omega\right) = \frac{4M}{\pi x_1}\sqrt{1 - \left(\frac{\delta}{x_1}\right)^2}$$

The limit cycle is given by:

$$-4\omega^2 + j\omega\left(3 - \omega^2\right) + \frac{4kM}{\pi x_1}\sqrt{1 - \left(\frac{\delta}{x_1}\right)^2} = 0,$$

which leads to $\omega = \sqrt{3}$ and:

$$\frac{4kM}{\pi x_1}\sqrt{1 - \left(\frac{\delta}{x_1}\right)^2} - 4\omega^2 = \frac{4kM}{\pi x_1}\sqrt{1 - \left(\frac{\delta}{x_1}\right)^2} - 12 = 0. \tag{A.51}$$

For $M = \pi$, $k = 1$ and $\delta = 0.01$, we obtain the following solution:

$$x_1 = 0.33318.$$

In the case of a relay with hysteresis and without a dead zone, the describing function is obtained from Equation (A.47) by setting $\delta = 0$:

$$N\left(x_1, \omega\right) = \frac{2M}{\pi x_1}\left[\sqrt{1 - \left(\frac{\Delta}{x_1}\right)^2} + \sqrt{1 - \left(\frac{\Delta}{x_1}\right)^2}\right] - j\frac{4M}{\pi x_1^2}\Delta.$$

The limit cycle is given by:

$$-4\omega^2 + j\omega\left(3 - \omega^2\right) - j\frac{4kM}{\pi x_1^2}\Delta + \frac{4kM}{\pi x_1}\sqrt{1 - \left(\frac{\Delta}{x_1}\right)^2} = 0,$$

which implies:

$$\omega\left(3-\omega^2\right)-\frac{4kM}{\pi x_1^2}\Delta = 0 \quad \text{and} \quad \frac{kM}{\pi x_1}\sqrt{1-\left(\frac{\Delta}{x_1}\right)^2}-4\omega^2 = 0.$$

The solution of these equations may lead to several different values of ω and x_1, i.e., different limit cycles. These limit cycles can be stable (attracting), unstable (repelling) or non-stable (saddle). The stability analysis of feedback systems containing relays remains an open problem. As the describing-function approach is an approximation, it may happen that we may predict erroneous results on the basis of it, such as the existence of limit cycles even if the linear system act as low-pass filter.

REMARK A.3. From measurements of the amplitude and frequency of the limit cycle, we may identify the parameters of a system characterized by two parameters. It is sufficient to solve equation (A.48) for these parameters.

Bibliography

[AGU 99] AGUIRRE J., TULLY D., Lake pollution model, available at http://online.redwoods.cc.ca.us/instruct/darnold/deproj/Sp99/DarJoel/lakepollution.pdf, 1999.

[AND 68] ANDREWS J., "A mathematical model for the continuous culture of micro-organisms utilising inhibitory substrate", *Biotechnology and Bioengineering*, vol. 10, p. 707–723, 1968.

[AND 04] ANDREOIU A., On power system stabilizers: Genetic algorithm based tuning and economic worth as ancillary services, PhD thesis, Chalmers University of Technology, Gothenburg, Sweden, 2004.

[AST 70] ASTRÖM K. J., *Introduction to Stochastic Control Theory*, ch. 4, p. 128-139, Academic Press, New York, 1970.

[AST 84] ASTRÖM K. J., HÄGGLUND T., "Automatic tuning of simple regulators with specifications on phase and amplitude margins", *Automatica*, vol. 20, p. 645–648, 1984.

[AST 86] ASTRÖM K. J., *Applied and Computational Complex Analysis*, vol. 1-3, Wiley, New York, 1986.

[AST 95] ASTRÖM K. J., HÄGGLUND T., *PID Controllers: Theory, Design, and Tuning*, 2nd edition, Instrument Society of America, Research Triangle Park, NC, 1995.

[AST 02] ASTRÖM K. J., Control system design, ch. 7, specifications, available at http://www.cds.caltech.edu/ murray/courses/cds101/fa02/caltech/ astrom-ch7.pdf, 2002.

[BAK 00] BAK M., Control of systems with constraints, PhD thesis, Department of Automation, Technical University of Denmark, Lyngby, 2000, available at http://www.iau.dtu.dk/secretary/pdf/bak.pdf.

[BOR 90] BORDENEUVE J., BABARY J. P., ABATUT J. L., NAJIM K., "Extended horizon control of an epitaxy furnace", J. E. Rijnsdorp, J. F. Macgregor, B. D. Tyreus, T. Takamatsu, Ed., *IFAC Symposium on Modelling and Control of Chemical Reactors, Distillation Columns and Batch Processes, DYCORD+ '89*, vol. 10, Elsevier, p. 707–723, 1990.

[CAL 88] CALVET J. P., ARKUN Y., "Feedforward and feedback linearization of nonlinear systems and its implementation using internal model control (IMC)", *Industrial and Engineering Chemistry Research*, vol. 27, p. 1822–1831, 1988.

[CEN 02] CENTURELLI F., LUZZI R., SCOTTI G., TOMMASINO P., TRIFILETTI A., "An efficient synthesis-oriented CAD implementation of Nyquist stability criterion", *Electron Technology Internet Journal*, vol. 34, p. 1–5, 2002, ch. 7, specifications, available at http://www.ite.waw.pl/etij/.

[CES 97] CESSNA R., A simple microeconomics model, available at http://online.redwoods.cc.ca.us/instruct/darnold/deproj/Sp97/Randy/randproj.pdf, 1997.

[CHA 97] CHATTERJI S. D., *Cours danalyse, Tome 2: Analyse complexe*, Presse Polytechniques et Universitaires Romandes, Lausanne, 1997.

[CHE 05] CHEN Y., MOORE K. L., "Relay feedback tuning of robust PID controllers with iso-damping property", *IEEE Transactions on Systems, Man, and Cybernetics, Part B: Cybernetics*, vol. 35, p. 23–31, 2005.

[CHI 90] CHIEN I. L., FRUEHAUF P. S., "Consider IMC tuning to improve performance", *Chemical Engineering Progress*, vol. 10, p. 33–41, 1990.

[CHI 03] CHIDAMBARAM M., SREE R. P., "A simple method for tuning PID controllers for integrator/dead-time processes", *Computers and Chemical Engineering*, vol. 27, p. 211–2003, 2003.

[GOO 91] GOODWIN G. C., GRAEBE S. F., SALGADO M. E., *Control System Design*, Prentice Hall, Upper Sddle River, NJ, 1991.

[GOR 02] GORINEVSKY D., "Loop-shaping for iterative control of batch processes", *IEEE Control Systems Magazine*, vol. 22, p. 55–65, 2002.

[GRA 00] GRADSHTEYN I. S., RYZHIK I. M., *Table of Integrals, Series*, 6th edition, Academic Press, New York, 2000.

[HAL 43] HALL A. C., *The Analysis and Synthesis of Linear Servomechanisms*, Technology Press, MIT, Cambridge, MA, 1943.

[HAR 64] HARRIOTT P., *Process Control*, McGraw-Hill, New York, 1964.

[HUR 64] HURWITZ A., "Extended horizon control of an epitaxy furnace", in BELLEMAN R. E., KALABA R., Eds., *Selected Papers on Mathematical Trends in Control theory*, p. 72–82, Dover, 1964, Translated from the German by H. G. Bergman.

[IKO 02] IKONEN E., NAJIM K., *Advanced Process Identification and Control*, Marcel Dekker, New York, 2002.

[JAM 47] JAMES H. M., NICHOLS N. B., PHILLIPS R. S., "Theory of servomechanisms", *MIT Radiation Laboratory Series*, vol. 25, 1947.

[KOI 91] KOIVO H. N., TANTTU J. T., "Tuning PID controllers: survey of SISO and MIMO techniques", *Proceedings of the IFAC Intelligent Tuning and Adaptive Control Symposium*, p. 75–80, 1991.

[KRY 47] KRYLOV N., BOGOLIUBOV N., *Introduction to Nonlinear Mechanics*, Princeton University Press, Princeton, NJ, 1947.

[LAV 72] LAVRENTIEV M., CHABAT B., *Methods for Complex Analysis*, Mir, Moscow, 1972.

[LJU 83] LJUNG L., SODERSTRÖM T., *Theory and Practice of Recursive Identification*, MIT Press, Cambridge, MA, 1983.

[MIN 03] MING-TZU H., CHIA-YI L., "PID controller design for robust performance", *IEEE Transactions on Automatic Control*, vol. 48, p. 1404–1409, 2003.

[MON 42] MONOD J., *Recherche sur la croissance des cultures bactériennes,*, Herman, Paris, 1942.

[NAJ 83] NAJIM K., MURATET G., *Pratique de la régulation numérique des processus industriels*, Masson, Paris, 1983.

[NAJ 88] NAJIM K., *Control of Liquid-Liquid Extraction Columns*, Gordon and Breach, New York, 1988.

[NAJ 89] NAJIM K., *Process Modeling and Control in Chemical Engineering*, Marcel Dekker, New York, 1989.

[NAJ 96] NAJIM K., POZNYAK A. S., IKONEN E., "Calculation of residence time for nonlinear systems", *International Journal of Systems Science*, vol. 27, p. 661–667, 1996.

[NAJ 99] NAJIM K., MURATET G., *Outils mathématiques pour le génie des procédés*, Dunod, Paris, 1999.

[NAJ 04] NAJIM K., IKONEN E., AÏT-KADI D., *Stochastic Processes, Estimation, Optimisation and Analysis*, Kogan Page, London, 2004.

[NIJ 90] NIJMEIJER H., VAN DER SCHAFT A., *Nonlinear Dynamical Control Systems*, Springer-Verlag, Berlin, 1990.

[NOR 92] NORMAN S. A., "Optimization of Transient Temperature Uniformity in RTP Systems", *IEEE Transactions on Electron Devices*, vol. 39, p. 205–207, 1992.

[ODW 03] ODWYER A., *Handbook of PI and PID Controller Tuning Rules*, Imperial College Press, London, 2003.

[RAD 75] RADEMAKER O., RIJNSDORP J. E., MAAELEVELD A., *Dynamics and Control of Continuous Distillation Units*, Elsevier, Amsterdam, 1975.

[RAM 95] RAMSHAW C., *Proceedings of the 1st International Conference on Process Intensification for the Chemical Industry*, BHR Group, Publication 18, London, 1995.

[SIL 03] SILVA G. J., DATTA A., BHATTACHARYYA S. P., "On the stability and controller robustness of some popular PID tuning rules", *IEEE Transactions on Automatic Control*, vol. 48, p. 1638–1641, 2003.

[SKO 03] SKOGESTAD S., "Simple analytic rules for model reduction and PID controller", *Journal of Process Control*, vol. 13, p. 291–309, 2003.

[SMI 75] SMITH C. L., CORRIPIO A. B., JR. J. M., "Controller tuning from simple process models", *Instrumentation Technology*, vol. 22, p. 39–42, 1975.

[TAV 03] TAVAKOLI S., TAVAKOLI M., "Optimal tuning of PID controllers for first and second order plus time delay models using dimensional analysis", *4th IEEE International Conference on Control and Automation*, p. 942–946, 2003.

[TSY 74] TSYPKIN Y. Z., *Relay Automatic Systems*, Nauka, Moscow, 1974.

[TYR 92] TYREUS B. D., LUYBEN W. L., "Tuning PI controllers for integrator/dead time processes", *Industrial and Engineering Chemistry Research*, vol. 31, p. 2625–2628, 1992.

[VIS 01] VISIOLI A., "Optimal tuning of PID controllers for integral and unstable processes", *IEE Proceedings on Control Theory Applications*, vol. 148, p. 180–184, 2001.

[ZIE 42] ZIEGLER J. G., NICHOLS N. B., "Optimum settings for automatic controllers", *Transactions of the ASME*, vol. 62, p. 759–768, 1942.

Index